# Transport and Transformation of Contaminants Near the Sediment-Water Interface

*Edited by*

Joseph V. DePinto
Wilbert Lick
John F. Paul

LEWIS PUBLISHERS
Boca Raton   Ann Arbor   London   Tokyo

**Library of Congress Cataloging-in-Publication Data**

Transport and transformation of contaminants near the sediment-water
  interface / edited by Joseph V. DePinto, Wilbert Lick, John F. Paul.
     p.  cm.
  Includes bibliographical references and index.
  ISBN 0-87371-887-9
  1. Water—Pollution.   2. Sediment, Suspended.   3. Sediment
transport.   4. Environmental chemistry.   I. DePinto, Joseph V.
II. Lick, Wilbert.   III. Paul, John F.
TD423.T73   1993
628.1'68—dc20                                                                          93-6153
                                                                                          CIP

This book contains information obtained from authentic and highly regarded sources. Reprinted material is quoted with permission, and sources are indicated. A wide variety of references are listed. Reasonable efforts have been made to publish reliable data and information, but the author and the publisher cannot assume responsibility for the validity of all materials or for the consequences of their use.

Neither this book nor any part may be reproduced or transmitted in any form or by any means, electronic or mechanical, including photocopying, microfilming, and recording, or by any information storage or retrieval system, without prior permission in writing from the publisher.

All rights reserved. Authorization to photocopy items for internal or personal use, or the personal or internal use of specific clients, may be granted by CRC Press, Inc., provided that $.50 per page photocopied is paid directly to Copyright Clearance Center, 27 Congress Street, Salem, MA 01970 USA. The fee code for users of the Transactional Reporting Service is ISBN 0-87371-887-9/94 $0.00 + $.50. The fee is subject to change without notice. For organizations that have been granted a photocopy license by the CCC, a separate system of payment has been arranged.

CRC Press, Inc.'s consent does not extend to copying for general distribution, for promotion, for creating new works, or for resale. Specific permission must be obtained in writing from CRC Press for such copying.

Direct all inquiries to CRC Press, Inc., 2000 Corporate Blvd., N.W., Boca Raton, Florida 33431.

© 1994 by CRC Press, Inc.
Lewis Publishers is an imprint of CRC Press

No claim to original U.S. Government works
International Standard Book Number 0-87371-887-9
Library of Congress Card Number 93-6153
Printed in the United States of America   1 2 3 4 5 6 7 8 9 0
Printed on acid-free paper

# About the Editors

**Joseph V. DePinto** is currently Professor of Civil Engineering and Director of the Great Lakes Program at the State University of New York at Buffalo. Dr. DePinto received his B.A. degree in Physics from Miami University, Oxford, Ohio in 1967. He has an M.S. in Physics (1970), an M.S. in Environmental Engineering (1972), and a Ph.D. in Environmental Engineering (1975) from the University of Notre Dame. After spending sixteen years on the faculty in the Department of Civil and Environmental Engineering at Clarkson University, he joined the faculty at the University at Buffalo in 1991.

Dr. DePinto's current research interests are focused on process experimentation and mathematical modeling of fate and transport of contaminants in aquatic systems. Within this broad area his ongoing research includes toxic contaminant-particulate matter interactions in aquatic systems, transport of particulate-associated contaminants in aquatic systems, biological accumulation of contaminants, and exposure analysis of contaminants through deterministic modeling. Dr. DePinto is a member of the EPA-funded modeling team that developed and applied an integrated exposure model for PCBs in Green Bay, Lake Michigan. A new direction in Dr. DePinto's research has been leading a team of scientists and engineers at the University at Buffalo who are developing a Geographically based Watershed Analysis and Modeling System (GEO-WAMS) for application in analyzing contaminant fate, transport and effects in Great Lakes watersheds. This system will couple a GIS (Geographic Information System) with existing and newly developed water quality models in order to take advantage of GIS capabilities relative to input, storage, analysis, retrieval, and display (visualization) of spatial-related data.

Prior to focusing his efforts on toxic contaminants, Dr. DePinto studied the transport and fate of phosphorus in tributaries and basins of the

Great Lakes. He made significant contributions to our knowledge base in the areas of the death and decomposition of phytoplankton and the associated rate and extent of phosphorus regeneration, analysis of the rate and extent of release of bioavailable phosphorus from various types of particulate matter in the Great Lakes, and investigation of methods for measuring and computing total phosphorus loadings to the Great Lakes from tributary systems.

In the broad area of understanding and quantifying the impacts of pollutants on natural aquatic systems, Dr. DePinto has received over $4 million in grants and contracts over the past eighteen years. These studies have led to over 80 scientific publications in this area and the direction of 27 Master's theses and eight (8) Ph.D. dissertations.

Dr. DePinto is a recipient of the Chandler-Misner Award for the outstanding contribution to the Journal of Great Lakes Research in 1981. He is a past President of the International Association for Great Lakes Research and is an Associate Editor of the Journal of Great Lakes Research. He is a member of several Technical Advisory Committees, including the U.S. Environmental Protection Agency, Office of Exploratory Research, Peer Review Panel for Chemistry and Physics: Water.

# About the Editors

**Wilbert Lick** is Professor in the Department of Mechanical and Environmental Engineering at the University of California at Santa Barbara (UCSB). He joined UCSB in 1979. His principal fields of interest are environmental engineering and science, fluid mechanics, and numerical methods with specific interests in sediment and contaminant transport in aquatic systems. His work includes laboratory experiments on sediment resuspension, deposition, flocculation, and settling speeds, and the sorption of contaminants onto sediments; the development and application of field devices to measure sediment resuspension; and the numerical modeling and verification of the hydrodynamics and the sediment and contaminant transport and fate in rivers, lakes, estuaries, and near-shore areas of the oceans. He has been concerned with these problems in the Great Lakes, the Santa Barbara Channel, Long Island Sound, the Venice Lagoon in Italy, Korea, and the Arabian (Persian) Gulf.

Previous to UCSB, he taught at Harvard University and Case Western Reserve University with visiting appointments at the California Institute of Technology and Imperial College, University of London. His Ph.D. is from Rensselaer Polytechnic Institute.

# About the Editors

**John F. Paul** is a physical oceanographer. He received a Bachelor of Science degree in Engineering Science with high honors from Case Institute of Technology in 1969. Dr. Paul received a Master of Science degree and a Ph.D. in Engineering (Fluid Mechanics) from Case Western Reserve University in 1971 and 1974. He stayed at Case Western through 1975 conducting postdoctoral work. From 1975 through 1981, Dr. Paul was located at the U.S. Environmental Protection Agency Large Lakes Research Station in Grosse Ile, Michigan, joining the EPA staff in 1977. Since 1981, Dr. Paul has been working at the EPA Environmental Research Laboratory in Narragansett, Rhode Island. His current position is Acting Associate Director for Near Coastal Systems in the Environmental Monitoring and Assessment Program (EMAP).

Dr. Paul's early research interests focused on developing numerical hydrodynamic models applicable for the U.S. Great Lakes, with particular emphasis on the time-dependent, three-dimensional aspects of water movement. This modeling experience was then applied to understanding and predicting pollutant transport and fate in the Great Lakes. After transferring to the EPA Laboratory in Narragansett, Dr. Paul's research broadened to include pollutant transport, fate, and effects in marine systems. His positions included Chief of the Exposure Assessment Branch, with responsibility for the marine modeling activities and for the interfacing of these modeling activities with the biological research conducted at the Laboratory.

Since 1988, Dr. Paul has been involved with EMAP, a multidisciplinary, interagency federal program whose long-term goals are to monitor and report on the condition of the nation's ecological resources and to evaluate the success of our country's cumulative environmental policies and regulations. He is responsible for the planning, development, and implementation of EMAP in marine systems. Dr. Paul has received the EPA's highest award (gold medal for exceptional service) for his work in EMAP.

# List of Contributors

R. Hans Aalderink, Nature Conservation Department, Water Quality Management Section, Agricultural University, Wageningen, Netherlands.

Herbert E. Allen, Department of Civil Engineering, University of Delaware, Environmental Engineering Program, Newark, Delaware

Robin Autenrieth, Texas A&M University, Department of Civil Engineering, College Station, Texas

Keith W. Bedford, Department of Civil Engineering, Ohio State University, Hitchcock Hall, Columbus, Ohio

Nelson Belzile, Department of Chemistry, Laurentian University, Sudbury, Ontario, Canada

Victor J. Bierman, Jr., Limno-Tech, Inc., South Bend, Indiana

Gerard Blom, Nature Conservation Department, Water Quality Management Section, Agricultural University, Wageningen, Netherlands

James S. Bonner, Texas A&M University, Department of Civil Engineering, College Station, Texas

Richard Carigan, INRS-Eau, University of Quebec, Sainte-Foy, Quebec, Canada

Joseph V. DePinto, Director, Great Lakes Program, State University of New York at Buffalo, Buffalo, New York

Edward H. Dettmann, U.S. EPA, Environmental Research Laboratory, Narragansett, Rhode Island

Sharon L. Ducharme, Aries Engineering, Inc., Concord, New Hampshire

David N. Edgington, University of Wisconsin, Center for Great Lakes Studies, Milwaukee, Wisconsin

Andrew N. Ernest, Texas A&I University, Environmental Engineering Department, Kingsville, Texas

Susan Thompson Leach, EBASCO Environmental, Boston, Massachusetts

James Lick, Department of Mechanical and Environmental Engineering, University of California at Santa Barbara, California

Wilbert Lick, Department of Mechanical and Environmental Engineering, University of California, Santa Barbara, California

Lambertus Lijklema, Section Integral Water Quality Management, Department of Nature Conservation, Agricultural University, Wageningen, Netherlands

Charles R. Myers, Department of Pharmacology and Toxicology, Medical College of Wisconsin, Milwaukee, Wisconsin

Kenneth H. Nealson, University of Wisconsin, Center for Great Lakes Studies, Milwaukee, Wisconsin

John F. Paul, Associate Director, Environmental Monitoring and Assessment Program, U.S. EPA, Narragansett, Rhode Island

*Peter G. Sly*, Rawson Academy of Aquatic Science, Ottawa, Ontario, Canada

*Andre Tessier*, INRS-Eau, University of Quebec, Sainte-Foy, Quebec, Canada

*Thomas L. Theis*, Department of Civil and Environmental Engineering, Clarkson University, Potsdam, New York

*Nelson A. Thomas*, U.S. EPA, Environmental Research Laboratory, Duluth, Minnesota

*Elizabeth H.S. Van Duin*, Nature Conservation Department, Water Quality Management Section, Agricultural University, Wageningen, Netherlands

*Thomas C. Young*, Department of Civil and Environmental Engineering, Clarkson University, Potsdam, New York

*Carl Kirk Ziegler*, Hydroqual, Inc., Mahwah, New Jersey

# Acknowledgments

The editors wish to thank the U.S. Environmental Protection Agency, Office of Exploratory Research for supporting the workshop that led to publication of this book. We also wish to thank several individuals who worked very hard on the preparation of this camera-ready manuscript. Among the many who helped, we would especially like to recognize Janet Demers, Monica Moshenko, and Scott Rybarczyk.

# Table of Contents

| | | |
|---|---|---|
| Chapter 1 | Introduction...........................1 | |
| | J.V. DePinto, W. Lick, and J.F. Paul | |
| Chapter 2 | EPA/ORD Role and Perspective in Sediment Research..........................7 | |
| | N.A. Thomas | |
| Chapter 3 | Lessons Learned from Siting of Boston Harbor Sewage Outfall................17 | |
| | J.F. Paul and E.H. Dettmann | |
| Chapter 4 | The Flocculation, Deposition, and Resuspension of Fine-Grained Sediments....................35 | |
| | W. Lick | |
| Chapter 5 | In-Situ Measurement of Entrainment, Resuspension and Related Processes at the Sediment-Water Interface................59 | |
| | K.W. Bedford | |
| Chapter 6 | The Impact of Physical Processes at the Sediment/Water Interface in Large Lakes.........95 | |
| | P.G. Sly | |
| Chapter 7 | Partitioning of Toxic Metals in Natural Water-Sediment Systems...................115 | |
| | H.E. Allen | |
| Chapter 8 | Reactions of Trace Elements Near the Sediment-Water Interface in Lakes..........129 | |
| | A. Tessier, R. Carignan, and N. Belzile | |
| Chapter 9 | Partitioning of Organic Chemicals in Sediments: Estimation of Interstitial Concentrations Using Organism Body Burdens....153 | |
| | V.J. Bierman, Jr. | |
| Chapter 10 | Predicting Metals Partitioning During Resuspension Events...................177 | |
| | J.V. DePinto, T.L. Theis, T.C. Young, and Susan Thompson Leach | |
| Chapter 11 | Biological and Chemical Mechanisms of Manganese Reduction in Aquatic and Sediment Systems....................205 | |
| | C.R. Myers and K.H. Nealson | |
| Chapter 12 | The Transport of Fine-Grained Sediments in the Trenton Channel of the Detroit River.......225 | |
| | C.K. Ziegler, W. Lick, J. Lick | |

| | | |
|---|---|---|
| Chapter 13 | Sediment Transport in Shallow Lakes-Two Case Studies Related to Eutrophication...........253 *L. Lijklema, R.H. Aalderink, G. Blom and E.H.S. Van Duin* | |
| Chapter 14 | Parameterizing Models for Contaminated Sediment Transport......................281 *J.S. Bonner, A.N. Ernest, R.L. Autenrieth, and S.L. Ducharme* | |
| Chapter 15 | The Effects of Sediment Mixing on the Long-Term Behavior of Pollutants in Lakes........307 *D.N. Edgington* | |
| Chapter 16 | Research Needs and Summary................329 *J.F. Paul, J.V. DePinto, and W. Lick* | |

# CHAPTER 1
# Introduction

Joseph V. DePinto
Wilbert Lick
John F. Paul

In the past two decades great strides have been made in the reduction of direct discharges of contaminants to surface waters (both freshwater and marine systems). Also, bans on the production and use of known toxic substances (e.g., DDT and PCBs) have led to significant load reductions and subsequent decrease in aquatic biota body burdens of these contaminants. Aquatic systems that have been the beneficiary of such abatement have shown improvements in water quality and biotic community integrity. In virtually all instances, however, the system recovery has been incomplete; surveillance and monitoring efforts continue to demonstrate the presence and impact of toxic substances. Problems of toxicity to aquatic organisms, disruption of ecosystem structure and functioning, and bioaccumulation of toxic substances in aquatic food chains are still being identified for both metals and organic chemicals.

In their efforts to address continuing ecosystem and human health problems related to toxic substances in surface waters, water resource managers have had to focus their efforts beyond point source control of contaminants. One such example is contaminated bottom sediments. These "in-place pollutants" represent a potential source of toxicity and other biological effects, which manifest themselves as direct effects on benthic organisms or bottom feeders or, indirectly, through resuspension of contaminated sediments, as acute or chronic effects on water column organisms.

Bottom sediments of aquatic systems become contaminated with toxic substances through a series of fate and transport processes acting on the contaminants introduced to the water column from external sources. Unlike many of the more conventional pollutants, toxic chemicals tend to have a strong affinity for particulate matter in aquatic systems. Thus, depending on the chemical properties and characteristics of the receiving water body, much of the introduced contaminants are sorbed by biotic and abiotic suspended matter and settle from the water column. Historically, this long-term accumulation of contaminants in bottom sediments was considered a safe repository of these relatively insoluble toxic substances. However, recent studies have demonstrated that, when external loads to a water body have been eliminated, the recovery of the system is not governed strictly by washout from the water column. There is a much slower response controlling the long-term recovery of the system that is governed by the interaction of contaminated bottom sediments with the overlying water.

There are numerous examples of the above scenario in both freshwater (Great Lakes nearshore and connecting channels) and estuarine (New Bedford and Boston Harbors, Chesapeake Bay, Puget Sound) systems. In particular, in its 1985 report to the International Joint Commission, the Great Lakes Water Quality Board (1985) designated 42 specific locations in the Great Lakes Basin that were exhibiting significant environmental degradation and severe impairment of beneficial uses (Great Lakes "Areas of Concern"). Indeed, in all but one of these Areas Of Concern it was felt that contaminated sediments were a major contributor to the degraded conditions.

In response to the concern about contaminated sediments and to the latest revision of the Clean Water Act, the Environmental Protection Agency, through its Great Lakes National Program Office, has established a new demonstration program known as ARCS (Assessment and Remediation of Contaminated Sediments). ARCS and programs like it in other regions are intended to assess the relative importance of in-place contamination for each system of concern and to establish and implement a rational remediation program. For example, one might ask the question: "If all external sources of toxic substances were eliminated to the greatest extent possible, would the system continue to experience significant detrimental biological or bioaccumulation impacts, and for how long after external input abatement would these conditions persist?" Another concern is the extent to which a tributary or nearshore area with contaminated sediments is contributing to the total load to the downstream receiving water body. Because not all systems behave identically with respect to in-place contamination, addressing these and other management questions on a site-specific basis requires a fundamental and quantitative understanding of the processes governing the interaction of sediments with overlying water with respect to toxic chemical transport, transformation, and biological effects. Furthermore, these processes must be integrated into an overall assessment model that can account for their relative importance and linkages for a given set of site-specific environmental conditions.

The chapters in this book, which evolved from a U.S. EPA Office of Exploratory Research sponsored workshop held in Galilee, Rhode Island, in August, 1988, are intended to provide the state-of-the-science in addressing the above needs relative to questions of exposure and biological effects of in-place contaminants. In addition to spawning this book, the workshop contributed significantly to the issuance by the EPA

Office of Exploratory Research of a special Request for Applications in 1990, entitled "The Role of Sediments in the Transport and Fate of Pollutants in Fresh Water and Estuaries."

Contained in this book are papers covering four topic areas: regulatory management perspectives, physical processes, chemical/biological processes, and process synthesis/modeling. Taken together these papers summarize our level of understanding regarding the transport and fate of contaminants in aquatic sediments and how this understanding could be used in making informed regulatory and remediation decisions. They also serve to identify the significant gaps in our knowledge of this subject matter, thus developing a statement of research needs.

In the first chapter on regulatory perspectives (Chapter 2), Thomas describes EPA's Sediment Quality Criteria program from a regulatory perspective. He also specifies the knowledge base that managers need to address sediment quality problems effectively. In Chapter 3, Paul and Dettmann use a case study (siting of the Boston Harbor sewage outfall) as an example of the interaction between regulatory managers and researchers in addressing environmental problems. It emphasizes the need to incorporate more sophisticated models in the assessment of impacts of particulate-associated contaminants in aquatic systems.

The next section of the book deals with sediment transport processes and emphasizes the physical aspects of this problem. Lick (Chapter 4) discusses recent work on flocculation (aggregation and disaggregation), deposition and resuspension of fine-grained sediments with particular emphasis on a quantitative description of the flocculation process. In Chapter 5, Bedford discusses approaches for measuring the combined effect of these processes in natural systems, that is, vertical sediment flux near the sediment-water interface. He presents several frameworks by which appropriate field measurements can be used to infer sediment entrainment rates. The physical processes at work near the sediment-water interface of large lakes are the subject of the next chapter in which Sly demonstrates how the spatial and temporal importance of deposition and entrainment can be evaluated from bottom sediment particle-size data.

The chemical and biological processes in sediments are examined in the next section of this book. In the last decade considerable progress has been made in the development of models for the equilibrium partitioning of metals between solution and pure solid phases. However, the binding of metals and organic chemicals to natural, heterogeneous sediments is the subject of current research. In Chapter 7, Allen reviews equilibrium models for metal speciation in natural sediments and discusses the important considerations in applying these calculations. Tessier, *et al.* (Chapter 8) have made extensive measurements of sediment depth profiles of several heavy metals in lakes that cover a wide range of pH values. From sediment and porewater concentrations, they determine the effect of precipitation, sorption, and diffusive fluxes on establishing the sediment profiles. The partitioning of organic chemicals in bottom sediments is addressed in Chapter 9. Here Bierman presents a model for estimating interstitial chemical concentrations based on measurements of benthic organism body burdens.

While equilibrium models are quite adequate for assessing partitioning of toxic chemicals within in-place sediments, DePinto, et al. (Chapter 10) offer evidence from field measurements that a kinetic approach may be more appropriate for sediment resuspension because it is such a dynamic process. They present a preliminary application of a time-dependent sorption-desorption model and discuss the relative importance of physical processes in governing metals partitioning during a resuspension event. Finally, in Chapter 11 Myers and Nealson present data on microbially-mediated reduction processes in sediments, with particular emphasis on reduction of manganese oxides, and discuss the implications for characterizing observed sediment geochemistry.

Four chapters of the book are devoted to the development and application of models of sediment and associated contaminant transport in surface waters. Ziegler, et al. (Chapter 12) present a field application of their cohesive sediment transport model to simulate the result of resuspension events in the Trenton Channel of the Detroit River. Recent experimental work on flocculation, deposition, and entrainment have been incorporated in the model. Another modeling case study (Chapter 13, Lijklema, et al.) demonstrates the importance of sediment transport on the eutrophication of shallow lakes. The results of calculations using several different sediment transport models are compared with each other and with field observations. In Chapter 14, Bonner, et al. used the results of laboratory-based process experiments to parameterize a model for transport of particulate-associated contaminants; this model successfully predicted observed particle concentrations (from two different sources) in large, mixed settling columns. In Chapter 15, Edgington applies a time-dependent model for the long-term transport and fate of radio-isotopes ($^{239-240}$Pu and $^{137}$Cs) in Lake Michigan to assess the average sedimentation rate and bottom sediment mixing depth in this large lake. Differences between results for the two elements are explained in terms of continued resuspension and deposition of "old" sediment.

Finally, in the last chapter the editors present a summary of the resource management and technical information presented in this book and their recommendations for future research necessary for more informed management of contaminated sediments.

## References

Great Lakes Water Quality Board. 1985. <u>Great Lakes water quality</u>. Annual report of Great Lakes Water Quality Board to International Joint Commission. IJC, Windsor, Ontario.

# CHAPTER 2
# EPA/ORD Role and Perspective in Sediment Research

## Nelson A. Thomas

## Introduction

Most chemical contaminants and organic wastes in aquatic ecosystems eventually accumulate in sediments where they may adversely affect the benthic biota, become a source of contamination in the water column, accumulate in biological tissues, and enter pelagic and human food chains. Contaminated sediments now appear to be the main source of toxic contaminants in many bays, lakes, and rivers. Because of their potential adverse impacts, the long periods of time associated with natural assimilation of many in-place contaminants, and the high costs of mitigation, sediments have become a focus of concern for many of EPA's research and regulatory programs. Programs in which sediment quality assessments are currently required include ocean dumping, National Environmental Policy Act (NEPA) of 1969 review, Superfund, and estuarine protection.

Historically, EPA's Office of Research and Development (ORD) had been conducting research on characterization and impact of contaminated sediments. The program was disbanded in 1983 because it was perceived that it did not contribute to the mission of EPA's Office of Water.

## Regulatory Applications of Sediment Quality Criteria

In 1985 a new sediment quality research program was initiated at the Environmental Research Laboratory-Duluth and the Environmental Research Laboratory-Narragansett to develop sediment quality criteria (SQC) that define safe concentrations of individual chemicals in sediment. SQC are needed to provide a chemical-specific basis for evaluating sediment contamination in relation to Superfund remedial action plans, dredged material disposal decisions, sediment wasteload allocations, Clean Water Act Section 301(h) analysis of benthic degradation, compliance monitoring, and other regulatory schemes (Table 1).

Table 1. Applications of sediment criteria in implementing major environmenal laws.

|  | 1977 Clean Water Act | 1987 Amendments | Ocean Dump Act | RCRA | Superfund SARA |
|---|---|---|---|---|---|
| Dumpsite Designation | X |  | X | X | X |
| Discharge Siting | X | X |  |  |  |
| Permit Decisions | X | X | X | X |  |
| Dumpsite Monitoring | X | X | X | X |  |
| Discharge Monitoring | X | X |  | X |  |
| Clean Area Identification | X | X |  | X | X |
| Clean-up Area Selection | X | X | X | X | X |
| Clean-up Goal Setting | X | X |  | X | X |
| Site Restoration | X | X |  |  | X |
| EIS Preparation | X |  | X | X | X |

Chemical-specific sediment quality criteria provide the most efficient regulatory strategy for relating benthic ecosystem degradation to specific contaminants. Other approaches for evaluating sediments, including bioassays of total sediment toxicity and benthic community structure analyses, do not reveal the relationship between benthic impacts and causal factors. The present SQC research program needs to be expanded to develop chronic test methods; augment toxicological databases; compare equilibrium partitioning, apparent effects threshold, and bioassay/toxicological test methodologies for developing SQC; and integrate available methods into a comprehensive approach for assessing and regulating sediment pollution.

## Current Sediment Quality Criteria Research Program

ORD's existing SQC research program is oriented toward the equilibrium partitioning approach, and most toxicological research has been conducted to test predictions of that approach. In the equilibrium approach, interstitial water concentrations of individual chemicals are predicted from equilibrium partitioning theory and compared with water quality criteria. The equilibrium partitioning approach has been most successful with respect to criteria for neutral organic chemicals. Some research on the partitioning, toxicity, and bioavailability of metals in sediment has also been conducted. The apparent effects threshold (AET) approach has been developed primarily by EPA Region X (Seattle). Development of screening-level criteria has not been pursued to the same extent as the equilibrium partitioning approach. Modeling efforts include fundamental process studies, adsorption of ionizable organics, and field testing of a sorption model for metals on aquifer solids. Although the different approaches to development of SQC are complementary, there has been little integration of these efforts. There has been little laboratory, and no field verification of SQC approaches under the existing ORD program.

## Future Sediment Quality Criteria Research

Future research efforts have been developed in phases that are dependent upon funding levels. In Phase I, field and laboratory investigations will be conducted to compare approaches for developing numerical SQC, including equilibrium partitioning, AET, and toxicological techniques. Field and laboratory data will be developed from a variety of freshwater and marine sites to expand the geographic applicability of the AET approach. Comparative toxicological databases, generated through use of chronic full-life-cycle toxicological tests to be developed under the Monitoring of Sediment Quality research area, will be used to derive contaminant-specific SQC and to recommend minimum test requirements for non-contaminant-specific sediment quality evaluations. Interim guidance documents on regulatory applications of SQC derived from a variety of contaminant-specific methodologies will be developed for selected priority chemicals.

In Phase II, research on the equilibrium partitioning, bioassay, and AET approaches for development of sediment quality criteria will be expanded. The equilibrium partitioning research will focus on SQC for metals, minimum toxicological data sets for extrapolation from water quality criteria to SQC, and field validation of proposed criteria. Research on the bioassay method will develop toxicological and bioaccumulation estimates of safe sediment concentrations for a set of fifteen priority chemicals. Correlations between contaminant concentrations, toxicity, and benthos distributions along pollution gradients at representative field sites will be used to investigate the AET method and to estimate SQC empirically. The different approaches to development of SQC will be compared and integrated for the assessment of a greater number of priority contaminants than in Phase I.

There will be a more comprehensive integration of the equilibrium partitioning hypothesis with other toxicological methods and field observations. Estimates of safe contaminant concentrations through classical dose-response experiments will be conducted for representative sediment types. Research comparing the relationships between organism tissue residues and responses, and estimates of sediment-to-organism bioaccumulation factors will be used to develop and evaluate SQC applicable to site-specific conditions.

Predictive models will be developed to estimate the equilibrium partitioning of ionic organic chemicals to sediments. A benthic exchange model will be developed to quantify the effects of sorption, diffusion, bioturbation, and sediment transport. Final guidance documents on regulatory applications of SQC will be developed and transferred.

## Sediment Quality Evaluation (Non-Chemical-Specific)

Assessment methods for evaluating sediment quality cannot always be based on chemical-specific criteria because of the great number of potentially hazardous chemicals in sediments. Further, under some regulatory strategies it is necessary to directly examine sediment toxicity and carcinogenicity, the condition of benthic resources, and physical characteristics. There is some technical evidence that SQC cannot accurately be applied on a chemical-specific basis without some knowledge or estimate of the joint actions among contaminants, which are typically found in mixtures in polluted sediment. Non-chemical-specific methods are often needed for sediment quality evaluation. Evaluation of sediment quality on the basis of biological impacts, toxicological effects, or contaminant interactions requires a better understanding of how benthic ecosystems respond to sediment degradation.

## Current Sediment Quality Evaluation (Non-Chemical-Specific)

EPA/ORD is currently funding a number of studies to describe the degradation and recovery of benthic ecosystems. Sediment characterization studies are being conducted to assess the stresses placed on the benthic systems. These projects are often based on correlations among sediment physical characteristics, contamination, disease, toxicity, and biological effects.

The critical need that is not being adequately addressed is an examination of causal relations among these factors. Regulations that require sediment quality evaluation can be fully successful only if there is certain knowledge that ecosystem perturbations are causally related to sediment contamination. Effects of mixtures of sediment contaminants are almost completely ignored in present research efforts.

## Future Sediment Quality Evaluation Research

All future research in sediment quality evaluation will be conducted under Phase II. Several key problems will be examined, including:

- The assimilative capacity of benthic ecosystems for organic enrichment and sediment contamination will be examined in relation to the development of new regulatory strategies for controlling benthic conatmination. One objective of this project will be to determine chemical and enrichment effects on benthic communities.

- Sediment toxicity will be related to its functional significance, including the elimination of benthic populations and consequent loss of trophic resources for commercial and recreational fisheries. This examination of the relevance of sediment toxicity will provide a stronger technical basis for the use of sediment bioassays as regulatory tools. In addition to toxicity, comprehensive assessments will include mutagenicity, genotoxicity, carcinogenicity, and fish tumor incidence.

- Models of the joint action of mixtures of sediment contaminants will be used to predict the total toxicity of field and experimental sediment in terms of the effects of additivity on sediment quality criteria.

- Data from field correlations will be used to design definitive experiments to demonstrate causal relations between sediment contaminants and the incidence of fish disease.

- The assessment of key physical characteristics that determine resuspension and contaminant flux potential is needed to determine the vulnerability of ecosystems.

- Data from field correlations will be used to develop a comprehensive assessment protocol.

- A strategy will be developed to characterize and identify the specific chemicals causing toxicity in contaminated sediments.

## Mitigation/Prevention of Sediment Quality Problems

A key sediment issue faced by a regulatory agency is how to prevent or mitigate the ecological damage caused by benthic contamination. Regulatory decision makers need predictive methods to describe contaminant deposition and distribution, bioaccumulation, toxicity, and ecosystem recovery. An essential element in developing a remedial action plan is the ability to predict rates of natural burial of contaminated sediments. Problems associated with the removal and disposal of contaminated sediments make the "do nothing" option an important consideration. To use this option effectively, regulators must be able to predict the rate of sediment deposition and the frequency of resuspension and contaminant exposure.

Five kinds of predictive methods must be developed and validated through field measurements:

- Transport/deposition models to relate in-place sediment contaminants to specific sources.

- Wasteload allocation strategies for sediment contaminants.

- Models of benthic community degradation and recovery.

- Models of acute and chronic toxicity and bioaccumulation.

- Microcosms to simulate specific remedial actions, such as dredging and capping.

One of the most common sediment mitigation procedures is to dredge contaminated sediments and dispose of them in another location. For a variety of reasons, the disposal site is often a wetland or an area that runs off into a wetland. While filling wetlands has an obvious impact, the effect of contaminants associated with the disposal of dredged material on wetlands is generally unknown. Information regarding the impact of toxic chemicals on wetlands would provide additional guidance to regulators and permit writers to use in making decisions on the appropriateness of disposal options.

## Current Research in Mitigation/Prevention of Sediment Quality Problems

Some of the research necessary to develop predictive methods for the sediment quality program has already begun, though the level of effort is insufficient to meet all research goals. The one research area that has been well developed by the Corps of Engineers and EPA concerns the effects of dredging and dredged material disposal. ORD started an initial project on sediment wasteload allocation in FY '88. Two projects on particle deposition and resuspension models have been supported by the Clean Water Act Section 301(h) regulatory program. Uptake of pollutants from contaminated sediments through the food chain is an integral part of the mass balance model being developed for Green Bay, Lake Michigan. The current wetland research program includes efforts on the mitigation of damaged or lost wetlands, the impact of cumulative losses of wetlands, and the functions of wetlands in improving water quality. However, this research does not include empirical studies on the effects of sediment contaminants on the biological components and processes of wetlands.

Predictive models for toxicity and ecosystem alterations are not available. Conceptual models of the bioaccumulation of sediment contaminants have not yet been quantified. Microcosms have not been applied to simulate sediment capping or other remedial actions that might be implemented at aquatic Superfund sites.

## Future Research in Mitigation/Prevention of Sediment Quality Problems

Phase I will focus on models of particulate deposition and resuspension and on predictions of the biological impacts of contaminated sediments. The ability to predict the fate of sediment contaminants will be advanced by strengthening research on deposition/resuspension models. Such models can be applied to a diversity of regulatory issues including source identification, wasteload allocation, nutrient release, and the question asked with increasing frequency: "What will happen if nothing is done?" Research on predictions of biological impact will address toxicity, bioaccumulation, and community recovery.

Phase II will focus on simulation of remedial actions, predicting the effects of toxic chemicals in the Great Lakes and wetlands, and refining methods for predicting the biological impacts of contaminated sediments. Microcosm research will initially consider the efficacy of alternative sediment capping options in open waters.

Research on the effects of toxic chemicals on wetlands will be conducted with experimental wetlands that can be stressed under controlled conditions. Dose-response curves are necessary to provide input into risk assessments and management decision schemes. Experimental wetlands provide the opportunity for controlling chemical inputs and for replicated exposures. Experimental wetlands at the EPA/ORD Monticello Ecological Research Station (Minnesota) would be used to determine the impact of sediment related contaminants on the biological activity under known exposure situations. The effect of contaminant additions to wetlands could be evaluated to systematically evaluate the potential for impact from chemical contaminants of sediments and dredged materials. This information will then provide guidance for appropriate sediment mitigation options.

## Monitoring of Sediment Quality

Monitoring benthic ecosystems is required under a variety of current EPA regulatory programs, including ocean dumping, estuarine and wetland protection programs, and NEPA review. It is likely to increase in importance as EPA's interest in sediment quality expands. Some sediment monitoring methods are well established, but others require development or refinement. Monitoring requirements have changed as knowledge of benthic processes has increased. Some very basic monitoring needs remain unresolved. For example, there is a need for chronic full-life-cycle toxicity tests to monitor sediment quality cost effectively. In addition, there is no standard method for collecting and processing interstitial water.

## Current Monitoring of Sediment Quality

Present research on sediment monitoring methods is not well coordinated. There are projects on optimal benthic community survey techniques in progress in EPA Regions V, IX, and X (Chicago, San Francisco, and Seattle). Monitoring of sediment toxicity and bioavailability is not an active research area, and the use of biochemical markers and interstitial water quality is rare in benthic monitoring. Integration and comparison of sediment survey methods is virtually non-existent.

## Future Monitoring Research

Phase I research in this area will concentrate on the development of chronic, full, and partial life-cycle toxicity tests for benthic freshwater and marine species. There is a critical need for these tests, which are more sensitive than the existing acute toxicity tests, for assessing sediment toxicity and for evaluating the various methods for developing sediment quality criteria described in the Sediment Quality Criteria sections.

Phase II research will develop and evaluate several additional kinds of sediment quality monitoring techniques, including benthic community structure, interstitial water quality, bioavailability, and biochemical markers. The interstitial water quality techniques will allow direct comparison of field conditions with sediment quality criteria based on the equilibrium partitioning theory. Biochemical markers will be adapted to sediment monitoring for the first time. Individual monitoring methods will be developed and tested against specific sediment contaminant gradients, and the different methods will be compared at selected freshwater and marine sites representing different kinds of sediment contamination.

In summary, EPA/ORD research efforts will continue to support the application of the various approaches of addressing contaminated sediments. The degree and timing to which ORD can undertake all the required tasks will depend on availability of resources. Current indications are that additional researchers and extramural dollars will be assigned to the sediment quality research program.

## Acknowledgments

I would like to acknowledge the contributions of Rick Swartz, David Hansen, and Sam Williams to this research effort.

# CHAPTER 3
# Lessons Learned from Siting of Boston Harbor Sewage Outfall

John F. Paul
Edward H. Dettmann

## Introduction

Control of the discharge of wastewater into coastal waters of the United States is regulated by the U.S. Environmental Protection Agency (EPA) under legislative mandates of the Federal Water Pollution Control Act (FWPCA) Amendments of 1972 (P.L. 92-532) and the Clean Water Act (CWA) of 1977 (P.L. 95-217). Volumetrically, the largest discharges into coastal waters have been municipal wastewater from publicly owned treatment works (POTWs). FWPCA mandated uniform national effluent limitations based upon secondary treatment [section 301(b)(1)(B)], defined in terms of biochemical oxygen demand (BOD), suspended solids, and hydrogen ion activity (pH) (40 CFR Part 133). Congress, incorporating section 301(h) in CWA, provided for a variance in the secondary treatment requirements for POTW discharges into certain ocean or estuarine water if the POTW adequately demonstrated that the discharge would not impair the integrity of the receiving waters and biota.

EPA issued regulations implementing section 301(h) on 15 June 1979. Shortly thereafter, these regulations were challenged in court, with certain provisions subsequently struck down. Section 301(h) was amended by Congress in 1981 (P.L. 97-11) following the court decision, stipulating seven statutory requirements for issuance of a 301(h) variance. The amended section covered discharges into deep waters of the territorial sea, into waters of the contiguous zone, and into saline estuarine waters where there are strong tidal movement and other hydrological and geological characteristics which would allow compliance with CWA. Criteria for implementation of the revised 301(h) program were specified by regulations contained in 40 CFR Part 125, Subpart G (47 Federal Register 53666, 26 November 1982).

The intent of the wastewater, or effluent, control program in FWPCA was to reduce discharge of pollutants into coastal waters by requiring

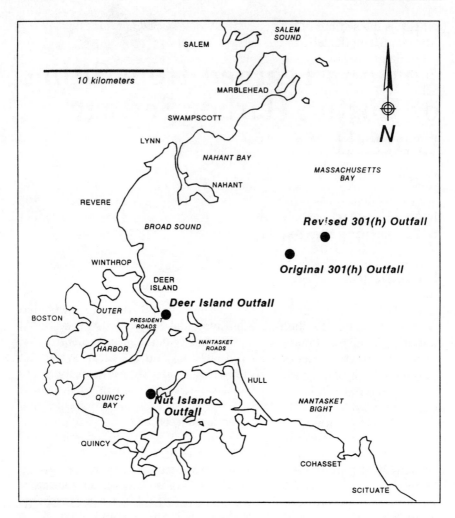

Figure 1. Boston Harbor and Massachusetts Bay, including existing and proposed original and revised 301(h) outfall sites.

treatment facilities to meet a minimum level of effluent quality in terms of the three parameters: BOD, suspended solids, and pH. The minimum level of quality for BOD and suspended solids is specified in terms of 7-day and 30-day average values not to be exceeded (45 mg/L and 30 mg/L, respectively), plus a 30-day average removal not to be less than 85 percent. The effluent values for pH are to be within the range of 6.0 to 9.0 (40 CFR 133.102). The common definition of secondary treatment is 85-90 percent removal of solids and BOD (Mueller and Anderson, 1983).

Control of BOD and solids does two things: (1) it reduces organic material, which directly consumes oxygen in the receiving waters and enriches the sediments, changing the biological communities around the discharge site (usually to the detriment of those communities), and (2) it eliminates the nutrients and toxic chemicals bound to the solids removed in the treatment process. It has been recognized that secondary treatment alone will not control all pollution problems associated with wastewater discharges. In particular, specific nutrient removal has been required over and above secondary treatment (advanced waste treatment), and pretreatment control has been necessary to reduce specific toxic chemicals (40 CFR Part 403; 40 CFR 125.64).

Even though a large fraction of particulate material is removed by wastewater treatment facilities, a sizable amount is still discharged into receiving waters. This particulate material that is discharged is of a fine-grain organic character. It is this type of material to which toxic chemicals have a tendency to bind (Baker, 1980; Dickson et al., 1987). Therefore, an understanding of the behavior of particulate material discharged into coastal waters is critical in understanding the transport pathways of toxic chemicals in the aquatic environment and the impacts on potentially affected biological species.

The intention of this chapter is to review briefly the recent history of the regulatory actions affecting the wastewater outfalls in Boston Harbor, with emphasis on the role of models for particulate behavior, in particular, sedimentation. This review of the Boston Harbor situation serves to illustrate where the state-of-the-science is in the application and use of modeling tools for addressing environmental problems in aquatic systems.

## Existing and Proposed Boston Harbor Outfalls

The greater Boston area [43 cities and towns served by Massachusetts Water Resources Authority (MWRA)] presently discharges its municipal wastewater into Boston Harbor through two discharges: a 300 MGD (13 m$^3$/sec) outfall located at Deer Island and a 160 MGD (6.9 m$^3$/sec) outfall on Nut Island (Figure 1). Operations at, and discharge through, the Deer Island facility commenced in 1895 and at Nut Island in 1904. In addition, sewage sludge is discharged through the Deer Island outfall during the outgoing tide. Deteriorated aquatic conditions in Boston Harbor are generally considered to be due to problems associated with the existing treatment of the wastewater: inadequate level of treatment, inadequate capacity, antiquated facilities, combined sewer overflows (CSOs), and discharge of sludge (Metcalf and Eddy, 1976).

In 1979, the Commonwealth of Massachusetts Metropolitan District Commission (MDC), the predecessor organization for MWRA, filed for a 301(h) variance (Metcalf and Eddy, 1979). The proposed outfall was to accommodate 520 MGD (23 m$^3$/sec) of primary treated wastewater and was to be located in 32 m of water in Massachusetts Bay (Figure 1). This

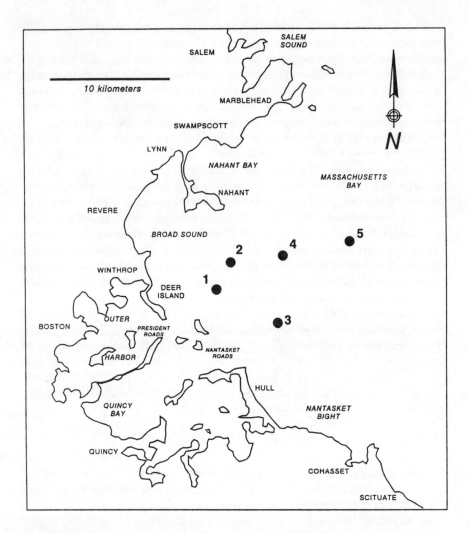

Figure 2. Boston Harbor and Massachusetts Bay, including proposed secondary outfall sites.

application was denied by EPA in 1983 because, in part, the proposed discharge would violate the state water quality standards for dissolved oxygen and would interfere with the protection and propagation of a balanced, indigenous population (BIP) of marine life (USEPA, 1983). A BIP is defined in section 301(h) regulations [40 CFR 125.58( f )] as an ecological community that: (1) exhibits characteristics similar to those of nearby healthy communities existing under comparable but unpolluted environmental conditions, or (2) may reasonably be expected to become reestablished in the polluted water body segment from adjacent waters if sources of pollution were removed.

In 1984, MDC filed a revised 301(h) waiver application for an outfall accommodating 490 MGD (21 m³/sec) of primary treated wastewater discharged into 36 m of water in Massachusetts Bay (Figure 1) (Metcalf and Eddy, 1984). This application was denied by EPA in 1985 in part because the proposed discharge would violate state water quality standards for dissolved oxygen during summer resuspension events, and would interfere with the protection and propagation of a balanced, indigenous population of marine life (USEPA, 1985).

The recourse was to apply for a National Pollution Discharge Evaluation System (NPDES) permit to discharge secondary treated wastewater. This process was initiated in 1985 by an act of the Massachusetts legislature creating MWRA to take over the sewer and waste operations of MDC. In 1987, a secondary outfall siting document was prepared by MWRA (1987), proposing to discharge 670 MGD (29 m³/sec) at one of five potential locations in Massachusetts Bay (Figure 2). In 1988 EPA issued a Supplemental Environmental Impact Statement (SEIS) (USEPA, 1988) evaluating the proposed MWRA outfall locations and recommending an area for the proposed outfall. An interesting note is that MWRA would first build a primary plant (to be operational in 1994), with the secondary plant to go on line in 1999. The MWRA and EPA documents thus had to evaluate the potential impacts from both primary and secondary outfalls.

## Methods Used to Evaluate Sedimentation

### MDC 301(h) Application

The original 301(h) application used the method described in the Revised Section 301(h) Technical Support Document (TSD) (Tetra Tech, 1982) to evaluate the potential distribution of particulate material from the proposed outfall. In the TSD, the following solids settling velocity ($V_s$) distribution is provided, based upon data from earlier section 301(h) applications for primary and advanced primary discharges:

    5 percent with $V_s > 0.1 cm/\text{sec}$,
    20 percent with $V_s > 0.01\ cm/\text{sec}$,
    30 percent with $V_s > 0.006\ cm/\text{sec}$, and
    50 percent with $V_s > 0.001\ cm/\text{sec}$.

It is assumed that the remaining solids settle so slowly that they remain suspended in the water column. Distances for solids deposition about the outfall location are determined using

where $D_a$ is the upcoast, downcoast, onshore, or offshore extent of the

$$D_a = V_a H_t / V_s ,$$

depositional area from the outfall; $V_a$ is the upcoast, downcoast, onshore, or offshore ambient current velocity; $H_t$ is the height of the average trapping level of the wastewater plume measured above the bottom [determined from a model for initial dilution (e.g., Muellenhoff et al., 1985)]; and $V_s$ is from above. The depositional areas are elliptical contours determined from the above computed distances. The settled solids are assumed to be uniformly deposited over each elliptical band. It is further assumed that 80 percent of the settled solids (for primary discharges) is organic material. This method provides estimates of accumulation of solids and organic material into the bottom sediment.

The estimates for the steady-state accumulation of solids around the proposed outfall were 160 gm/m² over an area of 6.3 km², and 25 gm/m² over 430 km² (USEPA, 1983). This high accumulation of sewage particulates with resultant organic enrichment was expected to cause large reductions in the number and density of pollution-sensitive species [in accordance with Pearson and Rosenberg (1978)] (USEPA, 1983).

The estimates of particulate accumulation and the location of the outfall site in a depositional area of Massachusetts Bay were used by the EPA to conclude that the proposed discharge location did not provide for sufficient transport and dispersion of diluted wastewater particulates to assure that water uses and areas of biological sensitivity would not be affected adversely (USEPA, 1983).

## MDC Revised 301(h) Application

The revised 301(h) waiver application used methods described in the Revised TSD (Tetra Tech, 1982) with solids settling velocity distribution determined from data for settling column experiments with 4:1 and 1:1 seawater effluent mixtures and with undiluted effluents (Metcalf and Eddy, 1984). The experiments were conducted with no, low, and moderate mixing in a settling column (provided by a rotating impeller). The settling column results were extrapolated to field conditions (dilutions of 100:1) using a second-order kinetic model for coagulation (Morel and Schiff, 1980; 1983). The coagulation model is described by

$$\frac{dC}{dt} = -aC^2 , \tag{1}$$

where C is solids concentration, $a$ is an empirical parameter, and t is time. The solution of equation (1) is

$$C/C_o = 1/(1+aC_o t), \qquad (2)$$

where $C_o$ is the initial concentration at t=0. MDC's results are shown as the curve labeled MDC in Figure 3. Note that this curve is substantially different from that suggested for use in the TSD, labeled as TSD in Figure 3.

In reviewing the MDC application for EPA, Tetra Tech (1984) noted that the appropriate depth scale was not used by MDC to extrapolate model results to field conditions. In particular, MDC used the depth scale of the laboratory settling column experiments instead of the entire plume trapping depth in Massachusetts Bay. The extrapolated settling velocity distribution, using the appropriate depth scale, agrees with the TSD settling velocity distribution. Therefore, Tetra Tech (1984) used the TSD distribution for their evaluation of sedimentation from the proposed

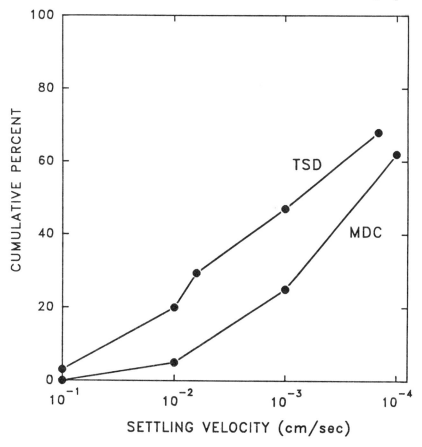

Figure 3. Settling velocity distributions in Revised Section 301(h) Technical Support Document (Tetra Tech, 1982) and MDC revised 301(h) application (Metcalf and Eddy, 1984).

outfall.

In reviewing MDC's data on the laboratory settling experiments, Paul (1984a) noted some interesting behavior in the data, and was able to fit the following generalized coagulation model to the data:

$$\frac{dC}{dt} = -abC_o^2 \left(\frac{C}{C_o}\right)^{1+1/b}, \qquad (3)$$

where $b$ is an additional empirical parameter. Equation (3) reduces to the second-order model for the specific case of $b=1$. The solution to equation (3) is (for $b \neq 0$)

$$C/C_o = 1/(1+aC_o t)^b. \qquad (4)$$

Paul noted that the sets of fitted parameters $(a,b)$ were directly related to mixing in the settling column. These results are listed in Table 1. The values of the fitted parameters for a second-order model ($b=1$) are also shown in this table. Note that the second-order model only provides a realistic representation for moderate mixing conditions. To explore this further, Paul (1984b) reanalyzed the quiscent settling data of Faisst (1980) for Los Angeles County Sewage District (LACSD) sludge using the model in equation (3). The results (Figure 4) clearly show a better fit to the data than a second-order model as applied by Morel and Schiff (1980) (Figure

Table 1

Results of Non-Linear Regression Analyses (Equation (3)) for MDC Sludge Settling Tests (from Paul, 1984a)

|  | best a,b | | | $b=1$ (second-order model) | |
| --- | --- | --- | --- | --- | --- |
|  | a | b | RMS deviation | a | RMS deviation |
| No Mixing | $5.8 \times 10^{-6}$ | .31 | 66.1 | $5.5 \times 10^{-7}$ | 117.1 |
| Low Mixing | $2.6 \times 10^{-6}$ | .43 | 85.1 | $6.3 \times 10^{-7}$ | 107.7 |
| Moderate Mixing | $1.9 \times 10^{-7}$ | .96 | 161.8 | $1.8 \times 10^{-7}$ | 159.7 |

RMS deviation is between predicted and observed values of suspended solids concentration (mg/l). Range of suspended solids data was 90-1000 mg/l.

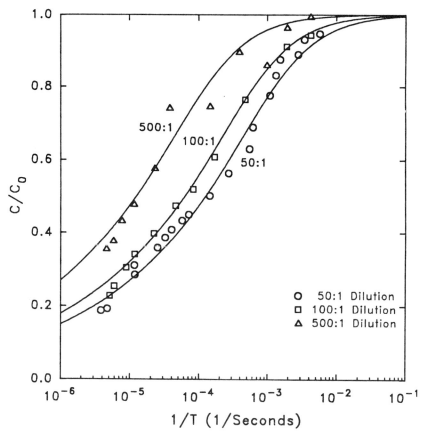

Figure 4. Data from Faistt (1980) for settling of LACSD sludge, with regression curves for fit to Equation (3)($a$=3.6E-06, $b$=0.25)(from Paul, 1984a).

5). The resulting fitted parameters for the LACSD sludge are consistent with the values from the MDC data for no mixing.

Application of Paul's model [equation (3)] to estimate sedimentation from the proposed MDC outfall provided estimates consistent with those of Tetra Tech (1984) using the TSD settling velocity distribution.

MDC, using their settling velocity distribution shown in Figure 3, predicted no impacts in the vicinity of the outfall due to solids deposition. Their estimates of sediment accumulation were low by almost two orders of magnitude (USEPA, 1985). EPA, using the information on settling velocity distribution provided by Tetra Tech and Paul, estimated that the maximum total deposition rate, for steady-state conditions, would be 100 g/m$^2$ over an 8.5 km$^2$ area (USEPA, 1985). These analyses estimated the distribution of initially deposited material and did not evaluate what would happen to the material that did not initially deposit around the outfall and to the material in the vicinity of the outfall that would subsequently be resuspended and transported elsewhere.

The information provided on the initial sedimentation of solids in the vicinity of the outfall was used, in part, to show that the proposed outfall location would not provide for sufficient transport and dispersion of the diluted wastewater and particulates to assure that water uses and that areas of biological sensitivity would be adequately protected (USEPA, 1985).

## MWRA Secondary Outfall Siting

The sedimentation analyses for the proposed MWRA primary and secondary outfalls incorporated a two-dimensional, convection-diffusion transport model for Massachusetts Bay, developed at Massachusetts Institute of Technology (Baptista et al., 1984; Westerink et al., 1985). The

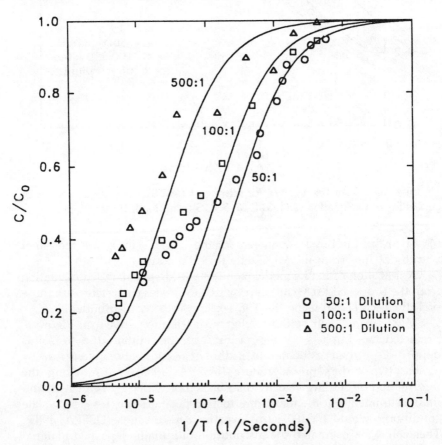

Figure 5. Data from Faistt (1980) for settling of LACSD sludge with regression curves for fit to second-order kinetic model (from Paul, 1984a).

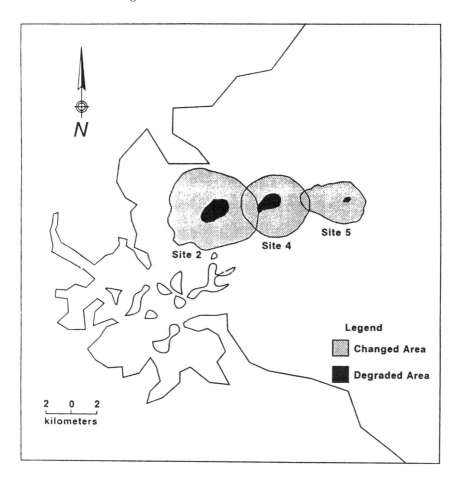

Figure 6. Areas of predicted changed and degraded benthic communities due to organic enrichment at MWRA sites with secondary effluent treatment (from USEPA, 1988).

removal of particulate material from the water column (sedimentation) was implemented as a first-order decay process, with decay times of 20 and 60 days to account for fast and slow removal, respectively (MWRA, 1988). Processes such as resuspension of deposited material and water column stratification were not incorporated in the model used by MWRA (1988). USEPA (1988) did incorporate a form of stratification in the transport model, albeit, in a very simplistic manner. The two-dimensional model was run as two separate layers, upper and lower water column, with no direct coupling between the layers. By using numerical modeling for predicting the farfield impacts, EPA made estimates for areas of solids accumulation in Massachusetts Bay from both primary and secondary outfalls at all of the proposed locations.

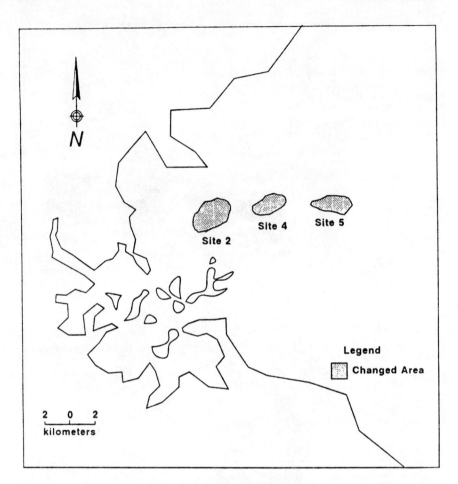

Figure 7. Areas of predicted changed benthic communities due to organic enrichment at MWRA sites with secondary effluent treatment (from USEPA, 1988).

Examples illustrating the predicted changed and degraded benthic communities are shown in Figures 6 and 7 for primary and secondary discharges, respectively. A degraded community was assumed to occur for a steady-state accumulation greater than 1.5 gm $C/m^2/day$ of organic material, while a changed community occurred for 0.1-1.5 gm $C/m^2/day$. The Supplemental Environmental Impact Statement (USEPA, 1988) recommended that the outfall diffuser location be in the area shown in Figure 8.

MWRA site 5 is comparable to the MDC revised 301(h) site. Results for the MWRA primary outfall at this location estimated accumulation of 75 $gm/m^2$ over a 12 $km^2$ area, comparable to 100 $gm/m^2$ over 8.5 $km^2$ area for the revised 301(h) discharge. That there are potential impacts

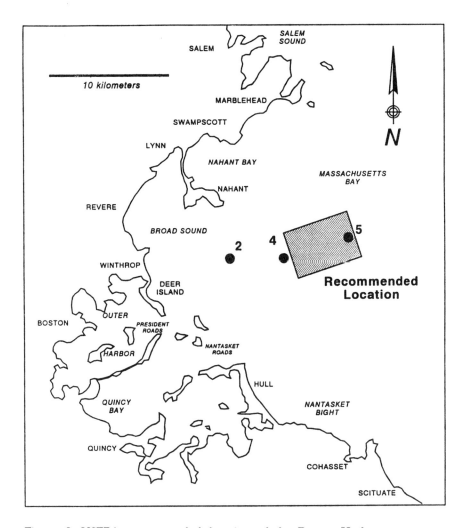

Figure 8. USEPA recommended location of the Boston Harbor wastewater treatment plant outfall (from USEPA, 1988).

from MWRA's five-year primary discharge is consistent with EPA's denial of the revised 301(h) application, indicating a need for secondary treatment to provide environmental protection in Massachusetts Bay. However, the MWRA primary discharge is temporary (five-year duration), and the potential impacts due to discharge of organic material should be reversible.

## Summary

The review of the recent history of the Boston area wastewater outfalls indicates that major environmental decisions are being made with relatively simplistic models for incorporating behavior of particulate material; the 301(h) decisions used models estimating the area over which the fastest settling fraction was deposited, and the MWRA siting decision used numerical models assuming first-order loss of particulates from the water column. We know that behavior of particulates in aquatic systems is more complex that this, as evidenced by the work of Lick (Chapter 4), Ziegler (Chapter 12), etc., in this volume.

What are some of the steps we need to take to incorporate the information presented by authors in this volume into models that can be used for environmental management decisions? We think the first steps have already begun by disseminating the information in this volume and by the interactions that were initiated at the workshop from which this volume was generated (DePinto *et al.*, Chapter 16).

The next step is to focus in on the two major types of problems that need addressing: (1) the initial discharge conditions and the impact in the vicinity of the discharge and (2) the farfield problem that looks at the broad area of impact, incorporating behavior of material that does not initially deposit in the vicinity of the outfall and of material initially deposited in the vicinity of the discharge that subsequently is resuspended and transported elsewhere. The models to address the first type need to incorporate the processes of particulate aggregation (coagulation) and disaggregation (e.g., Lick, Chapter 4; Lick and Lick, 1988) into models of plume dynamics (e.g., Muellenhoff *et al.*, 1985). Models to address the second type need to incorporate processes such as resuspension and deposition (e.g., Ziegler, Chapter 12), as well as dynamics of particle size distributions (e.g., Lick, Chapter 4).

Hopefully, the next technical volume that addresses the topic of transport and transformation of pollutants across the sediment-water interface will provide us with examples of these types of models and how they are being used in making environmental management decisions.

## Note

Although this work has been supported by the U.S. Environmental Protection Agency and reviewed according to procedures developed by the EPA Office of Research and Development, it should not be taken to reflect Agency policy.

# References

Baker, R.A. (editor). 1980. Contaminants and Sediments. Ann Arbor Science Publishers, Ann Arbor, Michigan.

Baptista, A.M., E.E. Adams and K.D. Stolzenbach. 1984. The solution of the 2-D unsteady, convective diffusion equation by the combined use of the FE method and the method of characteristics. Report No. 296, R.M. Parsons Laboratory, Massachusetts Institute of Technology, Cambridge, Massachusetts.

Dickson, K.L., M.A. Maki and W.A. Brungs (editors). 1987. Fate and Effects of Sediment Bound Chemicals in Aquatic Systems. Pergamon Press, New York.

Faisst, W.K. 1980. Characterization of particles in digested sewage sludge. Particulates in Water: Characterization, Fate, Effects, and Removal. M.C. Kavanaugh and J.O. Leckie (editors), American Chemical Society, Washington, D.C., pp. 259-282.

Lick, W. and J. Lick. 1988. Aggregation and disaggregation of fine-grained lake sediments. Journal of Great Lakes Research, 4(4):514-523.

Massachusetts Water Resources Authority. 1987. Secondary Treatment Facilities Plan, 7 Volumes plus Appendices, Draft Report, Boston, Massachusetts.

Massachusetts Water Resources Authority. 1988. Secondary Treatment Facilities Plan, Volume V, Appendix J., Final Report, Boston, Massachusetts.

Metcalf and Eddy, Inc. 1976. Wastewater engineering and management plan for Boston Harbor - Eastern Massachusetts Metropolitan Area: EMMA Study, Boston, Massachusetts.

Metcalf and Eddy, Inc. 1979. Application for Modification of Secondary Treatment Requirements for its Deer Island and Nut Island Effluent Discharges into Marine Waters. Volumes 1-5, Prepared for Commonwealth of Massachusetts Metropolitan District Commission, Boston, Massachusetts.

Metcalf and Eddy, Inc. 1984. Application for a Waiver of Secondary Treatment for the Nut Island and Deer Island Treatment Plants. Volumes 1-4, Prepared for Metropolitan District Commission, Boston, Massachusetts.

Morel, F.M.M. and S.L. Schiff. 1980. Geochemistry of municipal waste in coastal waters. Report No. 259, Massachusetts Institute of Technology, Cambridge, Massachusetts.

Morel, F.M.M. and S.L. Schiff. 1983. Geochemistry of municipal waste in coastal waters. Ocean Disposal of Municipal Wastewater: Impacts on the Coastal Environment, Vol. 1, E.P. Myers and E.T. Hading, editors, Sea Grant College Program, Massachusetts Institute of Technology, Cambridge, Massachusetts, pp. 249-421.

Muellenhoff, W.P., A.M. Soldate, Jr., D.J. Baumgartner, M.D. Schuldt, L.R. Davis and W.E. Frick. 1985. Initial mixing characteristics of municipal ocean outfall discharges: Vol. I - Procedures and applications, Vol. II - Computer programs. EPA/600/3-85/073a and b, U.S. Environmental Protection Agency, Narragansett, Rhode Island.

Mueller, J.A. and A.R. Anderson. 1983. Municipal sewage systems. Ocean Disposal of Municipal Wastewater: Impacts on the Coastal Environment, Vol. 1, E.P. Meyers and E.T. Hading, editors, Sea Grant College Program, Massachusetts Institute of Technology, Cambridge, Massachusetts, pp. 37-92.

Paul, J.F. 1984a. Empirically-based analysis of the Metropolitan District Commission's settling column data for their 1984 301(h) waiver application. Internal Technical Report, Environmental Research Laboratory, U.S. Environmental Protection Agency, Narragansett, Rhode Island.

Paul, J.F. 1984b. What model is valid for sedimentation estimates? Internal Technical Report, Environmental Research Laboratory, U.S. Environmental Protection Agency, Narragansett, Rhode Island.

Pearson, T.H. and R. Rosenberg. 1978. Macrobenthic succession in relation to organic enrichment and pollution of the marine environment. Oceanography and Marine Biology Annual. Review, 16:229-311.

Tetra Tech, Inc. 1982. Revised Section 301(h) Technical Support Document. EPA 430/9-82-011, U.S. Environmental Protection Agency, Office of Water Program Operation, Washington, D.C.

Tetra Tech, Inc. 1984. Technical Review Report Supplement. Prepared for U.S. Environmental Protection Agency, Washington, D.C., Contract Number 68-01-5906.

U.S. Environmental Protection Agency. 1983. Tentative Decision of the Administrator Pursuant to 40 CFR Part 125, Subpart G, Boston Metropolitan District Commission Deer Island and Nut Island Publicly Owned Treatment Works, Application for Section 301(h) Variance for the Secondary Treatment Requirements of the Clean Water Act, Washington, D.C.

U.S. Environmental Protection Agency. 1985. Tentative Decision Document. Metropolitan District Commission Boston, Massachusetts, Deer Island and Nut Island Publicly Owned Treatment Works, Application for Section 301(h) Modification of the Secondary Treatment Requirments of the Clean Water Act, Boston, Massachusetts.

U.S. Environmental Protection Agency. 1988. Boston Harbor Wastewater Conveyance System, Draft Supplemental Environmental Impact Statement, Volumes I and II, Boston, Massachusetts.

Westerink, J.J., K.D. Stolzenbach and J.J. Connor. 1985. A frequency domain finite element model for tidal circulation. Report No. MIT-EL 85-006, Energy Laboratory, Massachusetts Institute of Technology, Cambridge, Massachusetts.

# CHAPTER 4

# The Flocculation, Deposition, and Resuspension of Fine-Grained Sediments

## Wilbert Lick

## Introduction

Fine-grained sediments are here defined as particles whose diameters are less than 64 µm. These sediments are quite common in lakes, estuaries, and the nearshore areas of the oceans. For example, 90% of the surficial sediments in Lake Erie are fine-grained. Compared to coarse-grained sediments, fine-grained sediments have relatively large surface to mass ratios and therefore have relatively large chemical adsorptive capacities. For these reasons, fine-grained sediments are responsible for much of the contaminant transport in the Great Lakes, in estuaries, and the nearshore areas of the oceans.

Fundamental to a description of the transport of these fine-grained sediments is a quantitative analysis of their deposition and resuspension at the sediment-water interface. This analysis is complicated by two facts, both of which have to be considered in a realistic analysis. (1) Natural sediments are a mixture of particles with a wide distribution of mineralogical properties, shapes, sizes, effective densities, and hence settling speeds. In a typical sediment sample, particle sizes may vary by more than three orders of magnitude while settling speeds may vary by more than four orders of magnitude. (2) A large fraction of bottom sediments is cohesive. Because of this, the aggregation and disaggregation (flocculation) of these particles are important and continually affect the effective sizes, surface areas, densities, settling speeds, and deposition rates of the aggregated particles (flocs) as well as the chemical adsorption/desorption with these flocs.

These two facts must be considered in predicting the transport of particles in the overlying water as well as in understanding the properties of the sediment bed. In the overlying water, the flocculation of particles should be especially important near the sediment-water interface where high shear stresses and high sediment concentrations are present, both of which lead to high rates of aggregation and disaggregation. In addition, the properties of a sediment bed comprised of fine-grained, cohesive sediments will be greatly modified compared to a sediment bed made of uniform-size, non cohesive sediments. In particular, it has been shown by experiment that, at any particular stress, for fine-grained cohesive sediments, only a finite and relatively small amount of sediment can be resuspended while, for uniform-size, non-cohesive sediments (sands), the resuspension rate is approximately constant with time so that the amount resuspended increases with time.

It can be seen from the above that an understanding and quantitative description of the aggregation and disaggregation of fine-grained sediments in the overlying water and in the sediment bed are fundamental to a quantitative description of the sediment flux at the sediment-water interface. Our present understanding of flocculation in the overlying water will be reviewed in the following section. Flocculation obviously affects the effective densities and sizes of particles and hence their settling speeds and deposition rates. Recent work on settling speeds will be discussed in the third section. In the fourth section, recent laboratory investigations on the parameters affecting the properties of the sediment bed and resuspension from this bed will be discussed. Although laboratory investigations are necessary to determine the parameters upon which resuspension depends, it is also necessary for a complete analysis to know the properties of undisturbed sediments *in situ*. A device for measuring these properties and some recent measurements using this device will be described in the fifth section. A summary and concluding remarks are given in the final section.

## Flocculation

A picture of a typical medium-size (about 100 µm in diameter) floc is shown in Figure 1. It can be seen that these flocs are made up of individual particles (on the order of a few microns in diameter) that are relatively closely, and therefore strongly, bound together into clumps. Several of these clumps are then loosely, and therefore relatively weakly, bound to each other to form a floc. It can be shown that, as the flocs increase in size, the effective densities of the flocs decrease, and the resistance of the flocs to disaggregation also decreases.

As mentioned above, the effective sizes and densities of these flocs as well as the adsorption/desorption of contaminants from these flocs are continually being modified by aggregation and disaggregation. The rate of particle aggregation is determined by the rate at which collisions between particles occur and by the probability of cohesion after collision (the collision efficiency factor). Collisions are caused predominantly by three processes: Brownian motion, fluid shear, and differential settling.

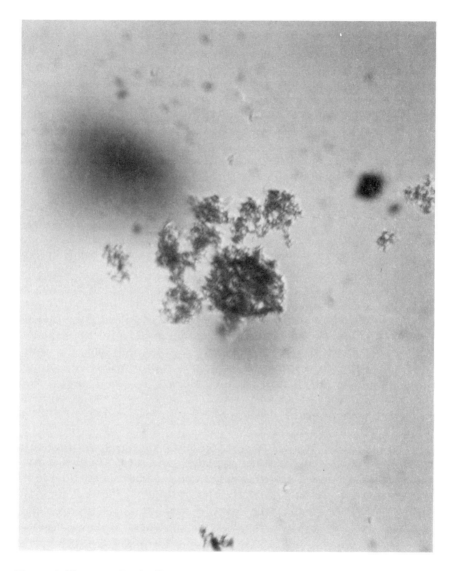

Figure 1. Photograph of a floc approximately 100 μm in diameter.

Brownian motion is due to the thermal energy of the fluid and is random in nature. Fluid shear will lead to collisions due to the relative motion between particles caused by this shear. Particles settling at different speeds will also cause collisions as large, faster-settling particles may collide with small, slower-settling particles.

As particles collide, only a fraction of the collisions will result in cohesion of the particles and the formation or increase in the size of flocs. While experimental determination of the collision efficiency factor for particles in natural aquatic systems is sparse, most available data indicate an approximate value between 0.001 and 0.1 for freshwater systems (O'Melia and Bowman, 1984; Bonner, 1983; Ali *et al.*, 1984; Lick and Lick, 1988) and a somewhat higher value by a factor of two or three for estuarine and sea waters (Edzwald *et al.*, 1974; Farley and Morel, 1986; Burban *et al.*, 1989).

The disaggregation of particles also occurs and is due to fluid shear and, more importantly, due to collisions between particles. The rates of disaggregation of fine-grained particles are not well known and must be determined through experiment. Even the processes that cause disaggregation are not well understood.

Experiments have recently been done on the aggregation and disaggregation of fine-grained sediments in freshwater, sea water, and estuarine water (Tsai *et al.*, 1987; Burban *et al.*, 1991; Xu, 1988). These experiments along with theoretical analyses of these experiments have quantified many of the processes governing the flocculation of fine-grained sediments and will be briefly reviewed here.

Although all of the tests were similar in character, to be specific, the tests on sediments in freshwater will be described. The basic experiments were performed in a horizontal Couette type flocculator. The sediments were natural bottom sediments from the Detroit River inlet of Lake Erie and were prepared by filtering and settling so that the initial (disaggregated) size distributions contained approximately 90% of their mass in particles less than 10 μm in diameter. The size distributions of the resulting suspensions were found to be rather uniform with the median diameters of the disaggregated particles varying from 3.5 to 5.4 μm.

The flocculator consists of two concentric cylinders with the outer cylinder rotating and the inner cylinder stationary. In this way, a reasonably uniform velocity gradient can be generated in the fluid in the annular gap between the cylinders. The width of the gap is 2 mm. The flow is uniform and laminar for shears up to about 900 s$^{-1}$ (approximately 9 dynes/cm$^2$), after which the flow is no longer spatially uniform. Particle size distributions were periodically measured during the tests using a Malvern particle size analyzer model 3600 E. The measurements are based on Fraunhofer diffraction and do not noticeably destroy or modify the flocs produced during the tests. These flocs can be extremely fragile and great care must be taken in the measurement and handling of these flocs so as not to modify their properties.

The sediments were initially disaggregated in a blender and then placed in the flocculator. The flocculator was then operated at a constant shear for about two hours. During this time, the sediments flocculated with the median particle size increasing with time until a steady state was reached. For the typical set of parameters, this always occurred in times less than two hours. Tests were run at shears of 100, 200, and 400 $s^{-1}$ and sediment concentrations of 50, 100, 400, and 800 mg/L.

For a sediment concentration of 100 mg/L and different shears, the results for the median floc diameter as a function of time are shown in Figure 2. At the beginning, there is a period of about 15 minutes during which the median sizes of the particles are small and change relatively slowly. The reason for this is that small particles are relatively inefficient in causing collisions compared with larger particles. As the particles increase in size, the collision rate increases, and the median size changes more rapidly. Still later, as the concentration of large flocs increases, disaggregation becomes more significant. As this happens, the net rate of change of the median diameter slows and eventually goes to zero in the steady-state, where the rate of disaggregation is equal to the rate of

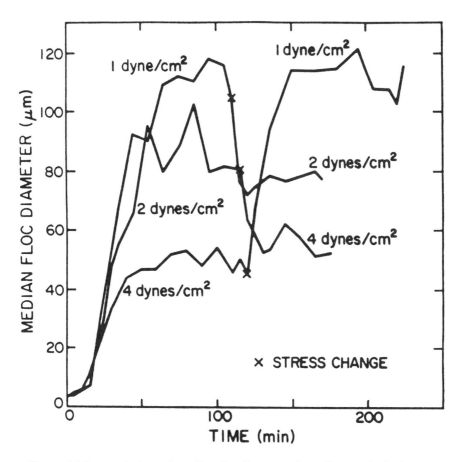

Figure 2. Time variations of median floc diameters for sediments in freshwater at a concentration of 100 mg/l.

aggregation. Fluctuations about this steady-state generally occur and these fluctuations are larger at smaller shears, i.e., at lower collision rates. The time to steady-state is about 50 minutes for a shear of 400 s$^{-1}$. This time increases as the shear decreases and is about 100 minutes for a shear of 100 s$^{-1}$.

After two hours, the applied shear was changed to a new value (100 s$^{-1}$ to 200 s$^{-1}$, 200 s$^{-1}$ to 400 s$^{-1}$, and 400 s$^{-1}$ to 100 s$^{-1}$). From Figure 2, it can be seen that, after the change in shear, a transient occurs and then a new steady-state is reached. For a particular shear, the median particle size for the second steady-state is approximately the same as the median particle size that was obtained at the same shear during the first part of the experiment.

Numerous experiments of this type have been done in freshwater, sea water, and estuarine waters for a range of shears from 100 s$^{-1}$ to 600 s$^{-1}$ and sediment concentrations from 10 mg/L to 800 mg/L. From these experiments, the following can be demonstrated: (1) The steady-state particle size distribution is independent of the manner in which the steady-state is approached. (2) The median floc size decreases as the shear stress increases. (3) The median floc size decreases as the suspended sediment concentration increases (see Figure 3). (4) Particles flocculate faster in sea water than in freshwater. (5) Flocs are smaller in sea water than in freshwater. (6) The sizes of flocs in estuarine waters (a mix of fresh and sea waters) seem to be weighted averages of the sizes of flocs in fresh and sea waters.

Theoretical analyses of these experimental results have also been made (Lick and Lick, 1988; Burban *et al.*, 1989). The analyses follow from a quite general formula for the time rate of change of the particle size distribution. Denote the number of particles per unit volume in size range k by $n_k$. The time rate of change of $n_k$ is then given by

$$\frac{dn_k}{dt} = \frac{1}{2} \sum_{i+j=k} A_{ij}\beta_{ij}n_i n_j - n_k \sum_{i=1}^{\infty} A_{ik}\beta_{ik}n_i$$
$$- B_k n_k + \sum_{j=k+1}^{\infty} \gamma_{jk} B_j n_j \qquad (1)$$
$$- \sum_{i=1}^{\infty} C_{ik}\beta_{ik}n_i + \sum_{j=k+1}^{\infty} \gamma_{jk} n_j \sum_{i=1}^{\infty} C_{ij}\beta_{ij}n_i .$$

The first term on the right-hand side of the above equation is the rate of formation of flocs by cohesive collisions between particles in size i and j. The second term represents the loss of flocs of size k due to cohesive collisions with all other particles. Binary collisions have been assumed. The quantities $A_{ij}$ are the probabilities of cohesion of particles i and j after

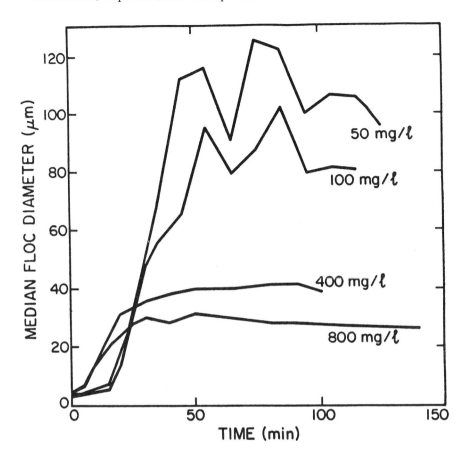

Figure 3. Time variations of median floc diameter for sediments in freshwater for an applied shear of 200 s$^{-1}$.

collision. They cannot be determined at present on the basis of theoretical arguments but have been determined from the experimental results. The quantities $\beta_{ij}$ are the collision frequency functions for collisions between particles i and j. They depend on the collision mechanisms of Brownian motion, fluid shear, and differential settling and are well known (Smoluchowski, 1917; Camp and Stein, 1943; Ives, 1978).

The third term on the right-hand side of equation (1) represents the loss of flocs of size k due to shear while the fourth term represents the gain of flocs of size k due to the disaggregation of flocs of size j>k due to shear. The coefficient $B_k$ should generally be a function of the shear, diameter, and effective density of the floc as well as dependent on the particular sediment. Numerous analyses (Argaman and Kaufman, 1970; Parker et al., 1972; Matsuo and Unno, 1981; Parker, 1982; Clark, 1982; Hunt, 1984) have attempted to determine this quantity from basic theoretical considerations. However, this quantity is still not well understood. The quantity $\gamma_{jk}$ is the probability that a particle of size k will be formed after the disaggregation of a particle of size j. Its value obviously depends on the manner of breakup of the particle of size j. If it is assumed that each floc can break up into two smaller flocs of size i and j such that i+j=k and that there is no preferred mode of breakup, then

$$\gamma_{jk} = \frac{2}{j-1}. \qquad (2)$$

The second to last term on the right-hand side of equation (1) represents the loss of flocs of size k due to collisions with all other particles. The last term represents the gain of flocs of size k after collisions between all particles i and j, where j is greater than k. Binary collisions have been assumed. The quantity $C_{ik}$ is the probability of disaggregation of a particle of size k after collision with a particle of size i. Again, these quantities cannot be determined on the basis of theoretical arguments but have been determined in a few cases from experiments.

It can be shown for fine-grained sediments that the above analysis as written is inadequate. In particular, for the theoretical analysis to approximate the experimental result that the median diameter decreases as the sediment concentration increases, it is necessary that three-body collisions be significant in the disaggregation process. Fluid shear and disaggregation due to binary collisions are insufficient to explain this well-documented experimental result. If three-body collisions are assumed to have a significant effect, then the resulting theory can accurately reproduce the experimental results of (a) mean diameter vs. time and (b) steady-state particle size distribution. In addition, it can be shown that, for the present range of parameters, the theory does not require disaggregation due to fluid shear. Fluid shear seems to have a negligible direct affect on disaggregation while collisions between particles (possibly due to shear but also due to differential settling and Brownian motion) are the dominant mechanism for disaggregation. At lower shears and sediment concentrations where larger and more fragile flocs may occur, this may no longer be true.

The fact that three-body collisions are significant in the disaggregation process may seem rather surprising since three-body collisions are generally rare by comparison with two-body collisions, especially in theoretical analyses. As an independent verification of this idea, it should be noted that large flocs are mostly water, and the largest ones generally are more than 99% water and less than 1% solid particles. Although the volume occupied by the disaggregated particles (in our experiments) is less than 0.1% of the total volume, because of the high water content of the flocculated particles, the volume occupied by the flocs (in the steady-state) is typically a few percent of the total volume. At this volume fraction, three-body collisions are quite probable.

From a comparison of the theory with the experimental results, the coefficients appearing in equation (1) were approximately determined. This allows the experimental and theoretical results to be applied to the prediction of floc sizes in the natural environment. This work, along with experimental work on the settling speeds of these flocs (see following section), then allows a more accurate description and prediction of sediment fluxes to the sediment-water interface.

## Settling Speeds

For a typical sediment, measured particle sizes may vary over three orders of magnitude while settling speeds may vary by more than four orders of magnitude. If a particle is spherical with known density, the diameter d and settling speed $w_s$ are related by Stokes' law (see Figure 4), which is

$$w_s = \frac{gd^2}{18\mu} (\rho_p - \rho), \tag{3}$$

where $g$ is the acceleration due to gravity, $\rho_p$ is the density of the particle, $\rho$ is the density of water, and $\mu$ is the molecular viscosity of water. The formula is valid for settling speeds such that the Reynolds number (Re = $\rho w_s d/\mu$) is less than about 0.5, i.e., for solid sedimentary particles with a diameter less than about 100 µm.

However, most particles are not spherical. More than that, because of flocculation, the average density of a floc is not the same as that of a solid particle, is quite often much less, and generally depends on how the floc was produced. In addition, the flow field is modified due to flow through the floc as well as around the floc. For these reasons, the settling speeds of flocculated particles are not uniquely related to particle sizes as in Stokes' law. Measurements of settling speeds of flocculated fine-grained sediments in fresh and sea waters have recently been made that demonstrate this (Burban et al., 1991), and preliminary results of these experiments will be reviewed here.

The measurements were made using a settling tube, a microscope to observe the flocs, and a camera to record their position as a function of time. The settling speed was obtained by photographing the floc at two different times, measuring the distance between the positions of the floc at these times, and then dividing by the time interval. The main

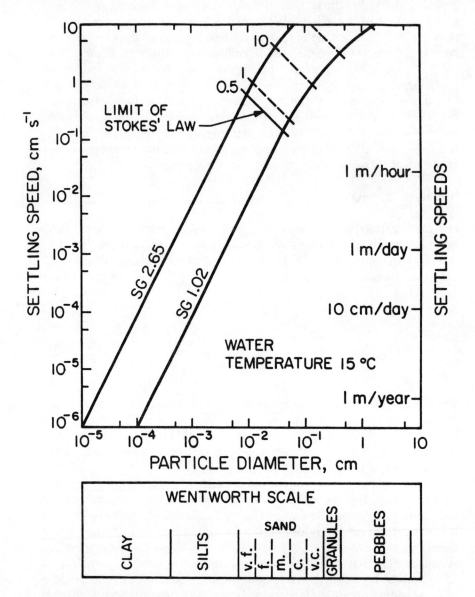

Figure 4. Settling speed as a function of particle diameter, Stokes law.

difficulty in the experiments (because of the low settling speeds) was the elimination of the natural convection currents in the settling tube caused by heat transfer at the sides and top of the tube and by the light source. These currents were essentially eliminated by careful insulation and by the use of a cool light source for only short periods of time. The accuracy of the measurements was checked by using uniform spherical glass beads of standard density and comparing the results with the predictions from Stokes' law. Good agreement was found, thus verifying the procedure.

The flocs whose settling speeds were to be measured were produced in the Couette type flocculator at shears from 100 to 400 s$^{-1}$ and sediment concentrations from 10 to 400 mg/L. Experiments were conducted in fresh and salt waters. A typical set of measurements of settling speeds for sediments in freshwater as a function of floc diameter for flocs produced at a sediment concentration of 100 mg/L and at shears of 100, 200, and 400 s$^{-1}$ is shown in Figure 5. It can be easily seen that almost all of the settling speeds are considerably less than that predicted by Stokes' law, some by more than a factor of 100. On closer examination, it is also clear that the settling speed of a floc depends on the conditions under

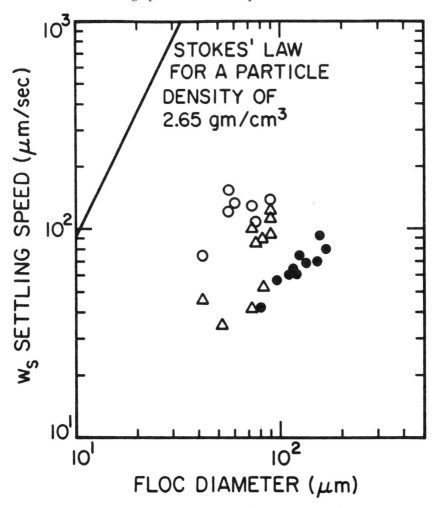

Figure 5. Experimental results for settling speed as a function of particle diameter. Freshwater and a sediment concentration of 100 mg/l. • 100 s$^{-1}$. ∆ 200 s$^{-1}$. O 400 s$^{-1}$.

which it was produced, i.e., flocs produced at low shears have lower settling speeds and therefore lower effective densities than do flocs produced at higher shears. Figure 6 shows a similar plot of settling speeds as a function of floc diameter but with flocs produced at a shear of $400 \text{ s}^{-1}$ and at sediment concentrations of 10, 100, and 400 mg/L. From this figure, it can be seen that flocs produced at low sediment concentrations have lower settling speeds, and hence lower effective densities, than do flocs produced at higher sediment concentrations. The variations shown in Figures 5 and 6 are characteristic of the entire range of parameters tested, both for fresh and salt waters.

From these measurements, the settling speed can be related to the floc

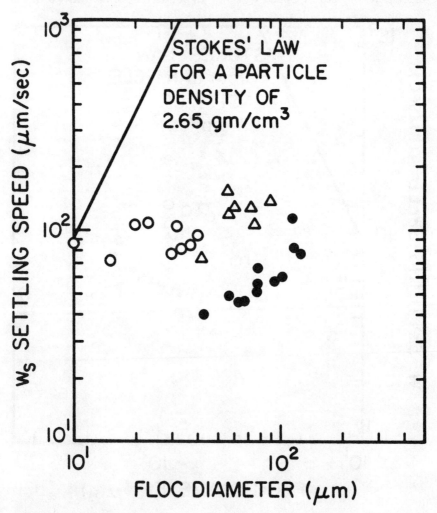

Figure 6. Experimental results for settling speed as a function of particle diameter. Freshwater and a shear of $400^{-1}$. • 10 mg/l. △ 100 mg/l. ○ 400 mg/l.

diameter with fluid shear and sediment concentration as parameters. An effective density can then be defined such that Stokes' law, equation (3), is satisfied. This relation between the density difference, $\rho_p - \rho$, and the diameter d can be approximated by

$$\rho_p - \rho = ad^m . \tag{4}$$

In previous studies by Hawley (1982) and Gibbs (1985), this relation was used and $a$ and $m$ were assumed to be constants independent of shear and concentration. For example, Hawley determined $m$ to be -0.97 while Gibbs found $m$ to be -1.22. In our studies, $m$ ranged from -1.01 (for a concentration of 400 mg/L and a shear of 400 s$^{-1}$) to -1.82 (for a concentration of 10 mg/L and a shear of 100 s$^{-1}$), i.e., $m$ (and $a$) depended on fluid shear and sediment concentration. In tests so far, these parameters seemed to be only weakly dependent on salinity with the values for $m$ slightly higher for salt water than for freshwater.

## Laboratory Measurements of Resuspension

The resuspension rate is significantly affected by particle size variations and also by cohesion between particles. At low stresses, only the finer particles on the surface of the bed can be resuspended while the larger particles are left behind and armor the bed. In addition, cohesion of particles and the resulting compaction of the bed cause the resuspension rate to vary with time after deposition and with depth. Because of this, sediments near the surface are less compacted and are relatively easy to resuspend, while sediments further down are more compacted and more difficult to resuspend. The result of all this is that, for fine-grained sediments at any particular stress, only a finite and relatively small amount of sediment can be resuspended as opposed to noncohesive, uniform-size sediments, which have a uniform rate of resuspension (Massion, 1982; Tsai and Lick, 1987).

Experimental work (Partheniades, 1972; Krone, 1962; Metha, 1973; Lee et al., 1981; Lick and Kang, 1987; MacIntyre et al., 1990; Tsai and Lick, 1987) has determined the dependence of the resuspension rate, E, and the total amount of sediment, $\varepsilon$, that can be resuspended at a particular stress as a function of (a) the turbulent stress at the sediment-water interface, and (b) the water content of the deposited sediments (or the time after deposition) for various sediments from both lakes and oceans. Some results for a sediment from the western basin of Lake Erie are shown in Figure 7. Shown is the net amount of sediment resuspended per unit area of bottom as a function of bottom stress, with time after deposition as a parameter. It can be seen that resuspension is a strong

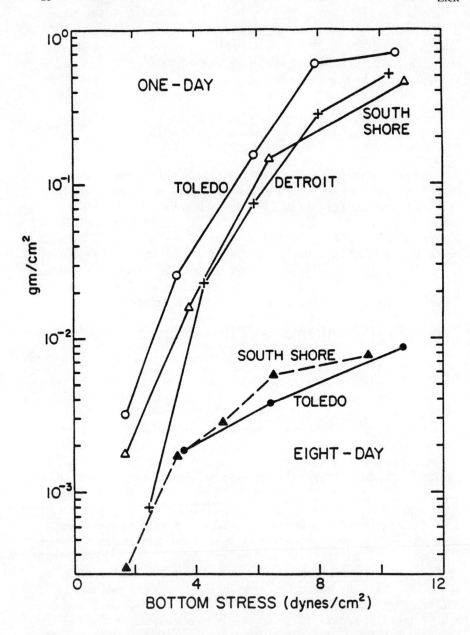

Figure 7. Net sediment resuspended per unit area of bottom sediment as a function of bottom stress with time after deposition of a parameter.

function of bottom stress and time after deposition. From other experiments of this type, it can be demonstrated that bed compaction (and the associated reduction in resuspension) is most rapid immediately after the bottom sediments are deposited. After 8 days, the compaction is rather slow and the results for 14 days, although lower, do not differ greatly from those for 8 days.

A formula for $\varepsilon$ which approximates the experimental data can be written as

$$\varepsilon = \frac{a_0}{t_d^n}(\tau - \tau_0)^m \quad , \quad \text{for } \tau \geq \tau_0 , \tag{5}$$
$$= 0 \quad , \quad \text{for } \tau < \tau_0$$

where $\varepsilon$ is the net amount of resuspended sediment per unit surface area in gm/cm$^2$, $a_0$ is approximately equal to $8 \times 10^{-3}$, both $n$ and $m$ are approximately equal to 2, $t_d$ is the time after deposition in days, $\tau$ is the shear stress (dynes/cm$^2$) produced by wave action and currents, and $\tau_0$ is an effective critical stress generally on the order of 1 dyne/cm$^2$. Each of the parameters $\tau_0$, $a_0$, $n$, and $m$ is dependent on the particular sediment (and the effects of benthic organisms) and needs to be determined experimentally. However, for fine-grained sediments, the values given above are a reasonably accurate first approximation.

The above formula is for the net resuspension. The total amount of sediment is not resuspended instantaneously but over a period of time on the order of an hour. In numerical computations, a reasonable approximation to the resuspension rate is that it is constant and equal to its initial value until all available sediment is resuspended and is then zero until further sediment is deposited and is available for resuspension.

The properties of the sediment bed are further modified by bacteria and benthic organisms. Their effects are quite diverse. Both bacteria and benthic organisms can bind the sediments due to mucus production within the sediments, at the sediment-water interface, and on burrow walls. Benthic organisms form fecal pellets that are generally larger than the nonpelletized sediments particles, but benthic organisms also burrow, stirring the sediments, disrupting the existing sediment fabric, and changing the water content. Preliminary work (Schink and Guinasso, 1977; Robbins et al., 1977; Rhoads et al., 1978; Fisher et al., 1980) has been conducted on these problems, but quantitative information on the effects of organisms on resuspension is not yet available. Recent work (Tsai and Lick, 1987) has demonstrated the effects of Nucula clams on the resuspension properties of sediments from Long Island Sound. In particular, it has been shown that the resuspension properties of sediments with clams are relatively constant with time after an initial transient of three to four days. In contrast, sediments without benthic organisms will continue to compact with time so that their resuspension properties change continuously with time. It is believed that other benthic organisms can affect sediment properties in a similar manner.

## *In Situ* Measurements of Resuspension

Laboratory experiments have determined the dependence of resuspension on various governing parameters, such as the applied shear stress, water content or time after deposition, particle size distribution, mineralogy, and number and types of benthic organisms. However, the state of the sediments *in situ* is not well known, and so the art of extrapolating laboratory results to the field is not well developed. A further disadvantage of the usual laboratory resuspension experiments is that they are cumbersome and very time-consuming.

In order to alleviate some of these difficulties, a portable device for the rapid measurement of sediment resuspension (called shaker hereafter) was developed (Tsai and Lick, 1987). It can be used in the laboratory for quick and reasonably accurate measurements of resuspension but, more importantly, it can be used on board a ship for rapid surveys of the resuspension of relatively undisturbed sediments.

The basic shaker consists of a cylindrical chamber inside of which a horizontal grid oscillates vertically (see Figure 8). The sediments whose properties are to be determined are placed at the bottom with water overlying these sediments. The grid oscillates in the water and creates turbulence that penetrates down to the sediment-water interface and causes the sediment to resuspend. The turbulence, and hence the amount of sediment resuspended, is proportional to the frequency of the grid oscillation.

The equivalent shear stresses created by the oscillatory grid were determined by comparison of results of resuspension tests in the shaker and a calibrated laboratory flume. The basic idea of the calibration is that, when the flume and shaker produce the same concentration of resuspended sediments under the same environmental conditions, the stresses needed to produce these resuspended sediments are equivalent.

The results of the calibration are shown in Figure 9, which shows a plot of the equivalent shear stress as a function of the period of oscillation of the shaker. Shown are the results of 49 tests for sediments from the western basin of Lake Erie, from Long Island Sound, and from the Venice Lagoon in Italy. Tests were done for two days and seven days after deposition. Although all of the sediments were fine-grained, they were quite different in mineralogy, water content, and resuspension rates. Nevertheless, all points fall reasonably close to a straight line in Figure 9. This demonstrates that the results are reproducible and, most importantly, that the equivalent shear stress produced by the shaker is independent of the sediments used for calibration. It can also be seen that all of the test results fall within 20% of the line denoting the average.

The shaker configuration shown in Figure 8 is the same as that used

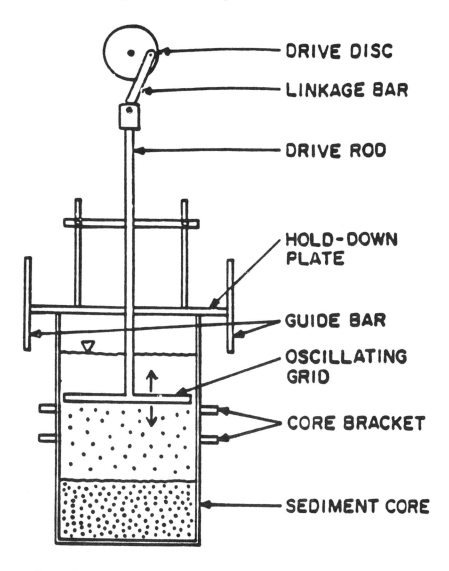

Figure 8. A schematic of the portable device for rapid measurement of sediment resuspension (shaker).

in the laboratory. Simple modifications allow it to be used in the field, where a diver can obtain undisturbed sediments. The basic modification is to replace the cylinder, which has a glued bottom plate, with an open cylinder. In order to obtain a sediment sample, the diver gently pushes the open cylinder into the sediments, covers the top of the cylinder, raises the cylinder out of the sediment bed with the undisturbed sediments inside the tube, and then covers the bottom of the cylinder to prevent the

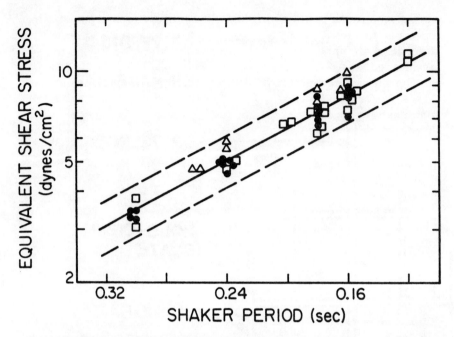

Figure 9. Equivalent shear stresses for the shaker as a function of the period of the grid oscillation.

sediments from falling out as the cylinder is raised to the surface. In field tests, the diver had no difficulty in obtaining samples and preserving an undisturbed sediment-water interface. For heavily polluted waters or when there is no diver available, the cylinder (with further modifications) can be attached to a pole and used to obtain undisturbed sediment cores from a ship without a diver (Tsai and Lick, 1986; Ziegler et al., 1987; Tsai and Lick, 1987).

## Summary and Concluding Remarks

Recent work on the flocculation, deposition, and resuspension of fine-grained sediments has been reviewed and discussed. These topics were chosen because they are basic to an understanding and prediction of the sediment and contaminant flux at the sediment-water interface, the topic of this volume.

Our present understanding of flocculation was reviewed first because it is fundamental to the processes of deposition and resuspension. From both experimental and theoretical work, a first approximation to the rates of aggregation and disaggregation of fine-grained sediments in fresh, ocean, and estuarine waters has been obtained. More importantly, a consistent and accurate technique for doing this has been developed and demonstrated. Only a limited range of fluid shears and sediment concentrations has been investigated. This range needs to be extended, especially to lower shears and sediment concentrations where much larger flocs are anticipated. Other types of sediments also need to be investigated.

The present model of flocculation is very inefficient since it treats each floc size independently. As a consequence, a large amount of computer time is needed to solve each problem. This model can be simplified by lumping flocs of approximately equal size together. This is especially necessary so that these flocculation processes can be included in transport calculations, where a flocculation calculation must be done at each of many grid points. This model simplification is presently being accomplished.

An important reason for understanding flocculation is because of its effect on settling speeds, a topic reviewed in the third section of the present paper. The settling speeds of flocs are typically one to two orders of magnitude lower than those prediced by Stokes' law for solid particles of the same size. Because of recent experimental work, a better understanding of settling speeds and the parameters on which they depend has now been obtained. It has been clearly demonstrated that the effective densities, and hence settling speeds, of flocs are affected by the conditions in which they are produced, in particular, by the shear stress and sediment concentration. Flocs produced at low concentrations and low shears have lower effective densities, and hence settling speeds, than do flocs produced at higher concentrations and shears. However, the particular dependence of settling speeds on diameter, shear, and concentration is not yet clear. A wider range of these parameters needs to be investigated before this dependence can be accurately determined.

Experimental work on resuspension has also been reviewed. An essential characteristic of fine-grained sediments is that, at any particular stress, only a finite and relatively small amount of sediment can be resuspended as opposed to non cohesive uniform-size sediments which have a uniform rate of resuspension. The parameters on which resuspension of fine-grained sediments depends have also been determined. A formula that approximates these data has been developed and is being used in modeling studies. However, the experimental data is only for low to moderate shear stresses and needs to be extended to higher stresses that occur during storms, for example.

Although the limiting cases of fine-grained, cohesive sediments and coarse-grained, non cohesive sediments are reasonably well understood, the transition between these two limiting cases and the parameters on which this transition depends are not understood. The transition obviously depends on particle size, but it seems that the distribution of particle size as well as the median size may be important. Resuspension parameters probably are not unique functions of median particle size. This needs much additional investigation before the resuspension of sediments can be adequately understood and predicted. Preliminary work on the effects of benthic organisms on resuspension has been done, but a more systematic and thorough investigation is necessary.

Although laboratory studies of resuspension are useful in order to determine the dependence of resuspension on various parameters, it is also necessary to know the state of the undisturbed bottom sediments. For this purpose, a portable device (shaker) has been developed and can be used in the laboratory as well as in the field to obtain reasonably accurate measurements of resuspension.

The processes described above are the basic components that are necessary for the understanding and prediction of the sediment and contaminant fluxes in lakes, oceans, and estuaries. It has been demonstrated that these processes are reasonably well understood. However, more work is needed to describe these processes accurately over a wider range of parameters. Of particular interest are the sediment and contaminant flux in regions where high rates of aggregation and disaggregation may occur, e.g., at the sediment-water interface and in the region of the thermocline. With the knowledge summarized above, these problems can now be more quantitatively understood.

Although additional laboratory experiments are necessary, it is also necessary to verify these processes by means of field tests. A systematic field test incorporating measurements of resuspension, deposition, and flocculation is needed. Modeling of these processes and of the overall transport of the sediments should be an inherent part of the field program.

Finally, our understanding of the processes of resuspension and deposition should be used to better understand the chemical release of contaminants as affected by resuspension and deposition. This should have a significant effect on the flux of contaminants at the sediment-water interface.

## Acknowledgments

This research was supported by the United States Environmental Protection Agency. Dr. Louis S. Swaby and Dr. Anthony Kizlauskas were the project officers.

## References

Ali, W., C.R. O'Melia and J.K. Edzwald. 1984. Colloidal stability of particles in lakes: measurement and significance. Water Sci. and Tech., 17:701-712.

Argamann, Y. and W.J. Kaufman. 1970. Turbulence and flocculation. J. Sanitary Eng. Div., 96:223-241.

Bonner, J.S. 1983. The vertical transport and aggregation tendency of freshwater phytoplankton. Ph.D. Dissertation, Clarkson University, Potsdam, NY.

Burban, P.Y., W. Lick and J. Lick. 1989. The flocculation of fine-grained sediments in estuarine waters. J. Geophys. Res., 94(C6):8323-8330.

Burban, P.Y., Y. Xu, J. McNeil and W. Lick. 1991. Settling speeds of flocs in fresh and sea waters. J. Geophys. Res., 95(C10):18213-18220.

Camp, T.R. and P.C. Stein. 1943. Velocity gradients and internal work in fluid motion. J. Boston Soc. Civil Eng., 30:219-237.

Clark, M.N. 1982. Discussion of forces acting on floc and strength of floc. J. Environ. Eng. Div., 108(EE3):592-594.

Edzwald, J.K., J.B. Upchurch and C.R. O'Melia. 1974. Coagulation in estuaries. Environ. Sci. Tech., 8:58-63.

Farley, K.J. and M.M. Morel. 1986. Role of coagulation in the kinetics of sedimentation. Environ. Sci. Tech., 20:2.

Fisher, J.B., W.J. Lick, P.L. McCall and A. Robbins. 1980. Vertical mixing of lake sediments by tubificid oligochaetes. J. Geophys. Res., 85:3997-4006.

Gibbs, R.J. 1985. Estuarine flocs: their size, settling velocity, and density. J. Geophys. Res., 40:3249-3251.

Hawley, N. 1982. Settling velocity distributions of natural aggregates. J. Geophys. Res., 87(C12):9489-9498.

Hunt, J.R. 1984. Particle aggregate breakup by fluid stress. Workshop on Cohesive Sediment Dynamics, Tampa, Florida.

Ives, K.J. 1978. Rate theories. In: The Scientific Basis of Flocculation, K.J. Ives (ed.), Sijthoff and Noordhoff International Publishers, B.V., Alphensan den Rijn, The Netherlands, pp. 37-61.

Krone, R.B. 1962. Flume studies of the transport of sediment in estuarial shoaling processes. Final Report to San Fransisco District, U.S. Army Corps of Engineers, Washington, DC.

Lee, D.Y., W. Lick and S.W. Kang. 1981. The entrainment and deposition of fine-grained sediments in Lake Erie. J. Great Lakes Res., 7:264-275.

Lick, W. and S.W. Kang. 1987. Entrainment of sediments and dredged materials in shallow lake waters. J. Great Lakes Res., 13(4):619-627.

Lick, W. and J. Lick. 1988. On the aggregation and disaggregation of fine-grained lake sediments. J. Great Lakes Res., 14(4):514-523.

MacIntyre, S., W. Lick and C.H. Tsai. 1990. Entrainment of cohesive riverine sediments. Biogeochemistry, 9:187-209.

Massion, G. 1982. The resuspension of uniform-size fine-grained sediments. M.S. Thesis, University of California, Santa Barbara, CA.

Matsuo, T. and H. Unno. 1981. Forces acting on floc and strength of floc. J. Environ. Eng. Div., 107(EE3):527-545.

Metha, A.J. 1973. Depositional behavior of cohesive sediments. Ph.D. Thesis, University of Florida, Gainesville, FL.

O'Melia, C.R. and K.S. Bowman. 1984. Origins and effects of coagulation in lakes. Schweizerische Zietschrift fur Hydrologie, 46:64-85.

Parker, D.S., W.J. Kaufman and D. Jenkins. 1972. Floc breakup in turbulent flocculation processes. J. Sanitary Eng. Div., 98:79-99.

Parker, D.S. 1982. Discussion of forces acting on floc and strength of floc. American Society of Civil Engineers. J. Environ. Eng. Div., 108(EE3):594-598.

Partheniades, E. 1972. Results of recent investigations on erosion and deposition of cohesive sediments. In: Sedimentation, H.W. Shen (ed.), Fort Collins, Colorado, pp. 20-1 to 20-39.

Rhoads, D.C., J.Y. Yingst and W.J. Ullman. 1978. Seafloor stability in central Long Island Sound. In: Estuarine Interactions, M. Wiley (ed.), Academic Press, New York, pp. 211-244.

Robbins, J.A., J.R. Krezoski and S.C. Moxley. 1977. Radioactivity in sediments of the Great Lakes: Post-depositional redistribution by deposit feeding organisms. Earth Planetary Sci. Lett., 36.

Schink, D.R. and N.L. Guinasso. 1977. Effects of bioturbation on sediment-seawater interaction. Marine Geol., 23.

Smoluchowski, M. 1917. Versuch einer mathematischen theorie der koagulations-kinetic kolloid losungen. Zietschrift fur Physikalische Chemi, 92:129-168.

Tsai, C.H. and W. Lick. 1986. A portable device for measuring sediment resuspension. J. Great Lakes Res., 12(4):314-321.

Tsai, C.H. and W. Lick. 1987. Resuspension of sediments from Long Island Sound. Proceedings of Conference on Coastal and Estuarine Pollution, Fukuoka, Japan.

Tsai, C.H., S. Iacobellis and W. Lick. 1987. Flocculation of fine-grained lake sediments due to a uniform shear stress. J. Great Lakes Res., 13(2):135-146.

Xu, Y.J. 1988. The flocculation of bentonite and barite in sea water. M.S. Thesis, University of California, Santa Barbara, CA.

Ziegler, C.K., C.H. Tsai and W. Lick. 1987. The resuspension, deposition, and transport of sediments in the Venice Lagoon. UCSB Report ME-87-3, University of California, Santa Barbara, CA.

# CHAPTER 5

# *In Situ* Measurement of Entrainment, Resuspension, and Related Processes at the Sediment-Water Interface

Keith W. Bedford

## Introduction

The resuspension and entrainment of bottom sediments is a difficult quantity to measure and an even more difficult quantity for managers to risk including in control or management strategies. Being driven basically by flow field physics, entrainment responds to a variety of different forcing functions and is quite dynamic. Recent instrumentation assemblages purport to measure resuspension and aspects of the associated transport. Yet the fact remains that there is no instrument available for making direct measurements of entrainment from the bed; it must be inferred.

The objective of this chapter is to assemble the frameworks for estimating entrainment and resuspension and, by drawing on various statistical measures for assessing error and variability, to evaluate the minimal instrumentation necessary to make each method operational. Of necessity, aspects of instrumentation suitability are discussed, but it is not the author's intent to evaluate the merits of individual transducers.

The material in this article is abstracted from a report (Bedford, U.S. Army Corps of Engrs. Wat. Exp. Sta., to appear) prepared by this author whose purpose is to present the frameworks and discuss the detailed aspects of the "field" implementation of each method. Space limitations prevent considerable detail on certain theoretical and implementation aspects from being presented here, and the reader is urged to obtain the report for these operational aspects.

## The Setting, Basic Equations, and Flux Definitions

Entrainment and deposition of bottom sediments are influenced primarily by fluid mechanical processes in the overlying water column and the composition and distribution of the material comprising the bottom. The purpose of this section is to summarize briefly their features and thereby describe the setting to which the instrumentation and subsequent measurements must apply.

## The Fluid Mechanical Setting

A considerable body of knowledge about the fluid mechanical processes at the bottom is available, particularly as regards the momentum field and corresponding boundary layers. Comprehensive papers about bottom flow processes include those by McCave (1976), Nihoul (1977), Nowell (1983), Nowell and Hollister (1985), Bechteler (1986), the works on the HEBBLE Project in *Marine Geology* (Hollister and Nowell, 1985), and the conference proceedings from the AGU Chapman conference in Argentina on cohesive sediment transport. A number of review articles have been written including Bowden (1978), Gust (1984), Grant and Madsen (1986), and Bedford and Abdelrhman (1987). An extremely important seminal article was written by Smith (1977) and in many ways his paper sets forth a number of the analytical concepts used to describe or predict bottom flow behavior to this day. Bioturbation and colloidal effects will not be included in this brief summary.

Characterization of bottom boundary layer (BBL) physics is quite challenging, principally due to the extremely complex interaction of the fluid mechanical forcing functions that potentially exist at each site. These forcing functions operate over a very large range of space and time scales and are generally classified (Bedford and Abdelrhman, 1987) as: (1) oscillatory motions composed of long- and short-period surface gravity waves and/or short- and long-period internal waves; (2) steady currents or mean flow; (3) rotation or Coriolis effects; (4) turbulence resulting from shear, separated flows, bottom roughness, wave breaking, etc.; and (5) stratification as possibly induced by salinity, temperature, or suspended sediment concentration. A schematic of the atmospheric and water column factors affecting the BBL appears in Figure 1 (Bedford and Abdelrhman, 1987).

In attempting to organize this complexity, analysis and interpretation have been commonly based upon three approaches: (1) empirical descriptions that are by and large problem and/or site specific; (2) the boundary layer approach, which is a predictive generalization based upon the filtering out of small scale effects; and (3) attempts based upon the laboratory experiments by, among others, Praturi and Brodkey (1978) Offen and Kline (1968), and Wallace et al. (1977) to understand the contributions of the intermittent small scale but intense effects on BBL structure.

<u>Scales of Activity.</u> It is important for each site/application to know

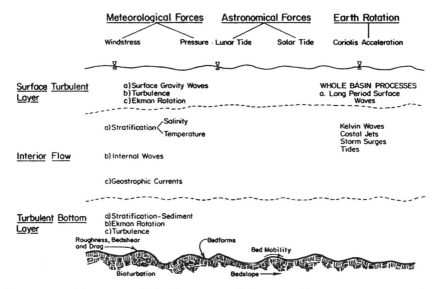

Figure 1. Schematic of the bottom and factors affecting transport and resuspension (Bedford and Abdelrhman, 1987).

the forcing function climatology. Table 1 (from Abdelrhman and Bedford, 1987 and Bedford, 1990, as adapted from the original version by Boyce, 1974) contains a summary of most of the important water column physical processes, a summary of the forces producing each process, and estimates of horizontal, vertical, and time scales. It is apparent in reviewing this table that the motions range from large to small in both spatial extent or time scale; it is also the case that the magnitude (amplitude) of larger scale processes is considerably larger than that for the small scale processes. A common procedure for showing the relationship between process variance ("magnitude squared") and process scale is via spectral analysis, and the power spectrum expected for typical surface water data is reasonably well known. Figure 2 contains a schematic of a typical spectral shape. It is noticed that at small wavenumber (low frequency), variability is quite powerful, anisotrophic, and controlled by geometry. The variability continuously extends over as much as eight decades of wavenumber space and is expended by viscosity at the molecular range. For isotropic homogeneous turbulence, there is an inertial subrange characterized by among several things a -5/3 slope dependency on wavenumber. In this region, the production and dissipation of turbulence are balanced. The appearance of this subrange in the data has considerable impact on data analysis as a number of verified theoretical results are known about fluxes, turbulence, and velocity and sediment profiles for these conditions.

As noted in the diagram, a spectral spike (or several spikes) might appear in the data. These spikes, should they be significant, represent

## Table 1

### Typical Bottom Boundary Layer Parameters (Bedford and Abdelrhman, 1987)

| Parameter | Deep Sea | Continental Shelf | Reference | Lake Erie Calm | Lake Erie Stormy |
|---|---|---|---|---|---|
| Bottom Ekman layer thickness (mid latitudes) | 10 m<br>10<br>5 | 50 m<br>100<br>50 | Gust (1982)<br>Bowden (1978)<br>Wimbush & Munk (1970) | ~60 m | ~20 m |
| Logarithmic layer thickness | 1-2 m<br>1<br>1 | 5 m<br>10<br>10 | Gust (1982)<br>Bowden (1978)<br>Wimbush & Munk (1970) | ~2 m | ~20 m |
| Viscous sublayer thickness | 1 cm<br>1<br>2 | 1 cm<br>0.1 | Gust (1982)<br>Bowden (1978)<br>Wimbush & Munk (1970) | 1 cm | 0.1 cm |
| Conductive sublayer thickness | 1 cm | | Wimbush & Munk (1970) | 1 cm | 1 cm |
| Diffusive sublayer thickness | 0.2 cm<br>0 (1 mm) | | Wimbush (1976)<br>Boudreau & Guinasso (1982) | 0 (1 mm) | 0 (1 mm) |
| Wave viscous boundary layer thickness | | 2 mm | Smith (1977) | --- | 1 mm |
| Time averaged horizontal velocity | 0.1-5 cm/sec<br>4<br>3 | 1-50 cm/sec<br>40<br>30 | Gust (1982)<br>Bowden (1978)<br>Wimbush & Munk (1970) | 1-5 cm/sec | 10-20 cm/sec |
| Free stream velocity | 3 cm/sec | | Wimbush & Munk (1970) | ~5 cm/sec | ~5 cm/sec |
| Friction velocity | 0.04-0.8 cm/sec | 0.4-2 cm/sec | Gust (1982)<br>Bowden (1978)<br>Wimbush & Munk (1970) | | |
| Roughness length | 1 cm | 5 cm | Gust (1982) | 5-10 cm | |
| Drag coefficient | $10^{-3}$<br>$3.1 \times 10^{-3}$ | $10^{-3}$ | Gust (1982)<br>Sternberg (1972) | $1.2 \times 10^{-3} - 2.5 \times 10^{-3}$ | |
| Bed shear stress | $0 (10^{-2}$ dyne/cm$^2)$ | $0 (1$ dyne/cm$^2)$ | | $0 (10^{-2}$ dyne/cm$^2)$ | $0 (1$ dyne/cm$^2)$ |
| Characteristic eddy viscosity | 2 cm$^2$/sec<br>1 | 200 cm$^2$/sec<br>100 | Wimbush & Munk (1970)<br>Boudreau & Guinasso (1982) | 40 cm$^2$/s | |
| Dissipation rate of turbulent kinetic energy (constant stress layer) | $0 (10^{-3}/z)$ watt/unit mass | $0 (1/z)$ watt/unit mass | | $0 (10^{-3}/z) 0(1/z)$ | |
| Molecular viscosity (Kinematic viscosity) | $1.8 \times 10^{-2}$ cm$^2$/sec<br>$1.5 \times 10^{-2}$ | | Boudreau & Guinasso (1982)<br>Chriss & Caldwell (1982) | SAME | |
| Von Karman's Constant | 0.41 | | Nowell (1983) | SAME | |
| Diffusion coefficient of salt | $2 \times 10^{-5}$ cm$^2$/sec | | Bowden (1978) | Does not apply | |
| Diffusion coefficient of heat | $1.4 \times 10^{-3}$ cm$^2$/sec | | Bowden (1978) | SAME | |
| Adiabatic temperature | $10^{-6}$ C°/cm | | Wimbush & Munk (1970) | Does not apply | |
| Bioturbation zone | 0 (10 cm) | | Gust (1982) | SAME | |
| Constant stress layer thickness | 10 cm | | Wimbush & Munk (1970) | 0 (10 cm) | 0 (1 m) |
| Kolmogorof length (logarithmic scale) | 0.5 cm | | Wimbush & Munk (1970) | 0.1-0.5 cm | |
| Monin Obukhov length scale | 100 m | | Wimbush & Munk (1970) | 10-20 m | |

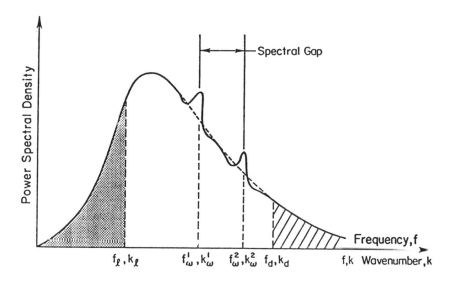

Figure 2. Schematic of variance spectrum and various sampling considerations.

organized wave motions by wave groups with a dominant frequency, $f_w$. The conditions by which Taylor's frozen turbulence hypothesis can be used to infer wavenumber spectra from wave-affected frequency spectra have been reviewed by Lumley and Terray (1983).

Each site in question is never affected by a persistant set of conditions; rather the suite or mix of physical processes changes at each site roughly in response to seasonal and local atmospheric phenomena (e.g., storms). Therefore, it is necessary to catalog the various processes and their successions. Bedford and Abdelrhman (1987) and Bedford (1990) present such an attempt with the intent being able to separate marine and freshwater systems principally by the presence of tidal activity and salinity-induced density gradients.

A First Look at Boundary Layers. As mentioned, there are three analytical approaches used to organize the BBL information. The boundary layer approach, a predictive methodology, is based upon Reynolds temporal average of the Navier Stokes equations and seeks solutions for simplified forcing functions. There are a considerable number and variety of such solutions, each distinguished by the relative importance of one or two such forcing functions over others. A method for categorizing the various layers was proposed by Bowden et al. (1976) and Bowden (1978) where, by use of separately defined length scales, segregation was achieved among rotation, stratification, and oscillatory forcings; Figure 3 contains a summary of the hypothesized layers along

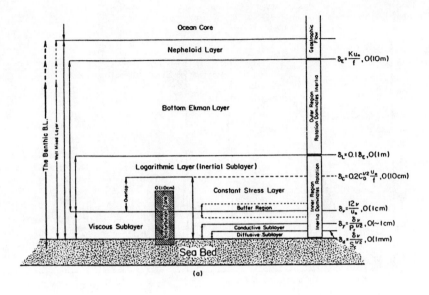

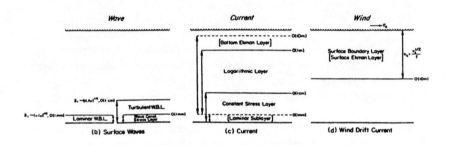

Figure 3. Boundary layer regimes and heights (Bedford and Abdelrhman, 1987).

with estimates of the thickness, while Table 1 contains a summary of observed values for the various measures of these layers. Obviously not all the layers in Figure 3 exist simultaneously and, therefore, a number of spatial cases are noted at the bottom of the figure. For the information in Table 1 and Figure 3, it is the case that depth plays a particularly important role in the various special cases of boundary layers. If the water is of great depth, then Wimbush and Munk (1970) suggest that the full bottom Ekman layer will develop to a height above bottom of $ku_*/f$, where k is von Karman's coefficient, $u_*$ is the shear velocity, and $f_c$ is the Coriolis parameter ($10^{-4}$ sec$^{-1}$). At such depths the boundary layer similarity scales are the free stream velocity or shear velocity and boundary layer thickness (Nowell and Church, 1979), In oscillatory flows with frequency, f, the similarity length scale is $u_*/f$. For stratified flows, Weatherly and Martin (1978) suggest that if N is the Brunt Vaisala frequency then $h = Au_*/f$ for $N<<f$ and $Au_*/\sqrt{(fN)}$ for $N>>f$. Often it is the case in estuaries, rivers, lakes, etc., that these depths are not achieved, in which case the full development of the BBL is limited. In such depth limited cases, Nowell and Church (1979) suggest that the local flow depth becomes the similarity length scale, while mentioning that the choice of velocity scale is not clear. Gieskes et al. (1976) suggests that in such flows the boundary layer height is the minimum elevation to which potential flow can be matched.

## Basic Equations

In order to understand the basis for entrainment measurement as well as to aid in understanding general BBL theory, it is necessary to summarize the basic equations. It is assumed that the flow is turbulent, but allowance for laminar flow for low Reynolds number is made. The bottom is assumed to be rough with four regimes defined: (1) a horizontal bed where roughness is due to sand grains; (2) weak bed forms (i.e., ripples) on a horizontal bed or roughness elements where roughness density (Morris, 1955; Nowell and Church, 1979) ranges over the three types of flow defined in these references (i.e., skimming, wake interaction, and isolated roughness); (3) a macro bedform regime (i.e., dunes), whose scale is quite large in contrast to the boundary layer sampling scale; and (4) a bed transport regime where the bedforms are washed out. Antidunes and peculiar nearshore bed features are not considered. Macrobedforms are considered, but only in so far as configuration information must be known, and momentum and sediment advection are accounted for in the basic equations.

Based upon these very general roughness categories, a coordinate system is defined (Figure 4), with the horizontal axes being placed on the bottom and the vertical axis positive up and aligned with gravity. For roughness cases (1) and (2) above, the horizontal location is the average location of the bottom in the manner of Yalin (1975). The bottom then is locally flat and, therefore, consideration of a "sloped" bottom such as a river is excluded; although it is possible to include this effect.

In the special case of boundary layer flows, it is common to perform a two-dimensional analysis with the vertical axis defined as above and the horizontal axis being defined as the axis coinciding with the velocity streamline. For a non flat bottom, particularly one containing a large undulation, the local streamline path must be parallel to the "bottom" for proper application of the boundary layer approximations. A two-dimensional coordinate system is then constructed with regard to this streamline. However, it is noted that the resulting local vertical axis is

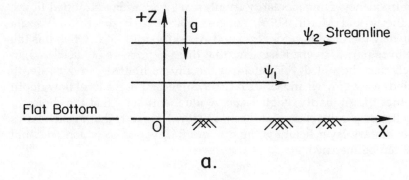

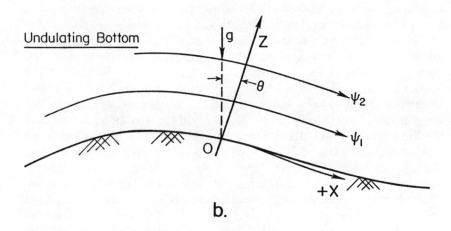

Figure 4. Coordinate system definitions for basic equations.

not coincident with gravity and, therefore, gravitational particle settling will contain both vertical and horizontal components in the local streamline coordinate system. Such difficulties place a considerable burden on the analyst and instrumentation. Local instrumentation "tower tilt" due to differential settling of the bottom can cause somewhat similar problems; therefore, this subject is addressed again later.

A number of authors have developed the basic equations of sediment transport from a multiphase or two-phase flow point of view: the works of Hunt (1954), Lumley (1976), Smith (1977), McTigue (1981), and van Rijn (1986) are a few relevant articles. The final equations governing the transport of a sediment with mass concentration, c $(M/L^3)$, are derived for low concentration conditions (van Rijn, 1986; Lumley, 1976) from the derivations based on mixture theory (Smith, 1977; van Rijn, 1986).

Turbulence is accouned for via a traditional Reynold's average uniform-weight-function temporal average following the rules of averaging allowed by this weight function (Bedford et al., 1987a; Dakhoul and Bedford, 1986). In the following, the overbar represents the temporal average as obtained over a finite time period, T, and the resulting correlation terms have not been "closed" by mean flow parameterizations. It is also assumed that the horizontal water and sediment velocities are identical and that the veritcal velocities differ only by the sediment settling with an average settling velocity $w_s$, relative to the water vertical velocity, $\bar{w}$. The high concentration equations are in Bedford (1990) as abstracted from the references mentioned. The final equations are:

Continuity

$$\frac{\partial}{\partial t} \rho + \frac{\partial}{\partial x_i} (\rho \overline{\mu_i}) = 0 \qquad [1]$$

Momentum Balance

$$\frac{\partial}{\partial t}(\rho \overline{\mu_i}) + \frac{\partial}{\partial t}(\rho \overline{\mu_i} \overline{\mu_j}) = -\frac{\partial \overline{p}}{\partial x_i} - \rho g \partial_{i3} + \frac{\partial}{\partial x_j}(\mu \frac{\partial \overline{\mu}}{\partial x_j} - \overline{\rho \mu_i' \mu_j'}). \qquad [2]$$

Conservation of Sediment Mass

$$\frac{\partial \overline{c}}{\partial t} + \frac{\partial}{\partial x_j}(\overline{c \mu_j}) - \frac{\partial}{\partial z} \delta_{i3} (\overline{w_s} \overline{c}) + \frac{\partial}{\partial x_j}(\overline{\mu_j' c'}) - \frac{\partial}{\partial x_j}(D \frac{\partial \overline{c}}{\partial x_j}) = 0. \qquad [3]$$

In these equations, indicial notation is assumed such that i (or j) = 1,2,3 corresponds to the x,y,z directions, respectively. The Kronecker delta function, $\delta_{i3}=1$ for i=3 and 0 for i=1,2. The molecular viscosity, u, and sediment molecular diffusion coefficient, D, have dimentsons $ML^{-2} t^{-2}$ and $L^2 t^{-1}$, respectively. Mixture density, $\rho$, has dimensions $ML^{-3}$.

## Definition of Flux Terms

Equation 3 may be written in flux form as

$$\frac{\partial \bar{c}}{\partial t} + \frac{\partial N_j}{\partial x_j} = 0. \qquad [4]$$

where the flux vector Nj is given by

$$N_j = \overline{\mu_j c} - \delta_{j3}\overline{w_s c} + \overline{\mu_j' c'} - D\frac{\partial \bar{c}}{\partial x_j}. \qquad [5]$$

If, for simplicity, x is considered to be the horizontal streamline coordinate direction, then a two-dimensional flux form follows:

Horizontal Flux

$$N_1 = N_x = \overline{uc} + \overline{u'c'} - D_x\frac{\partial \bar{c}}{\partial x}. \qquad [6]$$
$$\phantom{N_1 = N_x = }\;(1)\quad\;\;(3)\quad\;\;\;(4)$$

Vertical Flux

$$N_3 = N_z = \overline{wc} - \overline{w_s c} + \overline{w'c'} - D_z\frac{\partial \bar{c}}{\partial z}. \qquad [7]$$
$$\phantom{N_3 = N_z = }\;(1)\quad\;(2)\quad\;(3)\quad\;\;(4)$$

Term (1) in equations 6 and 7 is the advective flux, term (2) is the settling flux, term (3) is the turbulent flux, and term (4) is the molecular flux.

For the case where z=0, at the bottom, equation 7 simplifies as there is no advective flux; therefore

$$N_3(z=0) = N_z(z=0) = N_{z0} = \overline{w'c'} - \overline{w_s}\,\bar{c} - D\frac{\partial \bar{c}}{\partial z}. \qquad [8]$$
$$\phantom{N_3(z=0) = N_z(z=0) = N_{z0} = }\;(3)\quad\;\;(2)\quad\;\;(4)$$

In equations 7 and 8 it is possible to identify different expressions in Nz that can be used to define certain frequently used (and misused) verbal descriptions. For purposes of this chapter, three regimes are defined:

a. Depositional Flux  : $N_z < 0$   [9]

b. Equilibrium  : $N_z = 0$   [10]

c. Resuspension Flux : $N_z > 0$   [11]

The simultaneous presence of both settling (downward flux) and turbulent flux (either positive or negative) means that not only must the net or aggregate flux be labeled, which we have done as $N_z$ in Equations 7 and 8, but also the individual terms comprising this net flux must be defined. Therefore, for the flux at the bottom (equation 8), term (3) is the term responsible for turbulent transfer of material from the bottom to the water column and will be called the entrainment flux. Term (2) is the settling flux while term (4) is the molecular flux.

## Analytical and Computational Framework for Entrainment Flux

The expression for the time-varying vertical flux at the bottom is given in equation 8 and, were it possible to have a device that could measure these terms without destroying the physics of the near-bottom flow, then the data would be at hand. Such instruments do not exist at this time; therefore, the flux must be inferred with as much accuracy as possible. There are many approaches to an inference procedure, but in essence they are based upon measurements or analytical results that are easier to obtain. The data are then used in concert with a derived or assumed mathematical relationship between $N_{z0}$ and the more readily obtained data. The procedures for deriving relationships between $N_{z0}$ and more readily measured variables vary from those requiring few if any assumptions about the physics to those requiring several assumptions about the flow physics. As they all have been reported in the literature, the purpose of this section is to summarize where these inference procedures come from. The subsequent section summarizes measurement methods based upon these methods. The first two methods are computationally oriented and at first glance appear to require fewer assumptions, while the last two are more analytically based and require confirmation of the necessary physics in order to proceed.

1. Control Volume Approach: First used by Bedford et al. (1987b) to measure entrainment, the control volume or mass balance approach is an

]integral approach that is formulated by defining the integration volume as shown in Figure 5. If the bottom of the control volume, $z=\xi$, is placed at the bottom of the water column, and the top is located at $z=\eta$, then equation 4 can be integrated with the use of Leibnetz's theorem. The result is presented in Bedford et al. (1987b). The terms involving the time and space derivatives of the $\xi$ and $\eta$ positions possibly contribute to the mass balance for a variety of bedform changes during sampling and will be reviewed in the subsequent section. This approach is quite analogous to the Head (1960) entrainment formula for determining shear stress (momentum flux) at a solid boundary.

2. Differential/Computational Approach: Never tested, this procedure also begins with equation 4 and, invoking only minimal approximations, the derivatives of the fluxes are computed. Thereby, a second relation between bottom flux/entrainment and more readily measured variables is available. If the same control volume and two-dimensional flux equation are assumed as before and if the fluxes through each face are defined as area-average values, then a centered difference approximation might be employed to approximate equation 4. For a control volume with axes labeled as in Figure 5, symmetric or centered approximations in either finite difference or element form can be written for a coordinate shifted to the center of the control volume. If symmetric numerical operators for first-order differential terms are defined about the control volume centroid for time, $\delta_t$, and space, $\delta_x$ or $\delta_z$, then equation 4 may be approximated by

$$\delta_t \bar{c} + \delta_x \overline{N}_x + \delta_z \overline{N}_z + (E_t + E_x + E_z) = 0. \qquad [12]$$

where the E terms denote truncation errors for each operator, the size must be estimated for each calculation.

Equation 12 can be reconfigured and solved for the bottom flux when the grid point J is at $z=\xi$ and j+1 is located at $z=\eta$. The size of $\Delta t$ is strongly related to the averaging time used in the Reynolds average, a point to be discussed further ahead.

It should be noted that the choice of numerical approximations is not at all easy, and the selection could control the results. Symmetric operators have long been known (Oran and Boris, 1987) to manufacture parasite waves when applied to large first derivatives. Lagrangian methods (Oran and Boris, 1987) have been suggested as a remedy but seem inappropriate for use in the bottom as no mean flow exists. Suffice it to say that more accuracy requires more nodes, which means data collected at many more points in and around the control volume.

3. Assumed or Boundary Layer Profile Method: This approach is ubiquitous in the bottom momentum boundary layer field, where it is often used to estimate bottom shear stress. It is natural to ask if it could be extended to bottom concentration layers and entrainment fluxes. The procedure is relatively straightforward in concept. The boundary layer approximations are invoked (Schlicting, 1979) to derive a reduced set of two-dimensional equations valid along the streamline. A set of boundary conditions are employed that are tailored to reflect the physics being analyzed, and an exact solution is achieved that predicts the vertical profile of the dependent variable (i.e., velocity or concentration). Critical to the exact solution is the form of closure used to relate $\overline{w'c'}$ to the mean flow or its gradient. Eddy diffusivity coefficient preselects the type of turbulence and, therefore, the flow field physics.

There are a number of different boundary layer profile solutions, each appropriate to the specific set of assumptions necessary for that derivation. It is necessary to validate the existence of the profile before using information dervied from assuming its existence. Once existence has been ensured then, as is done for the momentum case, it is conceptually possible to differentiate the profile and estimate the components of the bottom flux, $N_{zo}$.

4. Phenomenological Turbulence Modeling Approaches: While not used too frequently, it is the case that *in situ* measurements might be made with sufficient accuracy that the second-order closure theories (see Launder, 1985) could provide another useful method for relating $\overline{w'c'}$ to more readily measured variables. These procedures account for history and non local effects in the turbulence and, in their most basic form, yield imposing differential transport equations for $\overline{w'c'}$. The work of Bedford (1987) appears to be among the first to suggest this approach, as adapted from the original works of Gibson and Launder (1976; 1978). Other forms based upon work by Mellor and Yamada (1982) have also been derived. Wall corrections will be required to account for the effect of a wall on the pressure strain correlation. Much simpler algebraic stress models (ASM) and differential stress models are derivable for boundary layer flows.

5. Empirical/Buckingham $\pi$ Methods: The final method for relating entrainment to more readily measurable quantities is simply to hypothesize the relation between a group of dimensionally consistent variables, perform a Buckingham $\pi$ analysis, and then perform the a procedure when attempting to predict $N_{zo}$ from groupings of bed material properties such as compaction, density, etc., as well as possible measures of fluid mechanics such as shear stress. Alternatively, instruments could be developed to measure the data, but calibration will

require the same empirical correction task.

In the following sections each of these methods is examined with particular attention devoted to implementation, errors, required instrumentation, and field experience to date.

## Sampling, Averaging, and Statistical Considerations for *In Situ* Measurement

Before proceeding to evaluate the frameworks presented in the previous section, it is necessary to consider aspects of how to translate the mathematical requirements and assumptions inherent in the equations into requirements for the field program. Of necessity then is a brief review of published experience on characterizing inherent sources of variability or error and how to minimize or control these errors such that the desired quantity can be obtained. The frameworks established above will be reviewed and evaluated in the next section with many of these concepts in mind.

Increasing emphasis on quantifying sources of variability has appeared in the literature, and much of this section has been collected from the works of Lumley and Panofsky (1964), Heathershaw and Simpson (1978), Soulsby (1980), Gross and Nowell (1983; 1985), Lumley and Terray (1983), Grant *et al.* (1984), Cacchione et al. (1987), and Grant and Madsen (1986).

Inspection of the various terms of the equations in the measurement frameworks indicates that a number of issues need to be addressed, including sensor size and accuracy, sampling rate, record length, the calculation of mean values and confidence intervals, various turbulence quantities and their confidence intervals, and the fluxes and their confidence intervals. Further, the boundary layer framework involves comparisons with exact solutions, which motivates questions of accuracy versus sampling resolution. All these items are to some extent discussed, with extensions, where possible, to the case at hand.

Before proceeding to questions involving the analysis of data, it is necessary to discuss several considerations in the physical circumstances by which these data are collected.

## Physical Considerations of Data Collection

Bottom flow and transport data are usually collected by bottom sitting towers with three legged pyramidal structures. Various types of pads are positioned on the legs to inhibit tower settling. It is extremely important to know the position of the tower relative to the bottom and the mean flow. While mean flow questions are dealt with a bit further on it is necessary to list the reasons why tower position knowledge is important.

Tower Tilt and Settling. With regard to the integral control volume equations (Method 1), Bedford et al. (1987a) have summarized the various factors that could lead to temporal or spatial changes in sampling volume position. These include: (1) changes in the location of the bottom due to time-varying depositional events; (2) a temporal change in bottom location or possible tower orientation due to the movement of large scale bedforms through the sampling region; and (3) uniform and/or differential tower settling due to unconsolidated bottom material. This last item could be a one-time activity at the time of deployment, or it could occur over time as the erosion and deposition patterns unfold. From the decomposition of the integrated horizontal flux term, it is apparent that the bedroom propagation and its effect on tower tilt is of consequence in the horizontal advection contribution. The effects of tower tilt on precision are discussed next.

Tower tilt has an additonal manifestation in that when the vertical axis does not coincide with gravity (i.e., it is rotated by some angle $\Theta$ from vertical in a two-dimensional coordinate system), then in the local-coordinate system gravitational settling will contribute to the observed local-coordinate horizontal flux. The settling velocity in the new local vertical coordinate ($z^*$) equals $w_s \cos\Theta$, which, in the limit of small angles, leaves $w_s$ essentially unchanged. The local positive horizontal ($x^*$) velocity is increased by an amount $w_s\Theta$ for negative rotations and is decreased a similar amount for positive tilts. All in all, for small angles, tilt does not affect the gravitational settling flux appreciably (Yalin, 1975). As will be seen later, its effects on mean flow and turbulent fluxes are quite another matter.

It is important to know the vertical location of the bottom relative to the instruments and the uncertainty in the measurements. As an example, Bedford et al. (1987b) use backscatter readings from a 3MHz echo sounder, which yields a ±1 cm location uncertainty, while Grant et al. (1984) report a 5 cm uncertainty from data collected with a 1MHz echo sounder and a mechanical gage. Vertical location also contributes to zero offset corrections, which are often necessary in analyzing momentum boundary layer data (Grant and Madsen, 1986; Cacchione et al., 1987).

Instrument Separation. The physical location of the instruments relative to one another can have an important effect, particularly when the measured variables are to be correlated and used for further interpretation of the physics. Typically the separation problem is not severe for the u and w measurement and its moment, $\overline{u'w'}$, because the same transducer is used. The problem is more troublesome for data collected by different instruments, say, for example, w and c and the

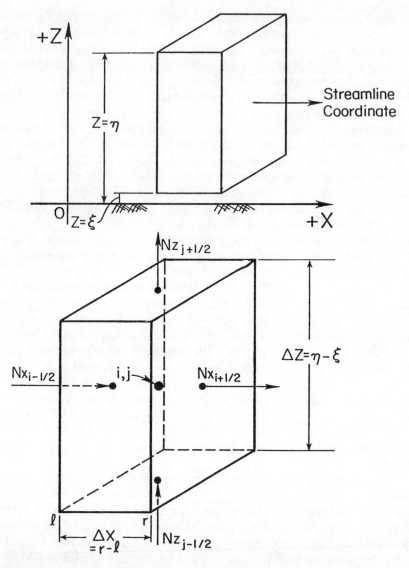

Figure 5. Control volume and grid definitions for control volume approach.

important turbulent/vertical flux $\overline{w'c'}$. Due to size and the necessity of non interference, instruments used to mesure these data are placed at a distance, h, apart. This separation basically means that organized flux activity with size 2h (wavenumber = $\pi/h$) cannot be resolved in the data. Since these separations are on the order of 0.5 m, this effectively precludes resolution of important flux activity in the bottom boundary layer where considerable variability occurs at these scales. The problem of near-bottom sampling of $\overline{w'c'}$ with large separated instruments is placed in jeopardy by simple separation arguments alone.

Instrumentation Size and Obstruction Considerations. The physical size of an instrument places limits on the degree of refinement in the data. Simple mindedly, assume the "sensor volume" is spherical with diameter d*. Then the minimum size of a physical of a physical process or its fluctuation that can be resolved directly (not inferred by Taylor's frozen turburlence hypothesis, TFTH, (Hinze, 1975) is 2d*, with a wave number of $\pi/d^*$. By more refined time sampling and TFTH, higher wave number activity can be inferred but aliasing is a potential problem in that the sampling volume might not be fully "refreshed."

The arms and legs of the instruments obstruct the flow and, at minimum, give an eddy filled wake. As a very rough rule of thumb (Schlicting, 1979), this wake would exist approximately 50 diameters downstream of the object causing the wake. Velocity pulses due to wakes yield data that are mostly unanalyzable. The shedding vortices could induce vibrations in the instruments and the tower, which are additional sources of unrealistic velocity measurements. Concentration measurements are in most cases much less affected by vibration, but since they are to be collected with velocity, errors are thereby introduced. The use of real time read-out strain gages from one or two critical positions on the tower is one method for determining vibration impacts.

## Instrument Sampling Effects

The instruments themselves introduce errors, and the record lengths collected and the frequency at which data are collected set limits on what part of the spectrum is resolved in the data. Consideration must then be given to parameterizing the ratio between the imposed variability from the instruments versus the natural variability introduced by the turbulence. The instrument-imposed variability must be quite small in contrast to the natural variability.

Instrument Noise and Calibration. It is extremely important that the noise levels and calibration uncertainty be parameterized in terms of the collected data. The standard deviation of the natural mean is defined as (Bendat and Piersol, 1971; Gross and Nowell, 1983)

$$\sigma_m = \pm \sigma/\sqrt{n}. \qquad [13]$$

A confidence band at the $\beta\%$ (e.g., $\beta=95\%$) level is determined by multiplying the standard deviation in equation 13 by the student t statistic (see for example, Gross and Nowell, 1983), that is

$$\sigma_m = t(n - 1,[(100 - \beta) \, 0.5])\sigma_m \qquad [14]$$

where $\sigma_m$ is measured during the record or averaging length T; $\sigma$ is the population standard deviation; and n is the number of statistically independent samples in the averaging interval (T) (n is also termed the number of degrees of freedom), which can be determined by the ratio T/d, where d is twice the duration of the autocorrelation time. This time is estimated by taking the autocorrelation plot of the data time series and finding the time corresponding to the first zero crossing (Gross and Nowell, 1983). Other procedures have been recommended by Soulsby (1980).

Instrument errors contribute to erros in the mean quantity (Gross and Nowell, 1983); therefore, knowledge of these limits for velocity and concentration must be known. The situation for velocity measurements is somewhat well in hand; for instance the ducted impellor current meters used by Smith and McClean (1977) have a 0.01 µ mean error; the acoustic tomography based BASS developed by A. J. Williams (Williams et al., 1987; Williams, 1985) has absolute noise and calibration uncertainties of ±0.03 and 0.3 cm/sec, respectively; and the frequently used Marsh McBirney electromagnetic (EM) current meters have (0.01-0.02) µ uncertainties for relatively steady flow. The EM current meter calibrations for combined wave current flows have recently been the subject of strenuous debate (Aubrey and Trowbridge, 1985; Guza, 1988; Aubrey and Trowbridge, 1988).

The state of concentration measurement accuracy is much poorer. Light and acoustic scattering devices are commonly used along with *in situ* data to measure concentration. Optical transmissometer accuracy has been reported to be 6.0% $\overline{c}$ (Zanevald et al., 1979; Bartz et al., 1978). Early acoustics-based transducers had accuracy and noise variability in the mean that was primarily due to configurational noise and was estimated as 45% $\overline{c}$ (Libicki et al., 1989). Second and third generation acoustic devices have decreased this variability to < 0.1 $\overline{c}$ (Libicki et al., 1987; 1989).

As many author have noted, while these errors in the mean are in some cases quite small, the error is magnified as the data are used to create perturbed or products of perturbed quantities. The example in Gross and Nowell (1983) shows that an error in the mean velocity of 0.01 $u$ magnifies to ±.05 $\overline{u}$ and ±0.1 $\overline{u'^2}$ for typical turbulent quantities. Here the overbar signifes the average quantity. Extensions to concentration measurement errors for $\overline{c'}, \overline{c'}$ and $\overline{c'^2}$ are analogous, while products uc and wc will contain additive errors. For example, if the error in w' equals 5% of the average fluctuation ($=.05\ \overline{u'}$) and the error in the concentration fluctuation is 10% ($=0.1\ \overline{c'}$), then the estimated first order measurement error is $0.15\ \overline{c'w'}$ or 15% of the mean. If the measurement error is relatively small compared to the natural variability, then the data can be used for higher order moment analysis.

Two other losses of information result from the length of the sampling interval, t*, and the digitization frequency, $f_d$. These effects are discussed next.

Record Length. The selection of the total length of record to be sampled affects the analyzable data in three ways (Soulsby, 1980): (a) lost low frequency variability can occur if the record length, t*, is equal to or less than the period of a large low-infrequency motion; (b) the record length must be long enough to allow the required stationary conditions to develop, a condition which often conflicts with constantly varying low frequency forcing function motions such as tides, seiches, etc.; and (c) the sampling variability in the estimate of the mean or variance and the size of the confidence bands depend on t*.

The first consideration is self-evident. If it is desirable to correlate a low frequency, large magnitude activity with bottom activity, then the record length must be long enough to resolve it. The implication being that all information sampled at $f_L$, $k_L$, or lower (Figure 2) is not resolvable. While the record length to the period of this motion, it is the case that (equations 13,14) detailed quantitative correlations can be done only when a number, n, of independent low frequency motions have been sampled. Confidence bands (equation 14) become accordingly large with fewer independent observations.

A corollary of this consideration is that the record length should be placed in a "special gap" between intense low frequency motions (Figure 2). Placing t* on a spectral speal, i.e., $k^1_w$, $k^2_w$, results in partial resolution of the coherent motion at those wave numbers, which results in biased calculations without physical meaning.

The consideration of stationarity in (b) above is significant as it provides the key by which data, statistical analyses, and proposed theories are prepared and compared consistently. All the statistical analyses rely upon the statistics of the fluctuations about the mean being stationary, they must not change with time. Furthermore, the mean flow definition used to calculate turbulence data for the basic turbulent flow equations and the resulting theoretically predicted profiles require, crudely, that the definition of an averaging interval, T, (Hinze, 1975; Monin and Yaglom, 1975) be coincident with the interval of time the statistics are stationary. Therefore, the time period of stationarity determines the upper bound on the averaging, interval, T. Stationarit y can be determined by run tests (Bendat and Piersol, 1971; Bedford et al., 1987a). For flows with steady mean flows stationarity can exist for some period of time. However, in flows where the mean is time varying, such as storm driven or tidally affected flows, the change in the mean flow is of the same order of magnitude as the turbulent fluctuations, and an averaging time T must be selected from the run tests such that the time varying means are reduced to a series of quasi-steady/stationary averaging intervals. Therefore, in order to select an averaging time that allows consistent comparison, it is prudent to select the smallest time that allows consistent comparison, it is prudent to select the smallest time scale found from the run tests.

Combined wind/internal wave and current flow fields, which are ubiquitous, confuse the issues discussed above. Quantitative guidelines do not exist for sampling in such conditions, although Soulsby (1980) suggests the use of phase averaging. Suffice it to say that, blended with the considerations above should be that averaging/stationarity times should be used that are much longer than the wind wave periods and smaller than internal wave periods (should these waves exist). Wave current sampling is a very difficult problem, especially when the waves sampled within T are nonlinear and give rise to drift as well as orbital contributions to the spectra (Lumley and Terray, 1983).

The final consideration of sampling variability is simply put, i.e., the random error in an estimate of a mean or variance is reduced for record length. The length of the record is now presumed to be T due to the stationary requirements for forming the statistical data and a confidence level, $\beta$, can be determined for this T. The entrainment flux and Reynolds stress measurements, $w'c'$ and $u'w'$, will contain the most uncertainty, have extemely wide confidence level bands and, therefore, require quite long records to reduce the uncertainty to acceptable levels. This requirement is due to the fact that event durations, d, in the flux terms $w'c'$, u, w are much smaller than those for the mean.

As an example, suppose the flux correlation time was approximately 3.0 seconds (see Bedford et al., 1987a) and the averaging time was 900 seconds (Bedford et al., 1987b); then these data contain 300 degrees of freedom. The variance of the $w'c'$ value at 67 cm above bottom was approximately $8\,\overline{w'c'}^2$ which gives a 95% confidence band of 32% of the mean for the flux $\overline{w'c'}$. If a 5% spread to the 95% confidence interval is required, then 10 hours of stationary data needs to be collected, while a 1% band to the 95% level requires 256 hours of such data. Clearly, stationarity requirements limit tidal and Great Lake flows to 8-20 minute averaging intervals (Soulsby, 1980; Gross and Nowell, 1983, 1985; Bedford et al., 1988). It is certainly the case that $w'c'$ and $u'w'$ will always contain high variability.

The procedure in equations 13 and 14 requires knowing the mean of the value being parameterized (e.g., the mean of the flux $w'c'$). Often this is circular logic as it is precisely these data for which we wish to design a sampling scheme before going to the field. Bendat and Piersol (1971) describe a procedure for evading this problem, which Soulsby (1980) has applied to the momentum boundary layer. The procedure for determining the error assumes the variability is normally distributed, then derives the error in terms of the mean flow variables instead of the mean of the variable being investigated.

The estimate for the sediment flux w'c' error, $\varepsilon_{wc'}$ proceeds in a similar fashion to the momentum derivation in the references (please see Bedford, 1990 for other details), that is

$$\varepsilon_{wc} = 2 \frac{\overline{w^2 c^2}}{\overline{cw^2} T} d^*_{cw}, \qquad [15]$$

where $d^*_{cw}$ is defined as

$$d^*_{cw} = \frac{1}{T-\tau} \int_0^{T-\tau} c(t) w(t+\tau) \, dt \qquad \tau \geq 0,$$
$$\phantom{d^*_{cw}} = \frac{1}{T-\tau} \int_0^T c(t) w(t+\tau) \, dt \qquad \tau < 0. \qquad [16]$$

It is noted from Soulsby (1980) that vertical scaling of these data can further reduce the necessity of individual calculations at vertical points (an implication of less required data collection). However, vertical scaling appears justified only under certain conditions and is not a universal collapsing. Gross and Nowell (1985) suggest vertical scaling is only possible when the ratio of the spectral strain rate to the mean shear is large; in essence, a requirement of separation of the wavenumber regions for the production and dissipation of turbulence. With the possibility of stratification due to sediment (Glenn and Grant, 1987) and the possible effects of bedload (Gust and Southard, 1983), the possibility of vertical scaling for entrainment and concentration data will need very careful examination for near-bottom conditions.

Sampling Rate. As seen in Figure 2, significant turbulent variations occur up to extremely high wave numbers and frequencies. If the instrument sampling rate is at a lower frequency or wavenumber ($f_d$, $k_d$), then some of this variability is lost, i.e., the shaded portion to the right of $f_d$, $k_d$, in Figure 2. A considerable amount is known both theoretically and experimentally for $f_d$ and $k_d$ in the inertial subrange. Kolmogorov (1941) first identified this range, and Monin and Yaglom (1975) and Townsend (1976) have shown that it is the portion of the spectrum where turburlence is neither created nor destroyed. If the sampling rate can be determined to fall into the intertial subrange, then a number of extremely powerful techniques for inferring the turbulent shear stresss and the bottom boundary layer physics, in general, can be brought to bear with a minimum of data collection required. It appears from the literature (Soulsby, 1980; Gross and Nowell, 1983, 1985; Grant et al., 1984; Bedford et al., 1988, etc.) that a sampling rate of 2-5 Hz will serve as a guideline for *in situ* sampling of the inertial subrange. Further review of the analysis methods available from inertial subrange physics appears in the next section.

## Regression and Curve Fitting

In processing the data for further analyses, two concerns arise that, if not addressed properly, add increased variability and uncertainty to the data.

These include: (a) the determination of the mean flow and the impacts of unaccounted for tower tilt and wave-current interactions and (b) the determination of profile fits from regression analyses and the inference of corresponding physics from this.

<u>Determination of the Mean</u>. Without doubt, it is essential to determine the mean value properly, as all turbulence variables and statistical summaries are defined relative to it. Stationarity conditions for the mean have already been discussed. To compare with all mechanics-based predictive theories, it is necessary to have the velocities rotated into a streamline coordinate system that permits no mean flow across streamlines. Local tower tilt relative to the flat bottom streamline or deformations due to combined bedforms and tower tilt can introduce errors in estimations of the mean. Though perhaps not too serious by themselves, the magnification of there errors in the correlated equantities $\overline{w'c'}$ and $\overline{w'c'}$ is severe.

As extended from Grant and Madsen (1986), it is possible to infer tower tilt errors and coordinate rotation errors by relating true to tilted errors as follows: for a coordinate system rotated through a small tilt angle, $\theta$, the tilted velocity ($u^*$, $w^*$) is related to the true velocity as follows:

$$\mu* = \mu + w\theta, \qquad [17]$$

$$w* = w - \mu\theta. \qquad [18]$$

No change in c (i.e., c*=c) is assumed unless the sampling volume is rotated completely away from the untitled volume, a requirement for large $\Theta$. With this in mind, the temporal average products become

$$\overline{uw} = \overline{u*w*}\left(1 + \left\{[\overline{w*^2} - \overline{u*^2}]\,\theta\sqrt{\overline{u*w*}}\right\}\right) \qquad [19]$$

and

$$\overline{cw} = \overline{c*w}\left(1 + \frac{u*\theta}{w*}\right). \qquad [20]$$

It is clearly apparent that, when attempting to use data collected at the bottom where w and w* are small, very small changes in the tilt can cause serious changes in the desired correlated quantities. It is, therefore, absolutely essential to account for or correct these tilt errors. Customarily, benthic boundary layer work has been done with current meters that measure only two axes of data. As Gross and Nowell (1983) point out, the correction of tower tilt and identification of the mean will require full three-dimensional data in order to avoid making severe assumptions about the directionality of the flow. Such current meters are only just emerging.

Mean Flow Conditions Affected by Waves. While space does not permit an exhaustive summary, it is the case that the problem of separating the mean flow from the turbulence is compounded considerably by the presence of gravity waves. Whether they are internal or surface waves, it is hypothesized that the net effect on the bottom boundary layer is the appearance of a thin (5-10 cm) nonlinear wavecurrent boundary layer at the bottom. The separation of wave and turbulence effects was substantively addressed by Lumley and Terray (1983) and Finnigan et al. (1984), and summarized for bottom boundary layer work by Grant and Madsen (1986). The presence of waves implies that observed variations in velocity will result from mean flow acceleration, turbulence, orbital velocities, and nonlinear drift. One way to achieve separation of wave and turbulence variability is by linear filtration techniques (Lumley and Terray, 1983).

Determination of Mean Profiles. It has been traditional in boundary layer theory to measure at a number of discrete points in the vertical dimension and, after time averaging, to use regression to find a "fit" to the vertical profile data. This is particularly the case for logarithmic linear fits to velocity data, wherein important shear stress calculations can be easily performed if the constant stress layer of the log law of the wall is present. If $\overline{u'}$ is the horizontal (streamline coordinate system) temporarily averaged velocity then the profile is described by u* [the friction velocity defined in terms of bottom shear, ($\tau_o$, as $\tau_o/\rho)^{1/2}$], k (von Karmen's coefficient taken here as 0.4), and $z_o$ (roughness). On a log-linear fit the slope is $(u*/k)^{-1}$ and the intercept is ln $z_o$. Confidence intervals for the slope and intercept can be derived (Gross and Nowell, 1983; Cacchione et al., 1987) from the student t distribution as a function of the number of measuring points in the vertical, m, and the correlation coefficient for the fit, $R^2$. The reader is referred to these references or Bedford (1990) for details.

These same confidence estimation procedures are applicable to sediment concentration profiles (Abdelrhman and Bedford, 1987; Bedford et al., 1987a). However, three classes of "fits" are possibly necessary, including

$$c \equiv \ln z \qquad [21a]$$

$$z \equiv \ln c, \text{ and} \qquad [21b]$$

$$\ln z \equiv \ln c. \qquad [21c]$$

Any or all of these may be observed in any one particular deployment.

As Grant and Madsen (1986) have displayed, all log-linear plots with large correlation coefficients are not necessarily log of the wall physics. Indeed, the data can be easily misinterpretated. For example, in the CODE paper by Grant et al., (1984), confidence bands of $u_*$ ±25% at the 95% level were selected, which implies that profiles from the four current meters were log-linear with an $R^2$ value of 0.993. Velocity profiles with $R^2$ values less than this were plagued with internal wave activity and rejected for "log of the wall" analysis. Clearly, the acceptance or rejection criteria based upon quantitative error measures in $u_*$ places the requirement for data collected at exceptionally high precision levels.

# Evaluation

With the sampling and variability considerations in hand, it is possible to make some comments about the relative merits of each procedure for estimating resuspension. After reviewing these procedures, a brief discussion of auxiliary shear stress measurements is presented.

With all these methods it appears that two items are universally required. First, it is quite clear that the inherent noise and instrument error when extended to the calculation of the turbulent moments must be small in comparison to the natural variability of the moments; therefore, instrument errors must be known. Secondly, in order to take advantage of a number of theoretically known and verified estimates, it is necessary that sampling frequencies be placed in the inertial subrange. This is when the sampling frequency is placed in the spectral gap. These two items will be assumed for the discussion that follows.

The remaining points to be evaluated are the errors involved in the numerical analysis required for each method and their size compared to existing variability in $\overline{w'c'}$, the number of sampling intervals versus this numerical error, and a summary of existing instrumentation capable of making these measurements. In all of the schemes, it is necessary to measure settling velocity, tower tilt and orientation, and grain size distribution; this paper will not address these measurements but rather assume that they are performed with equal precision or impression for each method. Tower tilt (±1°) and orientation are relatively easy to measure and need no further discussion while the other two data are particularly difficult to measure and should be the subject of focused symposia.

## Control Volume Method

In the control volume procedure, it is clear that the advective fluxes are being measured in both directions and compared to the time rate of change of mass in the control volume. It is important to note that only the fluxes and not their derivatives are measured in the vertical dimension, which is inherently more precise as derivative calculations are noisy (this discussion continues in the next subsection). To use this equation fully requires only knowledge of the vertical flux at one point off the bottom, the horizontal flux, and the time rate of change of mass in the control volume.

Insofar as horizontal advection is concerned, it has for the most part been ignored in all sediment transport theories, even the most sophisticated wave current sediment boundary layer theories (Glenn and Grant, 1987; Davis et al., 1988). This is, of course, no reason not to measure it, and it should be measured when sufficient instrumentation becomes available. Indeed, this term appears in all the measurement frameworks, and its incorporation should be a primary research goal for the future. Guidelines for its importance as regards the effects of bed deformation have been presented by McLean and Smith (1986).

With suitably frequent (5 Hz) data arranged over the boundary layer averaging time, the variability/confidence bands on $\overline{w'c'}$ ($z=\eta$) have been parameterized in the previous section; the remaining computational decision is how to evaluate numerically the total mass integral and its time rate of change. The accuracy of a numerical integration is a function of the number of vertical data points measured in the control volume over $\Delta z$ (Figure 5) between $\xi$ and $\eta$. Various numerical integration schemes are available including those for open and closed boundaries with equal and unequal spacing (Press et al., 1986). A method will not work if there are not, at absolute minimum, two concentration measurement points; one of which must be at $z=\eta$. Two-point or three-point measurements allow trapezoidal or Simpson's integration, respectively, but introduce error that is on the order of $\{\Delta z^3 \delta^3 c / \delta z^3\}$ which, for high settling near the bottom, would be larger than the integration itself as well as the other terms in the control volume sum. More measuring points in $\Delta z$ reduce this error; Simpson's three point and 3/8 (four point) rule could be used to reduce the error to $[(\Delta z/2)^4 \delta^4 c/\delta z^4]$ and $[(\Delta z/3)^5 \delta^4 c/\delta z^4]$ respectively. In general, the extended trapezoidal integration (Press et al., 1986) has a relative error that decreases as the square of the number of data/sampling points (N) in the vertical. For instance, a relative error of 25% occurs for two-point sampling while errors of 11% and 6% relative errors accrue for three and four point sampling. Clearly, four-point sampling is required in order to keep the numerical integration error the same size as the instrument error. In summary, this method is unsatisfactorily laden with computational noise if less than four concentration methods occur. Questions as to whether light-based concentration measurement devices can sample nondestructively in the thin wave-current boundary layer exist and cast doubt on their being able to fully collect the data required for the method. Accoustic devices have been used to estimate entrainment with this method (Bedford et al., 1988) in flows where horizontal advection is unimportant. In such situations, no noisy numerical spatial derviatives are computed, and, therefore, the controlling variability is in the measurement $\overline{w'c'}$ and $\overline{w}\,\overline{c}$ at $z=\eta$.

## Differential/Computational Method

This method has never been tried insofar as this author knows and is therefore, purely speculative. To make even the crudest use of this method, it is necessary to make numerical estimates of flux derivatives in both the x and z directions. The "leap frog" procedure or the staggered approach used for large gradients is nominally second order accurate in its numerical structure, due to the symmetrically centered finite differences (Press et al., 1986). For persistent horizontal advection, it might be necessary to consider Lagrangian schemes (Oran and Boris, 1987). If the data are collected at the faces of the control volume then, for example $\Delta z=(\eta-\xi)/2=0.5$ m and the relative error in the vertical flux derivative, the most important one, would be 25%. The minimum instrumentation required, even ignoring horizontal advection as before, is three points and includes a measurement of flux at $z=\eta$ and the time series of concentration at the centroid of the control volume and at $z=\xi$. However, the measurement of $\overline{c}$ $(z=\xi)$ is required to decompose $N_{zo}$ into a direct entrainment rate, and, once again, ultrasonic accoustic backscatter devices appear to be the only instruments capable of robustly and unobtrusively performing this measurement in the thin layer near the bottom. This three-point method can be extended to fourth order accuracy by use of a five-point tangent-secant derivative approximation (Oran and Boris, 1987), which required flux measurements at four of the five equi-spaced points within the control volume and $\overline{c}$ $(z=\xi,t)$. This reduces truncation errors to levels below instrument levels.

As mentioned, this flux estimation method is essentially untested and, were it possible to configure a non disruptive sampling, it would be worth comparing results with the integral approach. Magnification of $w'c'$ instrument errors by numerical differentiation is expected to be severe for the three point measurement (33-40% for the previously cited data) and therefore, of the same size as the natural variability whose measurements are to be made. Therefore, the five-point scheme appears to be nominally acceptable.

Of special note here are "one-point measurements" wherein a concentration and velocity measurement are made at only one vertical location. Data collected in this fashion can only measure the flux and concentration at the fixed point above bottom and unless the instruments are placed at the bottom, they cannot measure entrainment and resuspension. Additionally, if, as some do, the current meter only samples the horizontal velocities, then no data about sediment entrainment or vertical flux are possible whatsoever. Practically speaking, one-point measurements of velocity and concentration cannot measure entrainment and/or $N_{zo}$ and are of dubious value for continued use in attempting to measure these data. Two-point measurements are also deficient in that the derivatives collapse to first order correct in space or time, which essentially says that flux is constant, and, therefore, errors are potentially significant.

Clearly, the expectations as to "bottom sampling" must be fully known before selecting a sampling configuration with less than three vertical points.

## Inverse or Boundary Layer Methods

It is certainly the case that no greater attention has been paid to near bottom rationalizations than that accorded the boundary layer approach. The goal of the boundary layer approach is, as mentioned, an inverse methodology in that classes of exact solutions for the velocity and sediment profiles are sought for various and frequently observed boundary conditions. Ideally, a highly reduced suite of measurements can then be used to infer the detailed physics by reconstruction based upon the exact solution. This reconstruction can only be allowed if it is known that the exact solution employed fits the conditions being sampled. This last criterion is seldom checked.

The boundary layer method has been particularly succesful in analyzing for velocity, shear stress, and roughness data. Conditions necessary for checking the existence of various layers, particularly the log law of the wall have been thoroughly researched and summarized in the prior section. Methods for inferring shear stress are summarized later.

Bottom sediment boundary layer theories fall into three categories, distinguished by their forcing functions: current, wave, and wave-current sediment boundary layers. Each category has many proposed analytical solutions with fundamental work in each being by Smith (1977), Kennedy and Locher (1976), and Glenn and Grant (1987), respectively. For all three formulation categories, an equilibrium bottom boundary condition is imposed wherein the bottom flux equals zero; a statement that settling and entrainment fluxes are balanced. Therefore, since resuspension is a nonequilibrium process and since there is a lack of robust solutions for nonequilibrium conditions, it is not possible at this time to employ effectively the boundary layer solution approach to determine entrainment and resuspension. Given the relative success of this approach for momentum and shear stress, this is a discouraging state of affairs.

Certainly, the sediment boundary layer approach needs to be more fully and robustly developed to reach the level of utility achieved in momentum boundary layer analyses.

## Phenomenological Turbulence Model Approach

As yet this method is untested. It suffers many of the difficulties that the one-point measurement differential method shares, primarily that measurements cannot be made near enough to the bottom to qualify as a determination of $\overline{w'c'}$ at the bottom. The inherent wide-band variability in measuring fluxes also places this method somewhat at risk. The unique feature of this method is that it shows clearly a relationship between $\overline{w'c'}$ and turbulent kinetic energy (tke). As seen previously, tke is a musch easier variable to measure and is inherently a more accurate variable to correlate $\overline{w'c'}$ with. Further, for equilibrium conditions $\overline{u'w'}$, the turbulent shear stress, does not need to be measured as it does not affect w'c'. Of all the turbulence variables, Reynold's stress is the most difficult to measure and contains the wides error bands at a given confidence level.

Field data reported in Bedford et al. (1987b) support the development of this approach, but it remains an untested hypothesis in the traditional sense. Therefore, while it is not recommended as the preferable estimation framework, it remains a procedure that is testable with data already being collected for other frameworks.

## Empirical Buckingham π Approach

Two basic approaches to this type of estimation procedure exist in the literature, but both, in some sense, seek to relate what are perceived to be relevant variables. The necessary calibration coefficients usually cover axis offset, zero points, and curve shape for the regression relation and in general are highly data dependent.

One type of empirical procedure is the erosion flux, $S_E$, source sink term used to quantify nonequilibrium fluxes in numerical models. Procedures by Partheniades (1962, 1971), Scarlatos (1981), Hayter and

Mehta (1982), Parchure and Mehta (1985), and Ziegler and Lick (1986) (see Bedford, 1990; Table 10) all to a greater or lesser extent reflect fluid mechanical activity by use of the turbulent shear stress and bed characteristics through deposition time, consolidation history, bulk densities, and critical shear stresses for erosion. Nondisruptive *in situ* measurements of bottom conditions with the resolution (1 cm, 1 sec) required for accuracy commensurate with the available fluid mechanics measurements are only now being developed (see for example Libicki and Bedford, 1988). Therefore, the data necessary to verify these predictions *in situ* are at this time not collectable. These formulas will remain untested in the full scientific sense until instrument technology advances.

It is possible to construct instruments for inferring entrainment, based upon correlations between dimensionless groupings. Such is the case for example in the instrument adapted by Tsai and Lick (1987) from the early Rouse (1937) bell jar device that relates entrainment to the pumping frequency of a screen/grid contained in a cylinder. Such instruments also need considerable testing and comparison with other *in situ* data to check for inconsistency in the entrainment estimates.

## Auxiliary Variables

It is customary to relate the entrainment rate to the shear stress at the bottom, $\tau_b$. This is a momentum flux term that also is not directly measureable (Grant and Madsen, 1986) and, therefore, it must be inferred. The most common methods include (Bedford, 1990; Table 11) the one-point drag coefficient method (Sternberg, 1968, 1972), the two-point method (Caldwell and Chriss, 1979), the eddy correlation procedure (Heathershaw and Simpson, 1978; Gross and Nowell, 1985), the inertial dissipation method (Deacon, 1959), the log-profile method (Grant et al., 1984; Gross and Nowell, 1985) have the full three dimensional velocity data been collected with which to fully establish the log law and estimate shear stresses. This is particularly the case with the last three methods, all of which either directly or indirectly require an inertial subrange, separation of production and dissipation scales, and homogeneous or weakly stratified conditions for proper use. Under these conditions the estimates from all three methods are consistent in the values estimated for shear shress (Gross and Nowell, 1985). This verification is a notable achievement in bottom boundary layer research. The other methods require some independent method of verifying the existence of the proper conditions for use.

Again, however, difficulties arise when considering the estimate of shear stress for the nonlinear interacting wave-current boundary layer. At this time velocity measurements simply cannot resolve the bottom wave-current boundary layer (Grant *et al.*, 1984), and, therefore the three methods used for the log boundary layer calculations cannot be extended unambiguously to the wave-current situation.

## Final Comments

It is the case that statistically significant results now exist with which to test hypotheses about the momentum boundary layer. These results permit statements to be made about bottom shear. The statistical, error, and variability measures developed in the momentum bottom boundary layer straightforwardly adapt to the case of the sediment boundary layer, as seen here, and results from *in situ* experiments should be reported along with estimates of variability and precision. Unfortunately, the analogy to the momentum boundary cannot be further exploited as yet.

Various methods for estimating bottom entrainment are proposed here and, with one exception, are applicable to nonequilibrium conditons. The boundary layer approach, is due to be the least applicable method discussed. The control volume method appears to contain some advantages over all the others, in that it is an integral formulation that does not magnify errors by numerical differentiation. Yet the success of the control volume method appears to contain some advantages over all the others, in that it is an integral formulation that does not magnify errors by numerical differentation. Yet the success of the control volume method and all the other methods (except the empirical/Buckingham $\pi$) critically depends on having adequate sampling of the concentration and vertical and horizontal velocities at a sufficient number of points in the vertical. The control volume method requires that the least amount of velocity information, i.e., a one-point measurement of the three dimensional velocity at the top of the control volume. It also requires the concentration at the bottom, top, and sufficent internal points ($\geq 4$) to calculate accurately the concentration. The differential/computational method also requires multiple sampling in the vertical but requires additional instrumentation. Both methods are not accurate without at least four sampling intervals in the vertical. Turbulence models might provide a useful robust correlation between entrainment energy and turbulent kinetic energy, but one- and two-point measurements as discussed in the differential/computational method provide no quantitative information for entrainment estimates, especially when only horizontal velocities are measured.

All these methods contain error, and, with present instrumentation, it is possible to use several of these methods with the same data to see if any consistency can be obtained. Such should be the immediate goal of field deployments.

Finally, it is this author's opinion that the most challenging problem facing the measurement and analysis groups is how to adequately sample the wave-current boundary layers and near bottom sediments with enough precision and without disruption. At this point, the newly developed remote acoustic devices cited here hold promise and have been used to make the first such observations (Bedford *et al.*, 1988). The possible sensitivity of entrainment rates to the effects of waves is well documented and pronounced, and the problem of measurement in this crucial layer must be fully solved before entrainment estimation procedures are considered operational.

## Acknowledgments

The bulk of this work was supported by U.S. Army Corps of Engineers, Contract No. DACW-39-88-0040. Additional support was also made available via Office of Naval Research, Contract No. 00014-88-K-0252 and NOAA-Ohio Sea Grant Contract No. NA84AA-D-00079 R/EM-9. This support is very much appreciated.

## References

Abdelrhman, M. and K. Bedford. 1987. The recognition of spatial and ensemble averages in the benthic boundary layer. J. Geophys. Res., 92:11825-11844.

Aubrey, D. and J. Trowbridge. 1985. Kinematic and dynamic estimates from electromagnetic current meter data. J Geophys. Res., 90.

Aubrey, D. and J. Trowbridge. 1988. Reply. J. Geophys. Res., 93:1344-1346.

Bartz, R., R. Zanewald and H. Pak. 1978. A transmissiometer for profiling and mooring observations in water. Society Photo-Optical Inst. Engrs., 160:102-108.

Bechteler, W. (ed.). 1986. Transport of Suspended Solids in Open Channels. A. A. Balkema, Rotterdam.

Bedford, K. and M. Abdelrhman. 1987. Analytical and experimental studies of the benthic boundary layer and their applicability to the near-bottom transport in Lake Erie. J. Great Lake Res., 13:628-648.

Bedford, K., J. Dingman and W. Yeo. 1987a. Preparation of estuary and marine model equations by generalized filtering methods. In: Three-Dimensional Models of Marine and Estuarine Dynamics, J. Nihoul and B. Jamart (ed.), Elsevier Science Pub., Amsterdam, pp. 113-125.

Bedford, K., C. Libicki, O. Wai and R. Van Evra III. 1988. The structure of a sediment boundary layer in Central Long Island Sound. In: Physical

Processes in Estuaries and Shallow Bays, Wm. Van Leeuwan (ed.), Springer-Verlag, Heidelberg, pp. 446-462.

Bedford, K., O. Wai, C. Libicki and R. Van Evra III. 1987b. Sediment entrainment and deposition measurements in Long Island Sound. J. Hydraulic Engr., 113:1325-1342.

Bendat, J. and A. Piersol. 1971. Random Data: Analysis and Measurement Procedures. Wiley Interscience, New York.

Boudreau, B. and N. Guinasso. 1982. The influence of a diffusive sublayer on accretion, dissolution, and diagenesis at the sea floor. In: The Dynamic Environment of the Ocean Floor, K. Fanning and F. Manheim (ed.), F. T. Lexington Book Co., Lexington, pp. 115-142.

Bowden, K. 1978. Physical problems of the benthic boundary layer. Geophys. Surveys, 3:255-296.

Bowden, K., Y. Desaubies, A. Fuhrboter *et al.* 1976. Velocity variations, turbulence and stability. In: The Benthic Boundary Layer, I. McCave (ed.), Plenum Press, New York, pp. 231-245.

Cacchione, D., W. Grant, D. Drake and S. Glenn. 1987. Storm-dominated bottom boundary layer dynamics of the Northern California continental shelf: measurements and predictions. J. Geophys. Res., 92:1817-1827.

Caldwell, D. and T. Chriss. 1979. The viscous sublayer at the sea floor. Science, 205:1131-1132.

Chriss, T. and D. Caldwell. 1982. Evidence of the influence of form drag on bottom boundary layer flow. J. Geophys. Res., 87:4148-4154.

Dakhoul, Y. and K. Bedford, K. 1986. Improved averaging method for turbulent flow simulation, Part I: Theoretical development and application to Burger's transport equation. Int. J. Num. Methods in Fluids, 6:49-64.

Davis, A., R. Soulsby and H. King. 1988. A numerical model of the combined wave and current boundary layer. J. Geophys. Res., 93:491-508.

Deacon, E. 1959. The measurement of turbulent transfer in the lower atmosphere. In: Adv. in Geophysics, Vol. 6, Academic Press, New York, pp. 211-228.

Finnigan, J., F. Einaudi and D. Fau. 1984. The interaction between an internal gravity wave and turbulence in the stably-stratified noctual boundary layer. J. Atmos. Sci., 41:2409-2436.

Gibson, M. and B. Launder. 1976. On the calculation of horizontal, turbulent free shear flow under gravitational influence. J. Heat Transfer, 98C:81-87.

Gibson, M. and B. Launder 1978. Ground effects on pressure fluctuations in the atmospheric boundary layer. J. Fluid Mech., 86:491-499.

Gieskes, J., M. Brewers, S. Eittreim *et al.* 1976, Gradients of velocity and physical, biological and chemical properties. In: The Benthic Boundary Layer, I. McCave (ed.), Plenum Press, New York, pp. 211-229.

Glenn, S. and W. Grant. 1987. A suspended sediment stratification correction for combined wave and current flows. J. Geophys. Res., 92:8244-8264.

Grant, W. and O. Madsen. 1986. The continental-shelf bottom boundary layer. In: Annual Review of Fluid Mechanics, M. Van Dyke, J. Wehausen, and J. Lumley (ed.), Annual Review Inc., Palo Alto, pp. 265-306.

Grant, W., A. Williams, 3rd and S. Glenn. 1984. Bottom stresss estimates and their prediction on the Northern California continental-shelf during CODE-1: The importance of wave current interaction. J. Phys. Oceanog, 14:506-527.

Gross, T. and A. Nowell. 1983. Mean flow and turbulence scaling in a tidal boundary layer. Cont. Shelf Research, 2:109-126.

Gross, T. and A. Nowell. 1985. Spectral scaling in a tidal boundary layer. J. Phys. Oceanog, 15:496-508.

Gust, G. 1984. The benthic boundary layer. In: Oceanography, Vol. 3, Ch.8, Springer-Verlag, Berlin.

Gust, G. and J. Southard. 1983. Effects of weak bed load on the universal law of the wall. J. Geophys. Res., 88:5939-5952.

Guza, R. 1988. Comments on kinematic and dynamic estimates from electromagnetic current meter data by D. G. Aubrey an d J. H. Trowbridge. J. Geophys. Res., 93:1337-1343.

Hayter, E. and A. Mehta. 1985. Modeling of estuarial fine sediment transport for tracking pollutant movement. Rept. Univ. of Fla., UFL/COEL82/009, U.S. Env. Protection Agency.

Head, M. 1960. Entrainment in the turbulent boundary layer. ARC-RM 3152, London, United Kingdom.

Heathershaw, A. and J. Simpson. 1978. The sampling variabilitiy of the Reynolds stress and its relation to boundary shear stress and drag coefficient measurements. Estuarine Coastal Marine Sci, 6:263-274.

Hinz, J. 1975. Turbulence. McGraw-Hill Co., New York.

Hunt, J. 1954. The turbulent transport of suspended sediment in open channels.

Proc. Royal. Soc. London, 224A, pp. 322-335.

Kolomogorov, A. 1941. *Dokl Acad. Nau.* SSR 32, pp. 19-21.

Launder, B. 1985. Progress and prospects in phenomenological turbulence models. In: Theoretical Approaches to Turbulence, D. Dwoyer, M. Hussaini and R. Voigt (ed.), Springer-Verlag, New York, pp. 155-186.

Libicki, C. and K. Bedford. 1988. Remote and in-situ methods for sub-bottom sediment characterization. J. Coastal Science (to appear).

Libicki, C., K. Bedford. and J. Lynch. 1989. The interpretation and evaluation of a 3MHz acoustic device for measuring benthic boundary layer sediment dynamics. J. Acoustical Soc. of Amer., 85:1501-1511.

Libicki, C., K. Bedford, R. Van Evra III and J. Lynch, 1987. A 3 MHz acoustic sediment profiling system. In: Coastal Sediments '87, N. Kraus (ed.), Amer. Soc. Civil Engrs., New York, pp. 236-249.

Lumley, J. 1976. Two-phase and non-Newtonian flows. In: Turbulence, P. Bradshaw (ed.), Springer-Verlag, Berlin, pp. 290-324.

Lumley, J. and H. Panofsky. 1964. The Structure of Atmospheric Turbulence. Wiley-Interscience, New York.

Lumley, J. and E. Terray. 1983. Kinematics of turbulence convected by a random wave field. J. Phys. Oceanography, 13:2000-2007.

McCave, I. (ed.), 1976. The Benthic Boundary Layer. Plenum Press, New York.

McClean, S. and J. Smith. 1986. A model for flow over two-dimensional bed forms. J. Hydraulic Engrg., 112:300-317.

McTigue, D. 1981. Mixture theory for suspended sediment transport. J. Hydraulics Div., 107:659-674.

Mellor, G. and P. Yamada. 1982. Development of a turbulence-closure model for geophysical fluid problems. Rev. of Geophys. and Space Physics, 20:851-875.

Monin, A. and A. Yaglom. 1975. Statistical Hydromechanics. MIT Press, Cambridge, MA.

Morris, H. 1955. A new concept of flow in rough conduits. Trans. Amer. Soc. Civil Engr., 120:373-398.

Nihoul, J. (ed.). 1977. Bottom Turbulence. Elsevier Sci. Pub. Co., Amsterdam.

Nowell, A. 1983. The benthic boundary layer and sediment transport. Rev. of

Geophys. and Space Phys., 21:1181-1192.

Nowell, A. and M. Church. 1979. Turbulent flow in a depth-limited boundary layer. J. Geophys. Res., 84:4816-4824.

Nowell, A. and C. Hollister. 1985. Deep Ocean Sediment Transport. Elsevier, Amsterdam.

Offen, G. and S. Kline. 1968. A proposed model for the bursting process in turbulent boundary layers. J. Fluid Mech., 70:209-228.

Oran, E. and J. Boris. 1987. Numerical Simulation of Reactive Flow. Elsevier, New York.

Parchure, T. and A. Mehta. 1985. Erosion of soft cohesive sediment deposits. J. Hydraulic Engrg., 111:1308-1326.

Partheniades, E. 1962. A study of erosion and deposition of cohesive soils in salt water. Ph.D. Dissertation, U. of California, Berkeley, CA.

Partheniades, E. 1971. Erosion and deposition. In: River Mechanics, Vol. II, H. Shen (ed.), Water Resources Press, Fort Collins, CO.

Praturi, A. and R. Brodkey. 1978. A stereoscopic visual study of coherent structures in turbulent shear flow. J. Fluid Mech., 89:251-272.

Press, W., B. Flannery, S. Teukolsky and W. Vetterling. 1986. Numerical Recipes. Cambridge University Press, New York.

Scarlatos, P. 1981. On the numerical modeling of cohesive sediment transport. J. Hydraulic Res., 19:61-67.

Schlicting, H. 1979. Boundary Layer Theory. McGraw-Hill Co., New York.

Smith, J. 1977. Modeling of sediment transport on continental shelves. In: The Sea, Vol. 6, E. Goldbert, I. McCave, J. O'Brien and J. Steele (ed.), Wiley Interscience, New York, pp. 539-577.

Soulsby, R. 1980. Selecting record length and digitization rate for nearbed turbulence measurements. J. Phys. Oceanog, 10:208-219.

Sternberg, R. 1972. Predicting initial motion and bedload transport of sediment particles in the shallow marine environment. In: Shelf Sediment Transport: Process and Pattern, D. Swift, D. Duane and O. Pilkey (ed.), Dowden, Hutchinson and Ross, Stroudsburg, PA, pp. 61-82.

Townsend, A. 1976. The Structure of Turbulent Shear Flow. Cambridge University Press, Cambridge, United Kingdom.

Tsai, C-H and W. Lick. 1987. A portable service for measuring sediment resuspension. J. Great Lakes Res., 12:314-321.

van Rijn, L. 1986. Mathematical models for sediment concentration profiles in steady flow. In: Transport of Suspended Solids in Open Channels, W. Bechteler and A. A. Balkema (ed.), Rotterdam, pp. 49-68.

Wallace, J., R. Brodkey and H. Eckelman. 1977. Pattern-recognized structures in bounded turbulent shear flows. J. Fluid Mech., 83:673-693.

Weatherly, G. and P. Martin. 1978. On the structure and dynamics of the oceanic bottom boundary layer. J. Phys. Oceanog, 8:557-570.

Williams, III, A. 1985. BASS, an acoustic current meter array for benthic flowfield measurements. Marine Geol, 66:345-355.

Williams, III, A., J. Tochko, R. Koehler et al. 1987. Measurement of turbulence in the oceanic bottom boundary layer with an acoustic current meter array. J. Atmos. and Oceanic Tech., 4:312-327.

Wimbush, M. and W. Munk. 1970. The benthic boundary layer. In: The Sea, Vol. 4, A. Maxwell (ed.), Wiley Interscience, New York, pp. 731- 758.

Yalin, M. 1975. Mechanics of Sediment Transport. Pergamon Press, London.

Zanevald, J., R. Spinrad and R. Bartz. 1979. Optical properties of turbidity standards. Soc. Photo-Optical Inst. Engrs., 208:159-168.

Ziegler, K. and W. Lick. 1986. A numerical model of resuspension, deposition and transport of fine-grained sediments in shallow water. Rept. Dept. Mech. and Environmental Engrg., U. of California, Santa Barbara, CA.

# CHAPTER 6

# The Impact of Physical Processes at the Sediment/Water Interface in Large Lakes

P.G. Sly

## Abstract

This paper provides a summary of the more important physical process interactive at the sediment/water interface based on implications drawn from particle-size data. Examples are taken mostly from the Great Lakes, and from Lake Ontario in particular. The summary is largely based on previous work by the author and is intentionally limited in the depth of discussion. Aspects of sediment geochemistry and biological productivity and yield are also linked to physical processes active at the sediment/water interface.

## Introduction

This paper provides a summary of many of the more important physical processes active at the sediment/water interface that may be implied from particle-size data. Most of the examples in support of this summary are drawn from the North American Great Lakes, and from Lake Ontario in particular.

Relationships between particle-size and water depth are used to contrast the greater hydraulic energies of the marine environment relative to large lakes and to highlight anomalies, in particular the presence of relict deposits. Particle-size is used to differentiate among environments in which erosion, transportation, or deposition dominate the sand fraction. It is also used to discriminate among deposits that share a

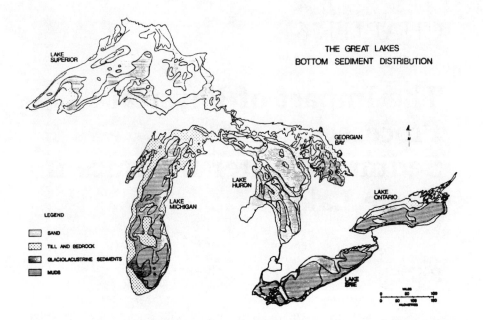

Figure 1. Distribution of sediment types in the Great Lakes, modified after Sly and Thomas (1974).

common mean particle-size but which may or may not be in hydraulic equilibrium. In the silt and clay size fractions, particle-size is used to discriminate between deposits that are likely formed in response to rather rapid settlement (as aggregates) and those that slowly accumulate in response to the settlement of individual particles. Particle-size and other characteristics of the nepheloid layer raise questions about the movement of very fine particles in deep water, and the process of sediment focusing.

## Particle-Size/Water Depth Relationships

Figure 1 shows the general distribution of sediments in the Great Lakes. The distribution is characterized by a very narrow peripheral zone of coarse fraction materials (sands and gravels), patchy exposures of glacial till and glaciolacustrine clay, and the dominance of silt and clay sedimentary basins. The distribution is strongly influenced by isostatic adjustment and bathymetry and, to some extent, material supply.

Rates of sedimentation in the Great Lakes are extremely variable. Much of the sand size fraction is easily reworked and therefore accumulations based on the thickness of such deposits are likely to be biased and unreliable unless monitored over extended time. Off the Niagara River mouth, rates of accumulations of sandy silts (at about 30

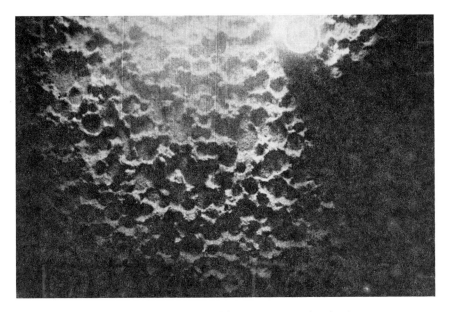

Figure 2. Sandy silt entrapped in pitted limestone at a depth of 24 m, Georgian Bay, after Sly and Sandilands (1988).

m depth) are about 1.8 mm/year, clayey silts (at depths between 70 and 90 m) range between 1.4 and 0.2 mm/year, and silty clays (more than 100 m depth) are about 0.08 mm/year (Sly, 1983b). Thus, although rising water levels provide conditions under which modern sediments will eventually cover relict deposits, the latter may continue to be exposed at sites of low modern sedimentation or in areas of non deposition. Figure 2 shows a pitted limestone surface (depth 24 m) in Georgian Bay. At this site, the micro-surface has remained within a non depositional zone for some time because of a lack of coarse and medium sand fractions in source materials. Only as water depths increased sufficiently to allow entrapment of silty materials has accumulation begun to occur. Despite the fact that submergence at this site may have begun over 6000 years before present(BP) (Lewis, 1969), the entrapped fines are of very much younger age.

Figure 3 provides a summary of the various forms of physical forcing functions that may influence water motions. Wind stress and, to a much lesser extent, riverine inflow are the principal causes of water motion in the Great Lakes. In the marine environment, tide induced flow may be of even greater significance than wind driven motions but this influence is negligible in the Great Lakes. Table 1 provides a comparison of hydraulic energies in each of the Great Lakes and from the Scotian Shelf (Atlantic seaboard), implied from the deposition depth of the

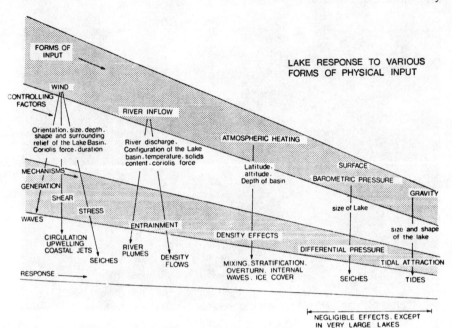

Figure 3. Physical stress-response effects in lake waters, after Sly (1978).

medium sand size fraction (2.5 phi) and the depth spread of the 2.0-3.0 phi sand size interval. These values are based on trend lines fitted over distribution data derived from numerous sources. Although the data are not complete for each lake and may reflect different sampling biases, they demonstrate a clear increase in hydraulic energy with lake size. Based on the general approximation that orbital velocities reduce by half at a depth of 1/9 of the wave length (Bascom, 1964), seabed environments such as the Scotian Shelf are at least twice as dynamic as those of Lake Superior which is, itself, two to three times more dynamic than Lake Ontario. Table 1 is consistent with wave frequency data from the Great Lakes, predicted from meterological records and supported by direct observations (Chesters, 1978).

Although sediment particle-size generally decreases with water depth, there are anomalies in this relationship. Figures 4 and 5 provide examples from the east and west ends of Lake Ontario. Using the formulation of Sheng and Lick (1979), near bed shear velocities have been calculated for wave conditions that approximate significant wave heights. The shear velocities have been related to particle-size to predict the size of material that could be set in motion at depth. The predicted size/depth relationships have been superimposed over actual sediment/depth sample data and generally show good agreement. Some anomalies, however, are noted on both figures. These anomalies indicate

Table 1

Comparative Hydraulic Energies in Large
Lakes and the Marine Environment

|  | 1,2,3 Ontario | 4 Erie | 5 Huron | 6 Michigan | 7,8,9 Superior | 10 Scotian Shelf |
|---|---|---|---|---|---|---|
| Mean water depth to 2.5 phi | 10-15 | 10-15 | 35-40 | 40-45 | 50-55 | 120-140m |
| Depth range for fraction 2.0 - 3.0 phi | 10 | 5-10 | 35 | 30 | 50 m | --- |

1) Sly (1983a)
2) Sly (1984)
3) Thomas et al. (1972)
4) Thomas et al. (1976)
5) Thomas et al. (1973)
6) Cahil (1981)
7) Mothersill (1969)
8) Mothersill (1971)
9) Thomas and Jaquet (1975)
10) Yorath (1967)

Note: Values approximate and rounded to nearest 5 m; areas of non deposition and anomalous deepwater deposits are excluded (see text).

that coarse sediments are present at depths significantly greater than would be expected from available hydraulic energies. Three possible explanations can be advanced: (1) local bathymetric control causing wave focusing or increased current velocity, (2) source control, and (3) the sediments are relict materials and derive from conditions unlike those presently existing in the lake.

Based on the evidence presented so far, it is not possible to be sure which of the explanations should apply. However, water levels in many of the Great Lakes have varied by more than +/- 100 m present level since postglacial times (Karrow and Calkin, 1985), and it is probable that such changes could exert considerable influence on partizle-size/depth relationships. Certainly, in Figure 5, the noted anomaly is coincident with the Niagara River outer bar deposits that were formed about 4000 years BP and that occur at a depth of about 20 m below present lake level (Sly, 1983a). The particle-size/depth relationships in Figures 4 and 5 also demonstrate that, with the dominance of SW winds across Lake Ontario (Phillips and McCulloch, 1972), wave motions are usually much greater in the Kingston area than at the Niagara River mouth. The mean depth of the 2.5 phi sand size is about 22 m at Kingston and only about 10 m off the Niagara River mouth.

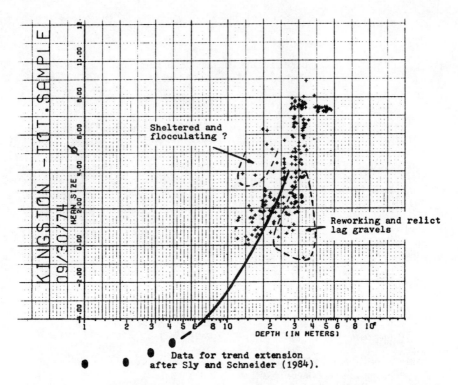

Figure 4. Particle-size trend lines predicted from significant wave height (Chesters, 1978) in the Kingston basin of Lake Ontario.

## Depositional Regime and Hydraulic Equilibrium

The form of the standard deviation envelope (Figure 6) provides a clear indication of particle-size end member population mixings (boulder/cobble - sand, and sand - clay). The sand minimum at about 2.5 phi is coincident with the lowest shear velocities required to induce particle motion (Hjulström, 1939). Materials in the fraction 2-3 phi may be considered those most easily mobilized at the minimum threshold of shear velocities. The presence of progressively coarser materials implies increasing water motion and greater erosional strength. Deposits characterized by finer particle-size (>2.5 phi) are more and more dominated by depositional regimes. That is, in the mid-sand fraction, particle-size indicates a balance between the dominance of erosional and depositional conditions, rather than some constant condition. Unfortunately, while mean particle-size is a general indicator of hydraulic energies it may not specifically relate to formative conditions (Sly, 1989b). For example, the minimum particle-size present in erosional deposits is most likely to relate to maximum shear velocities, and an occurrence of

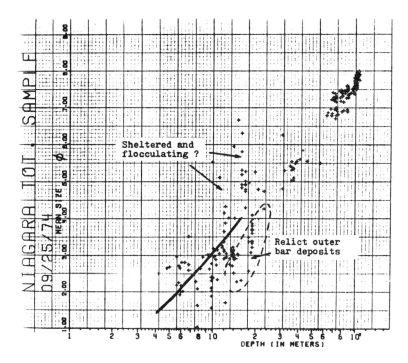

Figure 5. Particle-size trend lines predicted from significant wave height (Chesters, 1978) off the mouth of the Niagara River, Lake Ontario.

sands composed of the most easily mobilized fractions may be quite unrelated to any velocity in excess of the shear velocity threshold (effectively, the source material is the controlling factor and not the hydraulic energy).

The third and fourth moment measures (skewness and kurtosis) provide an important means of discriminating between sediments that are or are not in hydraulic equilibrium with their environment. Like the standard deviation envelope, there is a well pronounced feature in skewness/kurtosis relationships that marks the minimum threshold velocity for sand size particulates (Figure 7). Slight shifts towards erosional (high energy) or depositional (low energy) regimes, by the inclusion of small percentages of coarser or finer materials (respectively) cause major shifts in skewness and kurtosis values, which are particularly sensitive to end-member mixings with the sand fraction (Sly et al., 1983). Based on empirical relationships in marine and freshwater sediments, it has been found that those samples whose particle-size characteristics are most closely related to hydraulic shear velocities fall along the outer boundary curve of Figure 7. Sediment samples that plot progressively further away from this curve indicate an anamalous relationship with

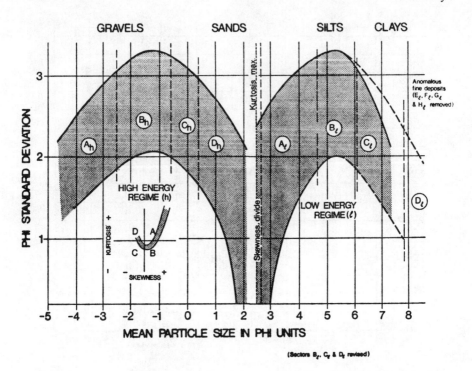

Figure 6. Standard deviation envelope based on empirical data, after Sly *et al.* (1983).

respect to hydraulic conditions. These empirical relationships are universal. Typical examples of such anomalies include: ice-drop debris (coarse particles in clayey silts), transitional interface materials (sandy silts in coarse gravels), and bimodal sands (mixed source populations).

Particle-size anomalies in sediment samples can be caused by several things: sampling at too great a depth in the substrate and thereby including more than one depositional layer; sediment mixing (biological or physical); fluctuating plume dispersions (or other source dispersions), which result in micro-variations within depositional units; or recovery of materials at a transitional interface (such as a relict lag gravel partly covered by modern silts), as a result of water level changes.

By using skewness/kurtosis values in addition to mean particle-size, it is possible to define the likely cause of anomalies that appear in relationships between mean particle-size and water depth. If the skewness/kurtosis relationships indicate that samples plot close to the boundary curve (Figure 7), mean size/depth anomalies are likely related to some feature of present-day hydraulic energy focusing or shelter, such as caused by bathymetric features. On the other hand, if the

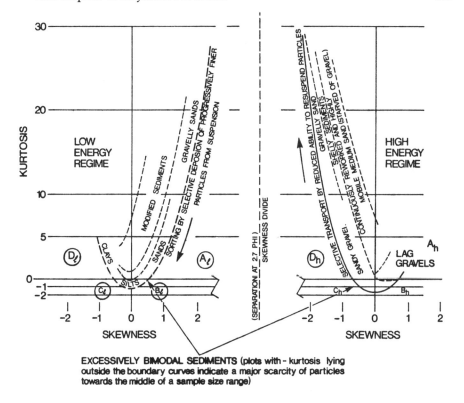

Figure 7. Skewness/kurtosis relationships for low and high energy regime sediments, after Sly et al. (1983).

skewness/kurtosis relationships are also anomalous, they can be used to define the cause of the anomaly. The presence of fines in otherwise medium-coarse sands (Figure 5 and Sly, 1983a) strongly supports the thesis that sediments of the Niagara outer bar are relict nearshore materials that now include small quantities of silty fines from modern plume dispersions of the river. In Figure 4, high energy regime relict lag gravels are modified by the presence of modern sandy silts (Sly, 1984); again, this is consistent with the influence of recent rising lake levels (Sly and Prior, 1984).

## Mean Particle-Size and the Silt/Clay Ratio

Figure 8 presents a remarkably well defined relationship between the percentage composition of silt and clay size fractions in predominantly depositional sediments. The numerical values noted at regular intercepts additionally define total sample mean phi size at these points. As samples fine from about 2.5 phi, the silt content increases to a maximum of nearly 70% at about 6.2 phi and thereafter it declines almost linearly to a minimum content of about 30% at 8 phi. The silt/clay ratio can be made unique only by plotting it against total sample mean particle-size. The relationships between the silt and clay fractions are, however, extremely consistent. In Figure 9, the silt/clay ratio is plotted against the total sample mean particle-size. Most of the data are derived from the Niagara site in Lake Ontario but, to extend the trend in data at the finest particle-sizes, a block of mid-lake samples have been added from a Tobermory data set (Georgian Bay, Lake Huron; Sly and Sandilands, 1988). This figure shows that the silt/clay ratios are particularly variable in samples with a total mean size between 1.5 and 6 phi. It also shows that at a total mean size finer than about 6 phi, there is almost no departure from the plotted trend line. It is implied that the combined effects of variable plume dispersion and flocculation cause the data scatter noted in samples of total mean size coarser than 6 phi. Likewise, it is implied that the lack of scatter in samples finer than about 6 phi reflects the fall of discrete particles at low concentrations and under conditions of minimum turbulence that allow sorting on the basis of subtle differences in particle-size. These interpretations are fully consistent with the field and experimental observations of Kranck (1975; 1980) and also with the distribution and interpretation of standard deviation values shown in Figure 3.

As a further guide to the particle-size structure of these sediment samples, the skewness/kurtosis values have been separated into two groups in the low energy regime sediments (total sample mean size finer than about 2.5 phi, and denoted by subscript e). Group 1 refers to normal samples that are in hydraulic equilibrium and lie close to the boundary curve of Figure 7. Group 2 refers to samples that depart from such equilibrium conditions and are anomalous. As a means of simple differentiation, group 1 contains no material coarser than 0 phi, but group 2 always contains some material of gravel size or larger. The sectors A, B, C, D (group 1) and E, F, G, H (group 2) refer to the + and - sign relationships of skewness/kurtosis values defined in Figure 7. Intercept plots indicate the total sample mean size (vertical axis) of sediments falling within each sector. High energy regime samples are denoted by subscript h and fall within only one skewness/kurtosis sector (D), as noted in Figure 7.

Plots of the spatial distributions of low energy sediments off the Niagara River mouth (Sly, 1983a) show that total sample mean particle-size is most variable in areas subject to plume excursions, and that samples containing sand, silt and clay have progressively less sand and clay as waters deepen over the nearshore zone (even at depths to 30 m or so; Sly, 1973). The rapid deposition that is necessary to retain much of

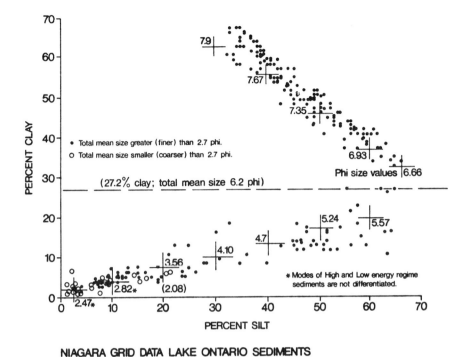

Figure 8. Percent silt and clay composition and mean particle-size of samples from the Niagara area of Lake Ontario, after Sly (1988a).

the silt fraction in this generally narrow zone (Sly, 1983a) can only occur as a result of flocculation and the formation of fine particle aggregates. This is consistent with the observations of Rosa (1985), regarding selective particle-size removal from the suspended load of adjacent lakewaters.

## Basin Deposits and the Nepheloid Layer

The accumulation of materials (mostly finer than about 6.5 phi) in mid-lake depositional basins is characterized by Stokean settlement, and it would be expected that particles pass directly across the water/sediment interface as they reach the bottom of the water column, providing that near bed shear velocities are negligible. However, during much of the stratified period, light transmission profiles and mid-lake water samples indicate that particle concentrations increase with water depth. The percent light transmission decreases steadily below the thermocline (Sandilands and Mudroch, 1983). In some cases, this nepheloid layer is particularly evident as a distinct layer above the lakebed (author, unpublished). At first, formation of the nepheloid layer appears to be

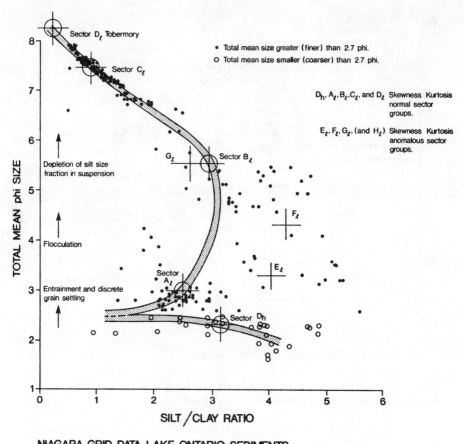

NIAGARA GRID DATA LAKE ONTARIO SEDIMENTS

Figure 9. The silt/clay ratio, in relation to mean particle-size, of sediments from the Niagara area of Lake Ontario, after Sly (1988a).

anomalous with respect to particle settlement; formation of the layer, however, indicates two effects: (1) A decrease in particle-size occurs with increasing water depth, coincident with increasing degradation of organic matter (Kemp and Lewis, 1968; Kemp, 1971). (2) There is a decrease in settling velocity with water depth, as particles degrade and become smaller. Particulates composed of calcium carbonate (Robertson and Scavia, 1983) also decrease in size with water depth, as they progressively dissolve in cold mid-lake waters. The net of these effects is to cause the concentration of particulates to increase but particle-size to decrease, as fine materials settle deeper through the mid-lake water column. The nepheloid layer persists throughout much of the stratified period but becomes dispersed with the onset of the fall overturn and formation of isothermal conditions. Based on calculations relating to organic carbon contents in the water column and surficial sediments, Sly (1987) estimated that about 50% of the particulate matter generated annually in the water column became entrained in the bottom sediments and that recycling in the water column would result in a maximum residence time of about 5 years for refractory organics and diatom fragments, which comprise most of the materials.

## Particle-Size, Geochemistry, and Sediment Focusing

Relationships between particle-size and geochemistry have been extensively described by numerous authors; and, for example, specific size fraction relationships with organic carbon, coprecipitates and clays are well known in the Great Lakes. On a more general basis, however, the spatial distributions of concentrations of geochemical parameters exhibit minimum variance in mid-lake basin sediments. Concentration gradients and spatial variability increase greatly around the basin margins (Sly, 1983a). Basin sediments reflect mid-lake water chemistry and primary production. The influence of shoreline contributions and point source discharge is most evident in the marginal sediments. Based on skewness/kurtosis relationships (Figure 7), sector B sediments (mean size about 5.5 phi) lie entirely within the influence of nearshore effects (Sly, 1983a). Sector C materials (mean size about 7.5 phi) mostly characterize mid-lake conditions but share some sector B characteristics near the C/B boundary. Sector D samples (mean size about 8.5 phi) are usually associated with glaciolacustrine clays that have very low organic carbon contents, although not exclusively so (Sly and Sandilands, 1988). These broad distributions, therefore, suggest partial separation of suspended loads in nearshore and offshore waters, based on major circulatory patterns (Sly, 1973).

Further, and at very large scales, lake dynamics may influence benthic productivity. Potentially this may be caused by restricting development of the oxidized surface micro-zone in deep water sediments during the stratified period, as a result of limited near bed water motion or stagnation (Sly, 1987). This effect implies that lake area to depth relationships could have subtle effects on fish yield.

The concept of sediment focusing suggests that fine particle-size materials follow some well developed pathway, from nearshore to mid-lake bottom. In many small lakes this may be true, but, in large lakes, with considerable areas of open water devoid of edge effects, the concept becomes less clear. As previously noted, mid-basin sediments strongly reflect open water chemistry and primary production. In some situations, however, this is modified by material contributed from marginal sources. Figure 10 provides dramatic evidence of this. Mirex was known to be derived from two major sources to Lake Ontario, the Niagara and Oswego Rivers (N and O, respectively). It is associated with fine particulates entrained in the nearshore circulation along the south shore of Lake Ontario. Between 1968 and 1977, the northern and westward extensions of the Oswego plume clearly follow the main circulatory gyre at the eastern end of the lake. Along much of the southern shore, however, there are northward extending "fingers" of high concentration that seem to break away from the pattern of eastward moving entrained flow. It is thought that these fingers represent "bursting" from the nearshore circulatory pattern under the influence of irregular northward bottom flows. These bottom flows provide a return for periodic upwellings along the north shore during the stratified period and fall overturn (Lee, 1972). Thus, mid-lake sediments are modified by a periodic influx from the nearshore zone but this is a considerably more complex phenomenon than suggested by the concept of simple focusing.

## Erosion

Particle-size, of course, may be used to define specific breaks in depositional regime, but it is difficult to be specific about the distribution of erosional zones unless particle-size is not a limiting factor. The removal of fine sands from a lag-gravel clearly defines an erosional condition but, without detailed analysis of near surface structure, it is hard to resolve the effects of erosion in modern muds (undefined mixtures of silt and clay). Based on Postma (1967) and the very low shear strengths of lake muds (sometimes less than 0.1 kPa; Sly, 1983b; 1984), these sediments may be as easily reworked as the most mobile sands. Thus redistribution of muds as a result of periodic high flow or turbulence events is quite possible (as noted in Figure 10 and discussed above). Based on hypolimnion flow velocities (Sly, 1973) and storm conditions (Chesters, 1978), it seems possible to rework muds at depths of 30-70 m in many parts of Lake Ontario.

In the Kingston basin of Lake Ontario, modern muds with relatively high contaminant levels form a veneer over rock and till surfaces (Sly, 1984). These muds do not accumulate to great thickness even though they may settle at rates >1 cm/year. Instead, they are re-mobilized under the influence of fall storms, and the particulates are entrained within the channel outflow of the St. Lawrence River. This effect is certainly present at depths of about 30 m (Figure 11) and may extend to depths of more than 70 m within the St. Lawrence trough (Sly, 1984).

In mid-basin areas, at depths of about 100 m and greater, hypolimnion shear velocities are usually far below the threshold required to erode modern muds (Sly, 1973), and this is largely substantiated by underwater photography (author, unpublished data) that shows no evidence of physical reworking at the sediment/water interface. However, there are zones of exposed glaciolacustrine clay that separate distinct sedimentary basins in Lake Ontario (Thomas *et al.*, 1972). At these locations, it is likely that bottom flows are sufficient to inhibit settlement of fines (non-deposition) rather than to erode materials.

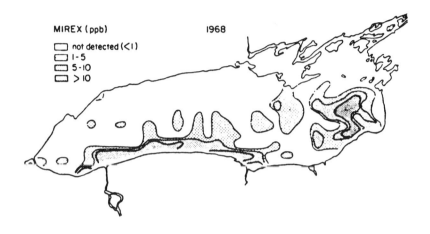

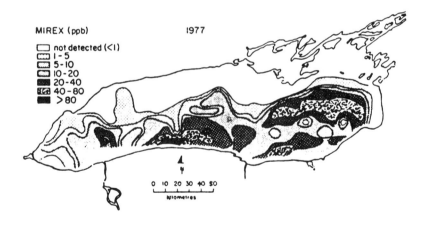

Figure 10. Distribution of mirex in the sediments of Lake Ontario, after Thomas and Frank (1987).

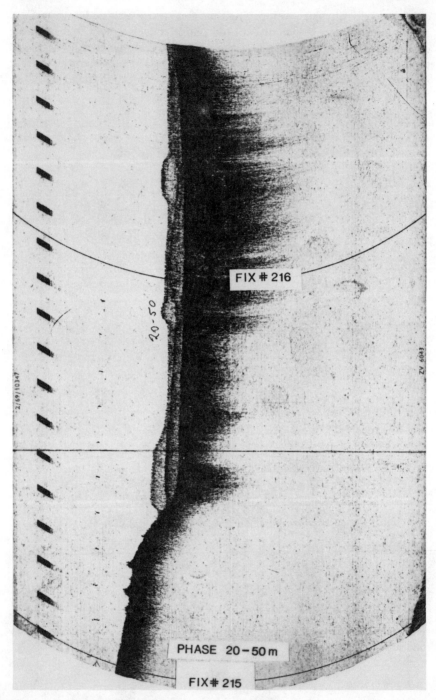

Figure 11. Partly eroded modern muds over glaciolacustrine clays at a depth of about 30 m in the Kingston basin of Lake Ontario.

# References

Bascom, W. 1964. Waves and Beaches - The Dynamics of the Ocean Surface. Doubleday, New York.

Cahill, R.A. 1981. Geochemistry of Recent Lake Michigan Sediments. Illinois State Geol. Surv., Circular 517, Champaign, Ill.

Chesters, G. 1978. Frequency and Extent of Wind-Induced Resuspension of Bottom Material in the US Great Lakes Nearshore Water. Great Lakes Basin Comm., Contract Report 77D1, Water Res. Center, Univ. Wisconsin, Madison, WI.

Hjulström, F. 1939. Transportation of Detritus by Moving Water. In: Trask, P.D. (ed.), <u>Recent Marine Sediments Symp.</u>, Amer. Assoc. Petrol. Geol., Tulsa, OK.

Karrow, P.F. and P.E. Calkin (eds.). 1985. Quaternary Evolution of the Great Lakes. Geol. Assoc. Can. Spec. Paper 30, Memorial Univ., St. John's, Nfd.

Kemp, A.L.W. 1971. Organic Carbon and Nitrogen in the Surface Sediments of Lake Ontario, Erie and Huron. *J. Sediment. Petrol.*, 41:537-548.

Kemp, A.L.W. and C.F.M. Lewis. 1968. A Preliminary Investigation of Chlorophyll Degradation Products in Lakes Erie and Ontario. *Proc. 11th Conf. Great Lakes Res.*, Internat. Assoc. Great Lakes Res., Ann Arbor. 206-229.

Kranck, K. 1975. Sediment Deposition from Flocculated Suspension. *Sedimentology*, 22:111-123.

Kranck, K. 1980. Experiments on the Significance of Flocculation in the Settling of Fine-Grained Sediment in Still Water. *Can. J. Earth Sci.*, 17:1517-1526.

Lee, A.H. 1972. Some Thermal and Chemical Characteristics of Lake Ontario in Relation to Space and Time. Report EG-6, Inst. Environ. Sci. and Engng., Great Lakes Inst. Univ. of Toronto.

Lewis, C.F.M. 1969. Late Quaternary History of Lake Levels in the Huron and Erie Basins. *Proc. 12th Conf. Great Lakes Res.*, Internat. Assoc. Great Lakes Res., 250-270.

Mothersill, J.S. 1969. A Grain Size Analysis of Longshore Bars and Troughs, Lake Superior. *J. Sediment. Petrol.*, 39:1317.

Mothersill, J.S. 1971. Limnological Studies of the Eastern Part of the Lake Superior Basin. *Can. J. Earth Sci.*, 8:1043-1055.

Phillips, D.W. and J.A.W. McCulloch. 1972. The Climate of the Great Lakes Basin. Climatological Studies No. 20, Environment Canada, Toronto.

Postma, H. 1967. Sediment Transport and Sedimentation in the Estuarine Environment. In: Lauff, G.H. (ed.), *Estuaries*, Amer. Assoc. Adv. Science. Publ. 83. Washington, DC, 158-179.

Robertson, A. and D. Scavia. 1983. North American Great Lakes. In: Taub, F. (ed.), *Lake and Reservoir Ecosystems*, Vol. 23, *Ecosystems of the World*, Elsevier, Amsterdam, 135-176.

Rosa, F. 1985. Sedimentation and Sediment Resuspension in Lake Ontario. *J. Great Lakes Res.*, 11:13-25.

Sandilands, R.G. and A. Mudroch. 1983. Nepheloid Layer in Lake Ontario. *J. Great Lakes Res.*, 9:190-200.

Sheng, P.Y. and Lick, W. 1979. The Transport and Resuspension of Sediments in a Shallow Lake. *J. Geophys. REs.*, 84(C4):1809-1826.

Sly, P.G. 1973. Sediment Processes in Great Lakes. In: Fluvial Processes and Sedimentation. *Proc. Hydrol. Symp. University of Alberta*, Calgary, May 1973, Nat. Res. Council, Ottawa, 465-492.

Sly, P.G. 1978. Sedimentary Processes in Lakes. In: Lerman, A. (ed.), *Lakes: Chemistry, Geology and Physics*, Springer Verlag, New York, 65-89.

Sly, P.G. 1983a. Sedimentology and Geochemistry of Recent Sediments Off the Mouth of the Niagara River, Lake Ontario. *J. Great Lakes REs.*, 9:134-159.

Sly, P.G. 1983b. Recent Sediment Stratigraphy and Geotechnical Characteristics of Foreset and Bottomset Beds of the Niagara Bar. *J. Great Lakes REs.*, 9:224-233.

Sly, P.G. 1984. Sedimentology and Geochemistry of Modern Sediments in the Kingston Basin of Lake Ontario. *J. Great Lakes Res.*, 10:358-374.

Sly, P.G. 1987. Benthos as it Relates to the Lake Ontario Food-Web and to Material Flows: Part I. *Proc. Food-Web II Workshop Internat. Joint Comm.*, Windsor, Ontario.

Sly, P.G. 1989a. Sediment Dispersion: Part 1-Fine Sediments and Significance of the Silt/Clay Ratio. *Hydrobiologia* 176/177:99-110.

Sly, P.G. 1989b. Sediment Dispersion: Part 2-Characterization by Size of Sand Fraction and Percent Mud. *Hydrobiologia* 176/177:111-124.

Sly, P.G. and J.W. Prior. 1984. Late Glacial and Postglacial Geology in the Lake Ontario Basin. *Can. J. Earth Sci.*, 21:802-821.

Sly, P.G. and R.G. Sandilands. 1988. Geology and Significance of Sediment Distributions in the Area of the Submerged Niagara Escarpment, Georgian Bay. *Hydrobiologia*, 163:47-76.

Sly, P.G. and C.P. Schneider. 1984. The Significance of Seasonal Changes on a Modern Cobble-Gravel Beach Used by Spawning Lake Trout, Lake Ontario. *J. Great Lakes Res.*, 10:78-84.

Sly, P.G. and R.L. Thomas. 1974. Review of Geological Research as it Relates to an Understanding of Great Lakes Limnology. *J. Fish. Res. Board Can.*, 31:795-825.

Sly, P.G., R.L. Thomas. and B.R. Pelletier. 1983. Interpretation of Moment Measures Derived from Water-Lain Sediments. *Sedimentology.*, 30:219-233.

Thomas, R.L. and R. Frank. 1987. Introduction. In: Thomas R., R. Evans, A. Hamilton, M. Munawar, T. Reynoldson, and H. Sadar. 1-4, Developments in Hydrobiology, DH 39, Junk, Amsterdam.

Thomas, R.L. and J.M. Jaquet. 1975. The Surficial Sediments of Lake Superior. Reprint, 9th Internat. Sedimentol. Conf., Nice.

Thomas, R.L., A.L.W. Kemp. and C.F.M. Lewis. 1972. Distribution, Composition and Characteristics of Surficial sediments of Lake Ontario. *J. Sediment. Petrol.*, 41:66-84.

Thomas, R.L., A.L.M. Kemp and D.F.M. Lewis. 1973. The Surficial Sediments of Lake Huron. *Can. J. Earth Sci.*, 10:266-271.

Thomas, R.L., J.M. Jaquet, A.L.M. Kemp and C.F.M. Lewis. 1976. Surficial Sediments of Lake Erie. *J. Fish. Res. Board Can.*, 33:385-403.

Yorath, C.J. 1967. Determination of Sediment Dispersal Patterns by Statistical and Factor Analyses, Northeast Scotian Shelf. Ph.D. Diss., Dept. Geol., Queen's Univ., Kingston, Ont.

# CHAPTER 7
# Partitioning of Toxic Metals in Natural Water - Sediment Systems

## Herbert E. Allen

Trace metals enter surface waters as a result of both natural processes and man's activities. Physical, chemical, and biological processes acting on these metals can result in transformation, translocation, accumulation, or dimunation of concentration.

Trace metal pollutants may be toxic to or bioaccumulated by aquatic organisms. However, the bioavailability of the metals is related to the concentration of specific chemical forms, not to the total. Therefore, one must be able to measure or to calculate the concentration of specific chemical forms in order to predict the biological effect from chemical analyses.

Fundamental to explaining the partitioning of metals between sediments and the associated pore water is knowledge of the chemistry of the sediment and of the metal in the solution phase. Studies of chemical speciation, both of natural materials and of model components, have provided some of the necessary information. However, while the fundamental principles governing partitioning to pure solid phases are understood, they have not been able to be applied to systems with the heterogeneity of sediments.

## Metal Speciation in the Aqueous Phase

There is an extensive literature that shows that metal toxicity to aquatic organisms is proportional to the concentration of free metal ions and perhaps that of some simple complexes (Sunda *et al.*, 1978; Anderson and Morel, 1978; Allen *et al.*, 1980; O'Donnell *et al.*, 1985; Luoma, 1983). Sunda *et al.* (1978) conducted bioassays using grass shrimp to assess the acute toxicity of cadmium. They found that either increased nitrilotriacetic acid (NTA) or salinity was able to diminish the toxicity. If the free cadmium ion concentration was considered, rather than total added cadmium, the free cadmium ion concentration was directly related to the toxicity irrespective of the concentration of added NTA or salinity.

Anderson and Morel (1978) reported similar results for the toxicity of copper, in the presence and absence of ethylene-diamine-tetra-acetic acid (EDTA), to a dinoflagellate. Allen et al. (1980) conducted algal assays using zinc as the toxicant. They tested five different chelators having stability constants for complexation with zinc, which varied over five orders-of-magnitude. They too found that the observed biological effect could be explained by considering the free metal ion concentration, not the concentration of total zinc. Thus, to be able to explain observed toxicity we must measure or predict these bioavailable forms.

Metals ions, such as $Cu^{2+}$ or $Ni^{2+}$, do not exist as such in aqueous systems. Rather such ions, referred to as free metal ions, are extensively hydrolyzed and have an inner sphere of water molecules. The general metal ion, $M^{2+}$, has a number of coordinated water molecules in its inner coordination sphere. For a metal with a coordination number of 6, it would be more proper to represent the free metal ion as $M(H_2O)_6^{2+}$. Complexation with a ligand such as hydroxide is then seen to be a displacement reaction:

$$M(H_2O)_6^{2+} + OH^- = M(H_2O)_5(OH)^+ + H_2O$$

or commonly:

$$M^{+2} + OH^- = MOH^+ + H_2O \qquad (1)$$

The relationship of the concentrations or, more correctly, the activities, of each of the species at equilibrium is described by the equilibrium constant for the reaction. In the case of the reaction in Eq. 1, this is

$$K = \frac{[MOH^+]}{[M^{2+}][OH^-]} \qquad (2)$$

Other monodentate ligands, such as $CN^-$, $Cl^-$, and $NH_3$ react similarly.

Organic ligands, such as EDTA, also form complexes that are described by chemical equilibria. Protonation of the ligand may have to be accounted for in determining the overall distribution of chemical species in the system. Protons compete with the metal ions for reaction with the ligand. Thus, for EDTA, which is a tetraprotic acid, we must consider the deprotonated species $Y^{4-}$ reacting with protons to form $HY^{3-}$, $H_2Y^{2-}$, $H_3Y^-$, and $H_4Y$ in addition to the metal complex MY.

In the multimetal - multiligand system, the computation of speciation requires use of chemical equilibrium computer programs. Several, such as MINEQL (Westall et al., 1976) and MINTEQ (Brown and Allison, 1987) are available. These programs can compute equilibria for all systems for which equilibrium constants are available.

There is a problem when natural organic complexing agents, such as humic and fulvic substances, are considered. These are decay products of natural organic matter that behave as polyelectrolytes and are important in complexation of metals in natural waters (Saar and Weber, 1982). Because they are decay products, they possess a wide variety of structures. Consequently their properties cannot be described in the same way as are properties of discrete compounds such as EDTA. The acid-base properties of humic substances are predominantly due to phenolic and carboxylic functional groups. These are also the sites to which metal will bind. Therefore, a description of the protonation properties of these materials is needed for an understanding of the metal-binding properties.

A typical acid-base titration curve for a humic extracted from an aquatic sediment is shown in Figure 1. This curve is similar to many others that have been run in this laboratory and elsewhere. It shows none of the strong inflection points characteristic of simple molecules. It has a featureless shape. Titrations with metals give similar curves.

Several approaches have been used to obtain parameters to fit these types of experimental curves (Dzombak et al., 1986). Most commonly the data are fit to models having one or two discrete classes of binding sites using the plotting technique of Scatchard (1949). More recently, a nonlinear least squares method to obtain a set of constants has been used in the computer program written by Westall (1982). Perdue and Lytle

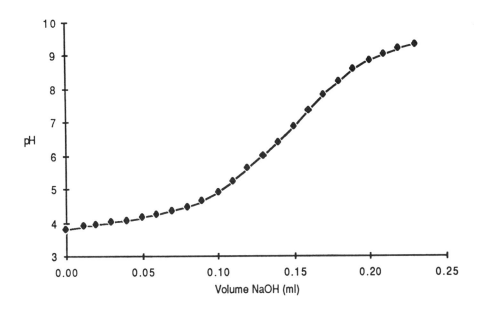

Figure 1. Acid-Base Titration of 50 mg/liter Humic Acid Isolated from Sediment. 0.1 moles NaOH/liter.

(1983) have described the data by a continuous multiligland Gaussian distribution model. The relative concentrations of each of the ligands are normally distributed relative to the $pK_i$ of the ligand

$$\frac{C_i}{C_L} = \frac{1}{\sigma(2\pi)^{1/2}} \exp\left[-\frac{1}{2}\left(\frac{\mu - pK_i}{\sigma}\right)^2\right] dpK \tag{3}$$

Shuman *et al.* (1983) have used an "affinity spectrum" method to evaluate the distribution of binding sites.

Fish *et al.* (1986) have compared the various models and discussed their limitations. They concluded that discrete ligand models are sufficient to describe metal-humate interactions. These constants can be directly incorporated into chemical equilibrium computer programs. However, such data are only now being developed, and the extent of their general applicability to a wide variety of environmental conditions has not been established.

## Metal Partitioning Between the Solid and Aqueous Phases

Sediments are highly heterogeneous solid phases including both mineral phases and detrital organic matter. Metals may be bound in a number of different manners including precipitation, coprecipitation, ion exchange, and adsorption (Förstner and Wittmann, 1979). If only total metal concentrations in sediments are considered, all forms are considered to have equal impact. Some forms of metals in sediment are resistant to exchange with water over long periods of time, and, therefore, they should not be considered in reactions involving partitioning between the sediment and water. For example, metals from rock debris and metals in the matrix of clay minerals are not subject to exchange over the approximately one decade time period of usual concern (Jenne, 1977). More refined consideration of sediment-water interaction requires knowledge of the metal species present.

There have been two principal approaches to the speciation of solid phases. One has been to subject the solids to a sequential selective fractionation, and the other has been a mechanistic approach in which the system is modeled as a series of simultaneous equilibria that are described by the pure component analogs.

### Selective Fractionation Approach

These approaches derive from the work of soil scientists who were attempting to determine the portion of the trace metal in soil that would be available for plant nutrition. Both single solution and sequential extraction approaches have been applied (Förstner and Wittmann, 1979; Salomons and Förstner, 1984). The single solutions employ complexing agents such as EDTA or diethylene-triamine-penta-acetic acid (DTPA) or weak acids to extract metals. Such extractants are nonselective and cannot impart information regarding the site in which the metal had been bound in the solid phase.

The sequential extraction methods have attempted to impart information about the site of metal binding by using a sequence of selective extractants to remove metals in weaker binding locations followed by those in stronger locations. Typically, authors have implied specificity with respect to the phases being extracted despite the operational nature of the method and consequently the interpretation of the results. A typical sequential extraction method, which has been employed in this laboratory (Allen and Unger, 1986), is shown in Table 1. In many studies the residual metals remaining after this, or a similar extraction series, are also determined.

These methods have been severely criticized (Rendell et al., 1980; Martin et al., 1987; Kheboian & Bauer, 1987; Tessier & Campbell, 1988). Some of the problems are readsorption of extracted metals onto residual or freshly exposed solids and a lack of being able to validate the extractions with known solid phases. It should also be noted that, even in the absence of these problems, the sequential extraction methods are not able to predict the distribution of a metal between the sediment and water phases.

## Mechanistic Approaches for Single Phases

Precipitation is the easiest of the mechanisms to be considered. Metal carbonates and metal sulfides are two classes of precipitates of importance in sediments. Precipitation of metals as their sulfides is likely the most important solubility controlling reaction for a number of metals in anoxic systems, while metal carbonates may be of importance in systems in which the alkalinity is high.

Sulfide and carbonate are the conjugate bases of diprotic acids. The solubility of cadmium carbonate, a typical case, is affected by pH due to the formation of bicarbonate ion and carbonic acid at pHs incurred in the environment. Soluble ligands will increase the solubility of trace metals. If relevant equilibrium constants and total concentrations of components are known, the solubility of the precipitated metal can be predicted in a straightforward manner (Allen and Unger, 1981).

Adsorption is probably the most important reaction for the control of metals in most oxic aquatic sediments. Important sorptive phases are iron and manganese oxides and particulate organic matter. Considerable work has been done to explain the behavior of metal ions at the interface

Table 1

Sequential Extraction Sequence for Sediment Fractionation

| Geochemical Interpretation | Extractant | pH | Temperature °C | Time Hr | References |
|---|---|---|---|---|---|
| adsorbed/ exchangeable | 1 M Mg(NO$_3$)$_2$ | 7 | 25 | 1 | Gibbs (19) |
| carbonate | 1 M NaOAc | 5 | 25 | 5 | Tessier et al. (20) |
| manganese oxides | 0.1 M NH$_2$OH-HCl 0.01 M HNO$_3$ | 2 | 25 | 1/3 | Chao (21) |
| oxidizables (organics) | 30% H$_2$O$_2$ | 2 | 85 | 5 | Gupta and Chen (22) |
| iron oxides | 1.0 M NH$_2$OH-HCl 25% HOAc | | 95 | 6 | Chester and Hughes (23) |

of metal oxides (Anderson and Rubin, 1981; Stumm, 1987; Hohl et al., 1980; Kramer and Allen, 1988). Oxide surfaces contain hydroxyl groups with ionizable hydrogen ions. These interfacial equilibria reactions can be treated in a similar manner as are solution equilibria. The reactions can be expressed as ionization or hydrolysis reactions:

$$\equiv SOH_2^+ + H_2O = \equiv SOH + H_3O^+ \qquad (4)$$

and

$$\equiv SOH + H_2O = \equiv SO^- + H_3O^+ \qquad (5)$$

where ≡SOH represents the surface hydroxyl. Conditional equilibrium constants are expressed in the similar manner to those for a solution phase reaction:

$$K_{cond,a1} = \frac{\{\equiv SOH\}[H^+]}{\{\equiv SOH_2^+\}} \qquad (6)$$

The concentration of charged sites can be determined from an alkalimetric titration (Hohl et al., 1980). The surface adsorption is the

$$K_{cond,a2} = \frac{\{\equiv SO^-\}[H^+]}{\{\equiv SOH\}} \quad (7)$$

sum of both chemical and electrostatic reactions:

$$\Delta G_{adsorption} = \Delta G_{intrinsic} + \Delta G_{coulombic} \quad (8)$$

$\Delta G_{adsorption}$ varies with the pH of the solution. $\Delta G_{coulombic}$ is the energy required to bring an ion from the bulk of the solution to a surface site at potential $\psi_o$:

$$\Delta G_{coulombic} = zF\psi_o \quad (9)$$

The surface potential can be from several electrostatic models. Westall and Hohl (1980) have show that they all are able to fit the experimental data. The intrinsic stability constant can then be expresses (Davis and Leckie, 1978):

$$pK_a^{int} = pH - \log \frac{\alpha}{1-\alpha} - \frac{zF}{RT \ln 10} \psi_o \quad (10)$$

where $\alpha$ is the fraction of ionized sites. This permits computation of the protonation state of the surface.

The binding of metals can then be expressed as surface complexation reactions:

$$\equiv SOH + M^{2+} = \equiv SOM^+ + H^+ \quad (11)$$

and

$$\equiv SOH + M^{2+} + H_2O = \equiv SOMOH + 2H^+ \quad (12)$$

We have recently studied the adsorption of copper onto manganese oxide because of the importance of this metal and of this sorbing phase in nature (Fu et al., 1991). The fit of adsorption data to the triple layer model is shown in Figure 2. In this study we were able to conclude that, for the Mn:Fe ratio found in typical aquatic sediments, adsorption of cadmium and copper by $\delta$-$MnO_2$ would be more important than the adsorption by iron oxides if the pH were less than about 7.0.

## Mechanistic Approaches for Multiples Phases

The previously discussed theoretical modeling frameworks do not operate as well for proton or metal exchange onto naturally occurring solids as they do for pure solid materials. Adediran and Kramer (1987) found that the adsorption onto the clay, organic, and iron plus manganese oxide phases of sediments were not additive. This suggests that they are not present in discrete phases, and simplifying assumptions required in models may not be valid. Oakley et al. (1981) reported that, in the linear region of adsorption isotherms, the contribution of the different components is additive. Honeyman (1984) also found that the contribution of sorptive phases in binary mixtures was additive.

Lion et al. (1982) studied the adsorption of lead and copper onto an estuarine salt marsh sediment and found that the adsorption was strongly pH dependent. The curves were typical strong adsorption edges, as for

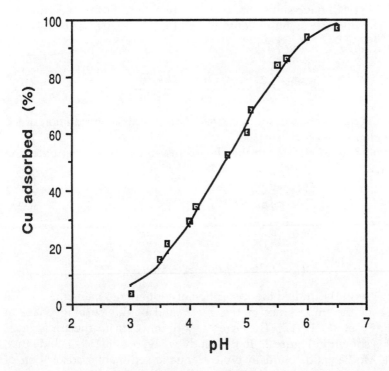

Figure 2. Comparison of Model Prediction with Experimental Data for Adsorption of Copper by $\delta$-MnO$_2$. Line Expresses the Model Prediction; Points are Experimental Data. I = 0.01 M NaNO$_3$; MnO$_2$ = 50 mg/L; Cu$_T$ = 8.0 x 10$^{-5}$ M; p*K$_{int,Cu^{2+}}$ = -0.36, p*K$_{int,CuOH^+}$ = 3.58.

the hydrous metal oxide sorbants. The Tessier (1979) sequential extraction procedure was used to attempt to remove various phases coating the sediments. They evaluated the changes in metal binding properties and found that over half of the metal was associated with the particulate organic matter. Davis (1982, 1984) and Laxen (1985) have also reported on the importance of natural organic matter in the adsorption of trace metals.

Metals that are complexed by organics in solution may either be more or less strongly adsorbed than the same metals in organic-free systems. A conceptual model has been presented to describe the effect of organics on adsorption (Benjamin and Leckie, 1981); no quantitative prediction of sorption is available.

Several papers have presented the fundamental requirements for modeling the interaction of a metal between sediment and the water with which it is in equilibrium (Oakley et al., 1981; Luoma and Davis, 1983; Davies-Colley et al., 1984; Jenne et al., 1986). The general formulation is shown in Eq. 13:

$$[M^{2+}] = \frac{[M]_{adsorbed}}{\sum_{sites} K_{intrinsic}[sites]} \tag{13}$$

This equation shows that three quantities must be known to predict the concentration of metal ion in equilibrium with a sediment:

1. The total concentration of metal adsorbed, or bound to any other type of site in which the metal is exchangeable. Some formulations use the quantity of metal bound in each of the sites. For example, Allen and Unger (1986) used a five step selective fractionation scheme to obtain the quantity of metal associated with each of the sites. They computed stability constants for each of the five types of sites. Jenne et al. (1986) have recommended that the total exchangeable quantity of metal be used because the fractionation methods do not sufficiently differentiate between the forms.

2. A stability constant is needed for each of the reactions of importance in the system. If a general model is desired, intrinsic stability constants are needed. If it is necessary to provide predictions only under a relatively limited set of environmental conditions of pH, ionic strength, and competing ions, conditional stability constants can be used.

3. The concentration of each of the available sites must be known. This provides the greatest limitation to application of such a model. Jenne et al. (1986) recommended that the concentration of the sites be determined by extractions. These are subject to the limitations described

previously.

Jenne et al. (1986) recommended that three sorbing phases be considered in the partitioning of metals. These are relative particulate organic matter, manganese oxide and iron oxide. Based on the literature, these were felt to have the greatest potential for metal binding.

Binding curves for the adsorption of protons or metal by sediment are generally featureless. As discussed with respect to humic materials, these curves can be described using a small set of equilibrium constants. For example, for the titration of the sediment shown in Figure 3, an almost perfect fit is obtained using pKs of 2.82, 5.67, 7.62, and 9.59. The titrations and determination of binding constants can be done with no *a priori* assumptions regarding the phases contributing to the sorption and without using stability constants determined for pure phases. Such an approach for describing binding by soils has been used by Nederlof et al. (1988). As pH is so important in controlling the partition of metals, it is clear that the proton binding characteristics of sediment must be understood before a comprehensive understanding of metal binding can be achieved.

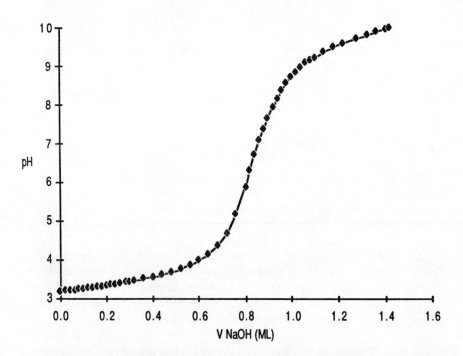

Figure 3. Acid-Base Titration of Sediment. Sediment Concentration = 1.5 g/liter, Ionic Strength = 0.01 Mole $NaNO_3$/liter, Base Concentration = 0.1 Moles NaOH/liter.

# Conclusion

Significant progress has been made in the last decade in development of models for partitioning of metals between solution and solid phases. These models have been used to explain the sorption of metals by pure oxide surfaces. Less advanced are approaches to prediction of metals by sediments. Frameworks for describing the binding of metals by heterogeneous sediment and soil systems have been advocated.

# References

Adediran, S.A. and J.R. Kramer. 1987. *Appl. Geochem.*, 2:213.

Allen, H.E., R.H. Hall and T.D. Brisbin. 1980. *Env. Sci. Technol.*, 14:441.

Allen, H.E. and M.T. Unger. 1986. In: *Chemicals in the Environment - Proceedings International Conference*, J.N. Lester, R. Perry and R.M. Sterritt (eds.), Lisbon.

Allen, H.E. and M.T. Unger. 1981. *Z.f. Wasser und Abwasser Forschung*, 13:124.

Anderson, D.M. and F.M.M. Morel. 1978. *Limnol. Oceanogr.*, 23:283.

Anderson, M.A. and A.J. Rubin (eds.). 1981. *Adsorption of Inorganics at Solid-Ligand Interfaces*, Ann Arbor Science.

Benjamin, M.M. and J.O. Leckie. 1981. *Environ. Sci. Technol.*, 15:1050.

Brown, D.S. and J.D. Allison. 1987. U.S. Environmental Protection Agency, Athens, Georgia, EPA-600/3-87/012.

Chao, L.L. 1972. *Soil Sci. Soc. Amer. Proc.*, 36:764.

Chester, R. and M.J. Hughes. 1967. *Chem. Geol.*, 2:249.

Davis, J.A.. 1984. *Geochim. Cosmochim. Acta*, 48:679.

Davis, J.A. 1982. *Geochim. Cosmochim. Acta*, 46:2391.

Davis, J.A. and J.O. Leckie. 1978. *J. Colloid Interface Sci.*, 12:1309.

Davis-Colley, R.J., P.O. Nelson and K.J. Williamson. 1984. *Environ. Sci. Technol.*, 18:491.

Dzombak, D.A., W. Fish and F.M.M. Morel. 1986. *Environ. Sci. Technol.*, 20:669.

Fish, W., D.A. Dzombak and F.M.M. Morel. 1986. *Environ. Sci. Technol.*, 20:676.

Förstner, U. and G.T.W. Wittmann. 1979. *Metal Pollution in the Aquatic Environment*, Springer Verlag, Berlin.

Fu, G., H.E. Allen and C.E. Cowan. 1991. Soil Science. 152: 72-81.

Gibbs, R.J. 1973. *Science*, 180:71.

Gupta, S.K. and K.Y. Chen. 1975. Environ. *Environ. Lett.*, 10:129.

Hohl, H., L. Sigg and W. Stumm. 1980. In: *Particulates in Water*, M.C. Kavanaugh and J.O. Leckie (eds.), Advances in Chemistry Series, 189.

Honeyman, B.D. 1984. Ph.D. Thesis, Stanford University.

Jenne, E. 1977. In: *Molybdenum in the Environment*, Vol. 2, W. Chappell and S.K. Petersen (eds.), Marcel Dekker, New York.

Jenne, E.A., D.M. DiToro, H.E. Allen and C.S. Zarba. 1986. In: *Chemical in the Environment - Proceedings International Conference*, J.N. Lester, R. Perry and R.M. Sterritt (eds.), Lisbon.

Kheboian, R. and C.F. Bauer. 1987. *Anal. Chem.*, 59:1417.

Kramer, J.R. and H.E. Allen (eds.). 1988. *Metal Speciation*, Lewis Publishers, Chelsea, Michigan.

Laxen, D.P.H. 1985. *Water Res.*, 19:1229.

Lion, L.W., R.S. Altman and J.O. Leckie. 1982. *Environ. Sci. Technol.*, 16:660.

Luoma, S.N. 1983. *The Science of the Total Environment*, 28:1.

Luoma, S.N. and J.A. Davis. 1983. *Mar. Chem.*, 12:159.

Martin, J.M., P. Nirel and A.J. Thomas. 1987. *Mar. Chem.*, 22:313.

Nederlof, M.M., W.H. Van Reimsdijk and L.K. Koopal. 1988. In: *Heavy Metals in the Hydrological Cycle*, M. Astruc and J.N. Lester (eds.), Selper, London.

Oakley, S.M., P.O. Newson, K.J. Williamson. 1981. *Environ. Sci. Technol.*, 15:474.

O'Donnell, J.R., B.M. Kaplan and H.E. Allen. 1985. *Proc. 7th Annual ASTM Sympos. Aquatic Toxicol.*, 485-501.

Perdue, E.M. and C.R. Lytle. 1983. *Environ. Sci. Technol.*, 17:654.

Rendell, P.S., G.E. Batley and A.J. Cameron. 1980. *Environ. Sci. Technol.*, 14:314.

Saar, R.A. and J.H. Weber. 1982. *Environ. Sci. Technol.*, 16:510A.

Salomons, W. and U. Förstner. 1984. *Metals in the Hydrocycle*, Springer Verlag, Berlin.

Scatchard, G. 1949. *Ann. N.Y. Acad. Sci.*, 51:660.

Shuman, M.S., B.J. Collins, P.J. Fitzgerald and D.L. Olson. 1983. In: *Aquatic and Terrestrial Humic Materials*, R.F. Christman and E.T. Gjessing (eds.), Ann Arbor Science Publ.

Stumm, W. (ed.). 1987. *Aquatic Surface Chemistry*, Wiley-Interscience, New York.

Sunda, W., D.W. Engel and R.M. Thuotte. 1978. *Env. Sci. Technol.*, 12:409.

Tessier, A. and P.G.C. Campbell. 1988. *Anal. Chem.*, 60:1475.

Tessier, A., P.G.C. Campbell and M. Bisson. 1979. *Anal. Chem.*, 51:844.

Westall, J.C. 1982. A computer program for determination of chemical equilibrium constants from experimental data. Oregon State University, Corvallis, Oregon.

Westall, J.C. and H. Hohl. 1980. *Adv. Colloid Interface Sci.*, 12:265.

Westall, J.C., J.L. Zachary and F.M.M. Morel. 1976. Tech. Note 18, Dept. Civil Engineering, Massachusetts Institute of Technology, Cambridge, Massachusetts.

# CHAPTER 8

# Reactions of Trace Elements Near the Sediment-Water Interface in Lakes

A. Tessier
R. Carignan
N. Belzile

## Introduction

Trace elements enter the aquatic environment from a variety of sources (natural weathering, atmospheric fallout, point sources.) Due to industrialization, their fluxes to the sediments have increased in last few decades (Hanson *et al.*, 1982 ; Nriagu *et al.*, 1979.) Models capable to predict how these increased inputs or other changes in environmental conditions (e.g., pH, redox potential, remedial action) affect the concentration of trace elements in the water column and sediments are needed for developing rational and effective environmental policies.

In order to predict adequately the concentration of trace elements in the water column, models of general application must take into account the most important processes that may affect the concentration of trace elements in lake waters (Armstrong *et al.*, 1987); these processes are either external (e.g., atmospheric fallout, weathering, anthropogenic point sources, water renewal rate) or internal to the lake (e.g., uptake by organisms, sedimentation, reactions at solid-liquid interfaces, fluxes of dissolved elements across the sediment-water interface.) Recent experiments involving the addition of trace elements to whole lakes or to enclosures have suggested that the transfer of trace elements at the sediment-water interface could play an important role in the removal of these elements from the water column (Hesslein *et al.*, 1980; Santschi *et al.*, 1986.) Several diagenetic reactions (oxidation of organic matter, formation of iron and manganese oxyhydroxides and of sulfides) that can influence trace element cycling are occurring in the upper layers of sediments. Despite the expected importance of sedimentary processes for trace elements, few studies have been conducted to understand them.

In this chapter, we use extensive measurements of trace element concentrations that we have performed in lake sediment solids and fluids to estimate the importance of some of the processes occurring in the upper sediment layers (fluxes of dissolved substances at the sediment-water interface; precipitation and sorption reactions in recent sediments.) The trace element data (concentrations in interstitial and overlying waters and in various layers of the sediments) were obtained from lakes chosen to represent a variety of lake pH values, trace element concentrations in the sediments and dissolved trace element concentrations in the overlying water.

## Dissolved Concentration Profiles

Examples of dissolved As, Cu, Ni, Zn, and Fe concentration profiles (comprising overlying water and upper sediment porewater data) collected with peepers (Hesslein, 1976; Carignan *et al.*, 1985) are given in Figure 1 for a site located in the littoral zone of an acid lake. It shows a number of interesting features that are common to many acid lakes where the sediment surface is oxidized.

### Iron

Figure 1 shows a peak in dissolved iron [Fe(II)], at about 3 cm below the sediment-water interface, which is presumably due to the reduction of Fe(III) by organic matter upon either burial of the sediments or seasonal fluctuations of the redox transition level. Reduction of Fe(III) occurs when stronger oxidants like dissolved oxygen, nitrate, and labile manganese oxides become exhausted in the upper layers of the sediments (Klinkhammer, 1980.) A portion of the Fe(II) that diffuses upward is reoxidized to Fe(III) and retained in the sediments as reactive authigenic Fe oxyhydroxides, whereas a varying portion can diffuse out of the sediments, to the overlying water; that latter portion should be relatively more important in acid lakes as the rate of Fe(II) oxidation decreases strongly with decreasing pH (Davison and Seed, 1983.) The downward diffusion of Fe(II) may be indicative of the presence of an Fe(II) sink, usually amorphous FeS(s), or crystalline sulfides (mackinawite, pyrite,) or may reflect non steady state conditions arising from seasonal oscillations of redox conditions.

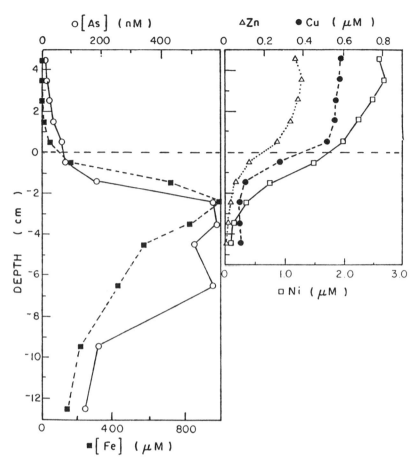

Figure 1. Porewater concentration profiles of As, Fe, Cu, Ni, Zn obtained in situ with porewater peepers (1cm vertical resolution) at a littoral station in Lake Clearwater (pH = 4.8; 46 22'N, 81 03'W; Sudbury area, Ontario). The horizontal broken line indicates the sediment-water interface.

## Arsenic

Dissolved arsenic concentration profiles also display a peak at about 3-4 cm below the sediment-water interface, which closely follows the dissolved Fe concentrations. Coincidence of the dissolved iron and arsenic peaks in the sediment column is commonly observed in lakes of various pH values and various arsenic contamination levels (Belzile and Tessier 1990.) If the profiles of Figure 1 were at steady state, the downward diffusion of As would imply the presence of an As sink, presumably a sulfide phase, below the peak. The upward diffusion of As could, in principle, lead either to its transfer to the overlying water or to its fixation by the authigenic Fe oxyhydroxides, close to the sediment-water interface. In this particular case, the persistence of a gradient above the sediment-water interface suggests some As diffusion to the overlying water. However, the enrichment of total As, which is consistently observed in the oxic zone of sediments (Crecelius, 1975; Holm et al., 1979; Nriagu, 1983; Johnson and Farmer, 1987) supports some fixation of diffusing As to the oxic layer.

The coupling of arsenic and iron is being increasingly documented. Hence, Edenborn et al. (1986) have added labile organic matter to surficial marine sediments, and found that arsenic was released from surficial sediments concurrently with iron, after the reductive dissolution of Mn was completed; this simultaneous release is consistent with the coincidence of As and Fe peaks shown in Figure 1. Several studies have also reported correlations between dissolved As and Fe concentrations in porewaters (Farmer and Lovell, 1986; Peterson and Carpenter, 1986; Edenborn et al., 1986; Belzile, 1988.) In addition, correlations have been generally observed between As and Fe concentrations in suspended and surficial bottom sediments (Crecelius, 1975; Langston, 1983); in agreement with these correlations, the solubilization of As with EDTA was reported to occur simultaneously with Fe(III) dissolution from oxic lake sediments (Aggett and Roberts, 1986). All these observations thus concur to suggest a close association of arsenic with the iron cycle.

## Trace Metals

In acid lakes, dissolved copper, nickel, and zinc concentrations (Figure 1) are typically high in the overlying water and decrease sharply below the sediment-water interface (Carignan and Nriagu, 1985; Carignan and Tessier, 1985; Tessier et al., 1989). The steepness of the profile, close to the sediment-water interface, tends to increase with decreasing pH. The downward fluxes of dissolved copper, nickel, and zinc generally observed near the interface of acid lakes lead to the obvious conclusion that their sediments can still trap dissolved trace metals. This conclusion contradicts the contention that progressive acidification of lake waters leads to a net release of trace metals from the top sediment layers to the overlying waters (Hanson and Norton, 1982); such a release would probably be observed only for a sudden drop in lake water pH (Anderson et al., 1987; Schindler et al., 1980).

## Solubility Equilibrium

It is one of the possible reactions that may affect the partitioning of trace elements among various compartments in a lake system.

## Anoxic Pore Waters

Solubility calculations performed on porewaters of lakes from the Sudbury area have shown saturation indexes (log $IAP/K_s$) close to 0 for amorphous iron sulfide in the anoxic layers of sediments, where sulfate is reduced (Carignan and Nriagu, 1985.) This calculation was reported to be consistent with the observation in these lakes of black deposits (1-2 cm below the oxic boundary) on the porewater peepers that were used to sample the interstitial water. At greater depths in the sediments, these researchers found that the ion activity product agrees more with the formation of siderite [$FeCO_3(s)$] than with that of iron sulfides. Sulfide formation may also be involved in the fixation of the trace metals below the sediment-water interface (e.g., Figure 1), since saturation indexes show saturation of the porewaters with respect to $ZnS(s)$ and $NiS(s)$, and oversaturation with respect to $CuS(s)$ (Carignan and Tessier, 1985; Carignan and Nriagu, 1985.) Copper is known to form relatively stable complexes with many ligands including humic and fulvic acids (Buffle, 1987); the apparent oversaturation observed for copper could be due to our inability to take into account complexation of this metal by natural organic matter in calculating its speciation. Similarly, the downward diffusion of As might be associated to the formation of $As_2S_3(s)$ or to a coprecipitation with iron sulfides (Aggett and O'Brien, 1985.)

## Overlying Waters

In contrast with anoxic porewaters, oxic lake waters do not appear to be saturated with respect to known pure solid phases of trace elements. For example, Figure 2a shows that dissolved zinc concentrations in the overlying waters of lakes of various pH and zinc contamination levels are largely undersaturated with respect to $Zn(OH)_2(s)$. Undersaturations are also observed for cadmium, nickel, and lead in the same lakes (Tessier *et al.*, unpublished results); the only possible exceptions to this general pattern are found for copper in a few lake waters with a pH above 8 that show a slight apparent over saturation with respect to copper oxide [$CuO(s)$] and malachite [$Cu_2CO_3(OH)_2(s)$] (Figure 2b). Since organic complexation could not be taken into account in the calculations, it is doubtful that the slight oversaturations of copper are real. The oxic lake

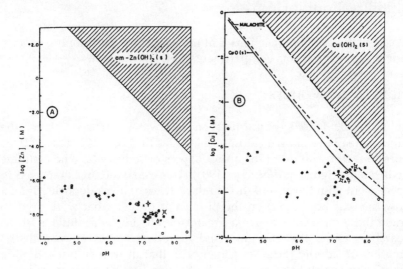

Figure 2. Solubility diagrams for zinc (a) and copper (b). Carbonate and oxide/hydroxide solid phases of these metals were considered. The solubility products and the formation constants of the inorganic complexes were obtained from Smith and Martell (1977) and Strumm and Morgan (1981). The highest concentration of total inorganic carbon (1.3mM) found in the lakewaters was used for the solubility calculations. Total dissolved zinc and copper concentrations obtained with porewater peepers at various sites of 26 lakes are shown as points with various symbols (Tessier et al. 1989). The sites were located in the littoral zone of the lakes.

waters are also largely undersaturated with respect to calcium, iron and aluminum arsenate, as well as to arsenic oxides such as $As_2O_5$ or $As_4O_6$ (Belzile and Tessier, 1990); these solids are far too soluble to exert any control on observed dissolved As concentrations.

## Benthic Fluxes

The concentrations of dissolved trace metals in the overlying waters generally decrease with increasing lake pH (e.g., see Figure 2) whereas those in the pore waters, where metal sulfides may precipitate, remain low for all lakes. For these reasons, the concentration gradients that develop at the sediment-water interface (and thus the downward fluxes) tend to decrease and even change sign with increasing pH as shown for zinc in Figure 3. The steep concentration gradients observed close to the sediment-water interface of acid lakes (e.g., Figure 1) indicate that

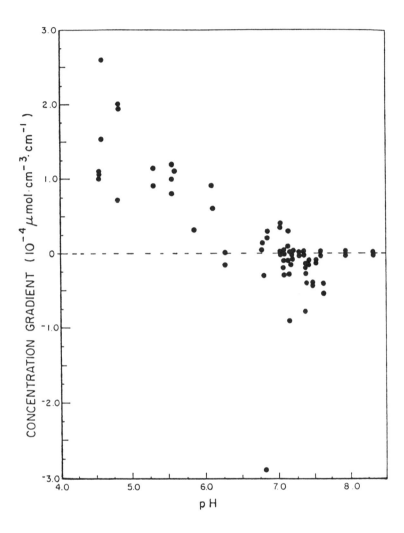

Figure 3. Concentration gradients of dissolved zinc across the sediment-water interface observed for lakes of various pH values. The values given correspond to the maximum gradients observed within the upper 2cm of the sediments. Positive values represent downward transport. From Tessier et al. 1989; reproduced with permission from Pergamon Press.

downward diffusion across the sediment-water interface may play an important role in the accumulation of certain trace metals in the sediments. For example, Carignan and Tessier (1985) have calculated that for lakes Clearwater (pH = 4.5; [Zn] gradient = $1.1 \times 10^{-4}$ µmol.cm$^{-3}$.cm$^{-1}$) and Tantaré (pH = 5.3; [Zn] gradient = $0.37 \times 10^{-4}$ µmol.cm$^{-3}$.cm$^{-1}$), at least 52 and 76%, respectively, of zinc deposition to the sediments could be accounted for by downward molecular diffusion. Similar calculations also indicated important diffusive contributions for copper (24-52%) and nickel (76-161%) in Lake Clearwater (Carignan and Nriagu, 1985). As mentioned in these studies, there are large uncertainties associated with the calculated fluxes due to the low vertical resolution of the concentration profiles (the concentration gradients might be steeper than assumed) and from neglecting metal transport across the interface by processes other than molecular diffusion (bioirrigation, physical mixing). In relation to this latter point, Carignan (unpublished results) has found that the transport coefficient for benthic irrigation in lake Tantaré is of approximately equal importance as the molecular diffusion coefficient. Owing to the paramount importance of measuring accurate concentration gradients and fluxes at (or close to) the sediment-water interface, the development of more suitable techniques for determining these quantities should be sought.

The appreciable contribution of diffusion to copper, nickel and zinc deposition in acid lakes creates subsurface peaks in total sedimentary trace metal concentrations at the depth where the trace metal fixation occurs; such peaks are observed usually for low pH lakes (e.g., pH < 5.5 for Zn) (Hanson et al., 1982.) This contribution of diffusion to trace metal deposition renders difficult the tracing of historical changes in lake water chemistry (e.g., acidification, increased or decreased metal deposition) from sedimentation rates and metal concentration profiles in sediment cores of acid lakes, as is currently practiced (Hanson et al., 1982; Ouellet and Jones, 1983a,b; Dillon and Smith, 1984; Nriagu and Rao, 1987.) Similarly, the intense post deposition remobilization of arsenic observed in all lakes complicates the historical reconstruction of its deposition.

## Sorption

Since thermodynamic calculations show undersaturation of oxic overlying waters with respect to trace element solid phases, sorption processes have been invoked to relate dissolved and sedimentary trace elements concentrations. Processes other than precipitation of a pure phase are usually included in the general term sorption; these processes include coprecipitation, adsorption, absorption, and surface precipitation

(Honeyman and Santschi, 1988). Both the distribution coefficient and the surface complexation concepts have been used to relate sorption of trace elements on suspended or bottom sediments to their dissolved concentrations in the overlying waters (Johnson, 1986; Tessier *et al.*, 1985, 1989; Belzile and Tessier, 1990).

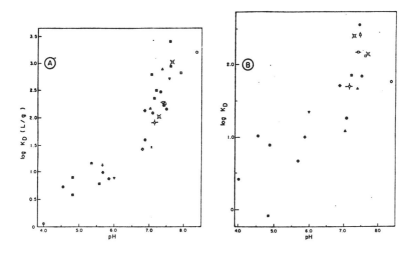

Figure 4. Distribution coefficients calculated with equation (1) for Zn (a) and Ni (b) at the same sites as in Figure 2. [Zn] and [Ni] were sampled with porewater peepers, whereas {Zn} and {Ni} were obtained from surficial (oxic) lake bottom sediments.

## Distribution Coefficient

Sorption of trace elements on particulate matter in aquatic environments is usually expressed (Sigg *et al.*, 1984, 1987; White and Driscoll, 1987; McIlroy *et al.*, 1986; Nyffeler *et al.*, 1984; Santschi *et al.*, 1986; Schindler 1981) as distribution coefficients, $K_D$ (L/g):

$$K_D = \frac{\{TE\}}{[TE]} \qquad (1)$$

where {} and [] stand for total trace element (TE) concentrations in the particulate (mole/g) and solution (mole/L) phases respectively. Figure 4

Table 1.
Linear regression equations describing the distribution coefficients or the apparent sorption constants.

### Cadmium

$\log K_D = 0.73\, pH - 3.29$ ; $r^2 = 0.81$; $N = 26$
$\log K_M = 1.03\, pH - 2.44$ ; $r^2 = 0.80$; $N = 26$

### Copper

$\log K_D = 0.43\, pH - 1.08$ ; $r^2 = 0.43$; $N = 31$
$\log K_M = 0.68\, pH + 0.34$ ; $r^2 = 0.75$; $N = 39$

### Nickel

$\log K_D = 0.48\, pH - 1.64$ ; $r^2 = 0.67$; $N = 21$
$\log K_M = 1.04\, pH - 2.29$ ; $r^2 = 0.87$; $N = 29$

### Lead

$\log K_D = 0.60\, pH - 1.14$ ; $r^2 = 0.81$; $N = 7$
$\log K_M = 0.81\, pH + 0.67$ ; $r^2 = 0.81$; $N = 7$

### Zinc

$\log K_D = 0.74\, pH - 3.03$ ; $r^2 = 0.83$; $N = 33$
$\log K_M = 1.21\, pH - 2.83$ ; $r^2 = 0.89$; $N = 41$

---

shows that the zinc and nickel distribution coefficients are strongly influenced by pH; similar results are found for Cd, Cu, and Pb (Tessier *et al.*, unpublished results). Table 1 gives the linear regression equations (log $K_D$ vs. pH) obtained for Cd, Cu, Ni, Pb, and Zn. This general behavior is in agreement with laboratory experiments that show increase with pH in the adsorption of these cations on oxides of iron (Benjamin and Leckie, 1981), manganese (Dempsey and Singer, 1980), silicon (Schindler *et al.*, 1976) and aluminum (Stumm *et al.*, 1976), as well as on humic material (Beveridge and Pickering, 1980) and clays (Farah and Pickering, 1977). The distribution coefficient is dependent upon solution and solid characteristics, which can vary from one lake to the other (Leckie and Tripathi, 1985). It can indeed be written:

$$K_D = \frac{\{S_1-M\}+\{S_2-M\}+\{S_3-M\}+...\{S_n-M\}+\{M_r\}}{[M^{z+}]+[ML(inorg)]+[ML(org)]} \quad (2)$$

where $\{S_1\text{-}M\}....\{S_n\text{-}M\}$ represent the concentration of trace metal M sorbed on the n sorbing substrates (organic matter, clays, iron, manganese, aluminium, silicon oxides, etc.) that are present in the sediment; $\{M_r\}$ represents the concentration of M tightly bound to the sediment components (e.g., in the crystal lattice of silicates, resistant oxides or sulfides); [ML(inorg)] and [ML(org)] stand for the concentrations of dissolved inorganic and organic complexes respectively. Examination of equation (2), shows that variations of sediment composition from one lake to the other should introduce variability in $K_D$ since the various substrates present different affinities for sorbing M; variability in $K_D$ values should also arise from varying $\{M_r\}$ and dissolved ligand concentrations.

## Surface Complexation

Sorption of trace elements in natural systems has also been expressed in terms of complexation by surface sites, using some of the concepts developped for the surface complexation models (e.g., Stumm et al., 1970; Davis and Leckie, 1979). The association of trace elements with Fe oxyhydroxides was studied using this approach (Johnson 1986; Tessier et al., 1985, 1989; Belzile and Tessier, 1990). Basic equations and definition of the various parameters used are given in Table 2. Combining equations (4) and (6)-(10) from this table leads to:

$$K_M = \frac{N_s * K_M}{[H^+]^{m+1}} = \frac{\{Fe-M\}}{\{Fe-ox\}[M^{z+}]} \quad (11)$$

and

$$K_A = N_s * K_A [H^+]^m = \frac{\{Fe-A\}}{\{Fe-ox\}[A^{n-}]} \quad (12)$$

for the sorption of cations ($M^{z+}$) and anions ($A^{n-}$), respectively. The concentrations of {Fe-ox}, {Fe-M}, and {Fe-A} have been estimated by

Table 2.
Basic equations for describing surface complexation of cations and anions on iron oxyhydroxide

---

For a cation ($M^{z+}$):

$$\equiv Fe-OH + M^{z+} + mH_2O \xrightleftharpoons{*K_M} \equiv Fe=OM(OH)_m^{z-m-1} + (m+1)H^+ \quad (3)$$

$$*K_M = \frac{\{\equiv Fe-OM(OH)_m^{z-m-1}\}[H^+]^{m+1}}{\{\equiv Fe-OH\}[M^{z+}]} \quad (4)$$

For an anion ($A^{n-}$):

$$\equiv Fe-OH + A^{n-} + mH^+ \xrightleftharpoons{*K_A} \equiv Fe=OM(OH)_{m+1}-A^{m-n} \quad (5)$$

$$*K_A = \frac{\{\equiv Fe-OH_{m+1}-A^{m-n}\}}{\{\equiv Fe-OH\}[A^{n-}][H^+]^m} \quad (6)$$

where $.K^M$ and $.K^A$ are apparent overall equilibrium constants for the adsorption of the cationic and anionic adsorbates, respectively; {} and [] refer to concentrations in the solid and solution phases, respectively, whereas "≡" refers to adsorption sites, either free ({≡Fe-OH}) or occupied by M or A.

At low density of adsorption:

$$\{\equiv Fe-OH\} \approx \{\equiv Fe-O-\}_T \quad (7)$$

where $\{\equiv Fe\text{-}o\text{-}\}_T$ is the total concentration of sites for the iron oxyhydroxides, which, in turn, can be expressed as;

$$\{\equiv Fe-O-\}_T = N_s \cdot \{Fe-ox\} \quad (8)$$

where $N_s$ is the density of sites of the Fe oxyhydroxides, and {Fe-ox} is the concentration of iron oxyhydroxides. The concentration of occupied sites can be related to the concentrations of M, {Fe-M} and A, {Fe-A}, associated with the Fe oxyhydroxides.

$$\{\equiv Fe-OM(OH)_m^{z-m-1}\} = \{Fe-M\} \quad (9)$$

$$\{\equiv Fe-OH^{m+1}-A^{m-n}\} = \{Fe-A\} \quad (10)$$

partial extraction of surficial oxic lake bottom sediments (Tessier et al., 1985, 1989; Belzile and Tessier, 1990); for a given site, these values have been used in equations (11) and (12), together with the estimated free ion concentrations of trace elements in the water overlying these sediments, to calculate $K_M$ and $K_A$ values derived from natural iron oxyhydroxides. It should be noted that $K_M$ and $K_A$ are apparent overall equilibrium constants, obtained from field measurements, that can vary with pH. Indeed, equation (11) predicts that a plot of log $K_M$ versus pH should lead to a straight line of slope m+1 and an intercept of log $(N_S \cdot {^*}K_M)$; similarly, a plot of log $K_A$ versus pH should yield, according to equation (12), a slope of -m and an intercept of log $(N_S \cdot {^*}K_A)$.

Figure 5a shows that a plot of log $K_M$ versus pH yields, for lakes of pH between 4 and 8.4, a straight line of slope 1.2 for zinc, in agreement with the simple model depicted by equation (11); these results, obtained for oxic lake bottom sediments, compare reasonably well with those obtained by Johnson (1986) for suspended sediments rich in iron in the Carnon River system (Figure 5a). Similar straight lines were obtained for Cd, Cu, Ni, and Pb in the various lakes (Table 1). The slopes of the regression lines, which represent the proton stoichiometry in the sorption reaction (number of protons released per metal sorbed), are 1 or slighty greater for Cd, Ni, and Zn (Table 1). These values are in agreement with the surface complexation model for adsorption of trace metals at the surface of oxide; according to this model, the proton stoichiometry should reflect a combination of surface complexation reactions involving the release of one or two protons per metal adsorbed (Schindler 1981). However, lead, and above all copper, yields slopes smaller than those observed for the other trace metals. These lower slopes can be attributed to our inability to take into account the organic complexes of these metals in our estimation of $[M^{z+}]$, which lead to an underestimation of $K_M$, according to equation (11); this effect will be greater at the higher pH values where organic complexation will be favored, thus leading to a decrease in the slope in a plot of log $K_M$ versus pH. Another characteristic of the field-derived $K_M$ values that can be examined is their sequence. Figure 6 shows that the sequence of increasing $K_M$ obtained from field measurements is Pb > Cu > Zn > Ni ≈ Cd, in agreement with those reported for adsorption of these metals on synthetic iron oxyhydroxides in well defined media (Leckie et al., 1983). If the pH dependence of the $K_M$ values obtained from field measurements and their sequence resemble those obtained in laboratory experiments, the field- and laboratory-derived values of $K_M$ themselves differ significantly in some cases (Johnson, 1986; Tessier et al., 1989).

In contrast to trace metals, a plot of log $K_A$ vs pH yields a slope

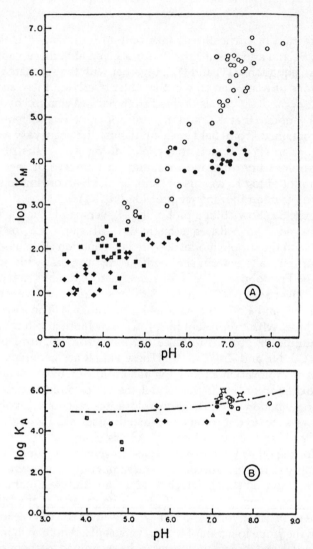

Figure 5. Apparent overall equilibrium constants for the sorption of zinc (a) and arsenic (b) on natural iron oxyhydroxides calculated with equations (11) and (12) from field measurements at the same sites as in Figure 2; they are shown as points ((0) for Zn and various aymbols for As). As(V) only was assumed to be present in the oxic overlying waters. For comparison, the values obtained by Johnson (1986) in the Caron river system for zinc in uncontaminated drainage waters, mine waters, and contaminated surface waters are present in Figure 5a. In Figure 5b, the curve was calculated with equation (12) from laboratory data reported by Pierce and Moore (1982) for the adsorption of As(V) on synthetic iron oxyhydroxides. Total iron (41.7 µM), total As(V) (0.667 µM), and the equilibrium partitioning of As(V) between the iron oxyhydroxides and the solution were used.

close to zero for arsenic (Figure 5b); this slope suggests that the surface species $=FeOAsO_3H^-$ is dominating in that pH range (Belzile and Tessier, 1990). As shown in Figure 5b, $K_A$ values derived for arsenic from field measurements agree well with those obtained in the laboratory for adsorption of arsenate on amorphous iron oxyhydroxides.

The treatment of the data according to a surface complexation model focuses on a given class of sorbing substrates, namely iron oxyhydroxides, and thus reduces the variability associated with the sediment composition; a weakness in the approach is, however, that the measurements of both iron oxyhydroxides and the associated metal concentrations rely on partial extraction reagents that are not completely selective. Additional variability in $K_D$ (which is partly corrected in $K_M$ calculations) arises from neglecting dissolved M speciation.

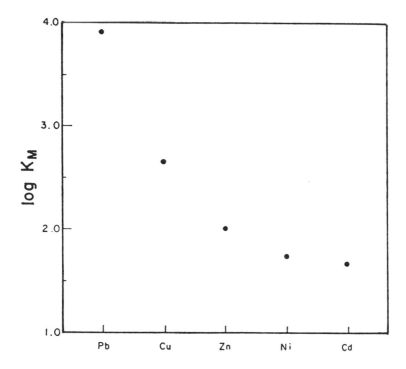

Figure 6. Sequence of $K_M$ values obtained from in situ measurements for the sorption of various trace metals on natural iron oxyhydroxides. The $K_M$ values were calculated with the linear regression equations given in Table I. A low pH value of 4 was chosen to minimize the effects of complexation by organic matter which was not taken into account in the calculation of $K_M$.

# Partitioning of Labile Trace Elements in Lakes of Various pH

The possible rate limiting steps for the exchange of sorbed trace elements between the surficial sediments (e.g., top first cm) and the water column could be either due to sorption-desorption reactions or transport of the trace elements from one compartment to the other. Examination of recent studies on the kinetics of sorption (Santschi et al., 1986; Balistrieri and Murray, 1984), desorption (Tipping et al., 1986), and sediment mixing due to bioturbation or physical processes (Santschi et al., 1986) suggests that these processes could all occur on time scales of a few weeks i.e., shorter than the water residence time of most lakes (Tessier et al., 1989). Owing to the expected mobility of sorbed trace metals, the regressions given in Table 1 between $K_M$ and pH can thus be used to describe the distribution of the quantities of the various trace metals between two pools of labile metal ($M_L$), i.e., dissolved and sorbed to iron oxyhydroxides of the sediments. It should be noted that other pools of labile metal (e.g., M associated with Mn oxides) are probably present in natural sediments; however, Fe-M is one of the most important pools, and the empirical regression equations relating sorbed and dissolved M (such as the ones given in Table 1 for the Fe oxyhydroxides) are not presently available for the other pools. Combining equation (11) with the following definition of the quantity of labile trace metal:

$$M_L = \{Fe-M\} \cdot a + [M] \cdot v + [ML(inorg)] \cdot v \qquad (13)$$

where a and v stand for the quantity of sediment and the volume of water leads to:

$$\frac{\{Fe-M\}}{M_L} = \frac{K_M \cdot \{Fe-ox\}}{v/a + K_M \cdot \{Fe-ox\} + v/a\,[ML(inorg)]} \qquad (14)$$

Possible complexation by dissolved organic ligands is not considered because of the lack of relevant equilibrium constants.

Figure 7 shows the plots of percent sorbed Cd, Cu, Ni, Pb, and Zn versus pH obtained with the regressions given in Table 1 and appropriate values for a, v, and {Fe-ox}; the values chosen for the

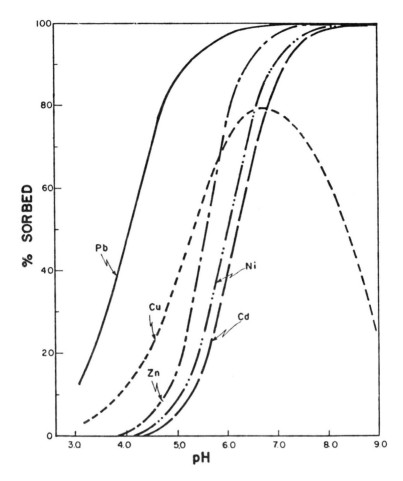

Figure 7. Percent of labile Cd, Cu, Ni, Pb, and Zn sorbed by natural iron oxyhydroxides as a function of pH calculated with equation (14). The values assumed for the calculations are: {Fe-ox} = 2.3% on a dry weight basis, v = 0.5L

typical of many shallow Canadian Shield lakes. Choosing greater lake depths or lower {Fe-ox} would shift the curves towards greater pH. These plots obtained from in situ measurements present the same characteristics as those reported for the adsorption of these trace metals on synthetic iron oxyhydroxides (e.g., Leckie *et al.* 1983) i.e. the sequence of sorption of the trace metals is the same and there is a strong pH dependence of sorption. Figure 7 shows that the relative importance of the two pools vary with pH. For lakes of low pH values (pH < 3 for Pb and < 5 for Zn, Cd and Ni), most of the trace metals are in the water column, and thus adsorption of these trace metals on synthetic iron oxyhydroxides (e.g., Leckie et al., 1983) i.e., the sequence of sorption of the trace metals is the same, and there is a strong pH dependence of sorption. Figure 7 shows that the relative importance of the two pools varies with pH. For lakes of low pH values (pH < 3 for Pb and < 5 for Zn, Cd, and Ni) most of the trace metals are in the water column, and thus sorption to iron oxyhydroxides should not play an important role in regulating their dissolved concentrations in the overlying waters; at higher pH values (pH > 7 for Cd, Ni, and Zn and > 6 for Pb) the trace metals are prominently associated with the sedimentary iron oxyhydroxides. As expected, Pb is strongly associated with the Fe oxyhydroxides and should be less mobile than the other trace metals. Changes in lake pH within the ranges 5 to 7 for Cd, Ni, and Zn and 3 to 5 for Pb will influence the partitioning of these trace metals between the water column and the sediments. Many poorly buffered lakes in the pH range 5.5-6.5 can be subjected to pH changes of about 1 pH unit fall during spring melt.

## Concluding Remarks

Considerable uncertainties still remain in our knowledge on the sedimentary processes that involve trace elements as well as on the relative importance of these processes in the regulation of the dissolved concentrations of the elements in lake water. Precipitation of trace elements seems to occur only in anoxic situations, where sulfide is present. With the exception of FeS(s), the exact solid compounds formed have, however, not been isolated or identified precisely; their nature has only been inferred from such indirect measurements as saturation indexes. There is some evidence in the literature that trace elements can be removed from the water column by direct reaction with the bottom sediments (Hesslein et al., 1980; Santschi et al., 1986.) In agreement with these findings, our results suggest that, in acid lakes, diffusive fluxes of some trace metals across the sediment-water interface to the zone of sulfide formation could be an important process influencing trace metal concentrations in lake waters. In circumneutral shallow lakes, most of the labile trace metals appear to be associated with iron oxyhydroxides present in the surficial oxidized sediments, and, according to the few kinetic data available, should be exchangeable with the water column within a relatively short period of time. This contention is supported by the rapid release of trace metals from the sediments to the overlying water that has been observed in enclosures upon sudden acidification (Santschi et al., 1986; Anderson et al., 1987.) Other field measurements present evidence that the scavenging by sinking particles (especially plankton) is mainly responsible for the removal of trace elements from the water column (e.g., Sigg *et al.*, 1987.) We have not studied this mechanism in our lakes, but it can be speculated that it would be responsible for most of the trace metal deposition in the circumneutral lakes. In brief, the relative importance of all the processes participating in the regulation of trace elements should vary among lakes and among trace elements; variables to consider in a systematic study of their relative importance should include pH, sedimentation, mean depth, flushing rate, and sediment characteristics.

## Acknowledgments

Financial support from the Québec Fond pour la Formation de Chercheurs et l'Aide à la Recherche, from the Wildlife Toxicology Fund, and from the Natural Sciences and Engineering Research Council of Canada is acknowledged. One of us (N.B.) was supported by a post doctoral fellowship from Institut National de la Recherche Scientifique. The able scuba diving skills of R. Beauchemin are gratefully acknowledged.

## References

Aggett, J., and G.A. O'Brien. 1985. Detailed model for the mobility of arsenic in lacustrine sediments based on measurements in Lake Ohakuri. *Environ. Sci. Technol.* 19:231-238.

Aggett, J., and L.S. Roberts. 1986. Insight into the mechanism of accumulation of arsenate and phosphate in hydrolake sediments by measuring the rate of dissolution with ethylenediaminetetraacetic acid (EDTA). *Environ. Sci. Technol.* 20:183-186.

Anderson, R.F., P.H. Santschi, U.P. Nyffeler, and S.L. Schiff. 1987. Validating the use of radiotracers as analogs of stable metal behaviour in enclosed aquatic ecosystem experiments. *Can. J. Fish. Aquat. Sci.* 44(Suppl. 1):251-259.

Armstrong, D.E., J.P. Hurley, D.L. Swackhamer, and M.M. Shafer. 1987. Cycles of nutrient elements, hydrophobic organic compounds, and metals in Crystal Lake. Role of particle-mediated processes in regulation. In *Sources and Fates of Aquatic Pollutants*, eds. R.A. Hites and S.J. Eisenreich, *Adv. Chem. Ser.* 216:491-518. Washington: American Chemical Society.

Balistrieri, L.S. and J.W. Murray. 1984. Marine scavenging: trace metal adsorption by interfacial sediment from MANOP site H. *Geochim. Cosmochim. Acta*

48:921-929.

Belzile, N. 1988. The fate of arsenic in sediments of the Laurentian Through. *Geochim. Cosmochim. Acta* 52:2293-2302.

Belzile, N. and A. Tessier. 1990. Interactions between arsenic and natural sedimentary iron oxyhydroxide. *Geochim. Cosmochim. Acta* 54:103-109.

Benjamin, M.M., and J.O. Leckie. 1981. Multiple-site adsorption of Cd, Cu, Zn, and Pb on amorphous iron oxyhydroxide. *J. Colloid Interface Sci.* 79:209-221.

Beveridge, A. and W.F. Pickering. 1980. Influence of humate-solute interactions on aqueous heavy metal ion levels. *Water Air Soil Pollut.* 14:171-185.

Buffle, J. 1987. *Complexation reactions in aquatic systems. An analytical approach.* Chichester: Ellis Horwood.

Carignan, R., F. Rapin, and A. Tessier. 1985. Sediment porewater sampling for metal analysis: a comparison of techniques. *Geochim. Cosmochim. Acta* 49:2493-2497.

Carignan, R., and J.O. Nriagu. 1985. Trace metal deposition and mobility in the sediments of two lakes near Sudbury, Ontario. *Geochim. Cosmochim. Acta* 49:1753-1764.

Carignan, R. and A. Tessier. 1985. Zinc deposition in acid lakes: the role of diffusion. *Science* 228:1524-1526.

Crecelius, E.A. 1975. The geochemical cycle of arsenic in Lake Washington and its relation to other elements. *Limnol. Oceanogr.* 20:441-451.

Davis, J.A., and J.O. Leckie. 1979. Speciation of adsorbed ions at the oxide/water interface. In Chemical Modeling in Aqueous Systems, ed. E.A. Jenne, *Adv. Chem. Ser.* 93:299-317. Washington, D.C.: American Chemical Society.

Davison, W. and G. Seed. 1983. The kinetics of the oxidation of ferrous iron in synthetic and natural waters. *Geochim. Cocmochim. Acta*, 47:67-79.

Dempsey, B.A. and P.C. Singer. 1980. The effects of calcium on the adsorption of zinc by $MnOx(s)$ and $Fe(OH)_3(am)$. *In Contaminants and Sediments*, Vol 2: *Analysis, Chemistry and Biology*, ed. R.A. Baker, pp. 333-352. Ann Arbor, Michigan: Ann Arbor Science Publishers.

Dillon, P.J. and P.J. Smith. 1984. Trace metal and nutrient accumulation in the sediments of lakes near Sudbury, Ontario. In Environmental Impact of Smelters, ed. J.O. Nriagu, pp. 375-416. New York: John Wiley and Sons.

Edenborn, H.M., N. Belzile, A. Mucci, J. Lebel, and N. Silverberg. 1986. Observations on the diagenetic behavior of arsenic in a deep coastal sediment. *Biogeochem.* 2:359-376.

Farmer, J.G. and M.A. Lovell. 1986. Natural enrichment of arsenic in Loch Lomond sediment. *Geochim. Cocmochim. Acta* 50:2059-2067.

Farrah, H. and W.F. Pickering. 1977. Influence of clay-solute interactions on aqueous heavy metals ion levels. *Water Air Soil Pollut.* 8:189-197.

Hanson, D.W., and Norton, S.A. 1982. Spatial and temporal trends in the chemistry of atmospheric deposition in New England. In International Symposium on Hydrometeorology, pp. 25-33. Urbana, Illinois: American Water Resources Association.

Hanson, D.W., S.A. Norton, and J.J. Williams. 1982. Modern and paleolimnological evidence for accelerated leaching and metal accumulation in soils in New England, caused by atmospheric deposition. *Water Air Soil Pollut.*, 18:227-239.

Hesslein, R.H. 1976. An *in situ* sampler for close interval pore water studies. *Limnol. Oceanogr.* 21:912-914.

Hesslein, R.H., W.S. Broecker and D.W. Schindler. 1980. Fates of metal radiotracers added to a whole lake: sediment-water interactions. *Can. J. Fish. Aquat. Sci.* 37:378-386.

Holm, T.R., M.A. Anderson, D.G. Iverson, and R.S. Stanforth. 1979. Heterogeneous interactions of arsenic in aquatic systems. In Chemical Modeling in Aqueous Systems, ed. E.A. Jenne, pp. 712-736. Washington: American Chemical Society.

Honeyman B.D. and P.H. Santschi. 1988. Metals in aquatic systems. *Environ. Sci. Technol.* 22:862-871.

Johnson C.A. 1986. The regulation of trace element concentrations in river and estuarine water contaminated with acid mine drainage: the adsorption of Cu and Zn on amorphous Fe oxyhydroxides. *Geochim. Cosmochim. Acta* 50, 2433-2438.

Johnson, L.R. and J.G. Farmer. 1987. Arsenic mobility and speciation in the sediments of Scottish inland and coastal waters. In *Proceedings of the International Conference on Heavy Metals in the Environment*, pp. 218-222.

Edinburgh: CEP Consultants.

Klinkhammer, G.P. 1980. Early diagenesis in sediments from the eastern equatorial Pacific. II. Pore water metal results. *Earth Planet. Sci. Lett.* 49:81-101.

Langston, W.J. 1983. The behavior of arsenic in selected United Kingdom estuaries. *Can. J. Fish. Aquat. Sci.* 40:143-150.

Leckie, J.O., D.T. Merrill, and W. Chow. 1983. Trace element removal from power plant wastestreams by adsorption/coprecipitation with amorphous iron oxyhydroxide. AICHE Symposium Series, pp. 28-42.

Leckie, J.O. and V.S. Tripathi. 1985. Effect of geochemical parameters on the distribution coefficient Kd. In *Proceedings of the International Conference on Heavy Metals in the Environment*, pp. 369-371. Edinburgh: CEP Consultants.

McIlroy, L.M., J.V. DePinto, T.C. Young, and S.C. Martin. 1986. Partitioning of heavy metals to suspended solids of the Flint river, Michigan. *Environm. Toxicol. Chem.* 5:609-623.

Nriagu, J.O. 1983. Arsenic enrichment in lakes near the smelters at Sudbury, Ontario. *Geochim. Cocmochim. Acta* 47:1523-1526.

Nriagu, J.O., A.L.W. Kemp, H.K.T. Wong, and N. Harper. 1979. Sedimentary record of heavy metal pollution in Lake Erie. *Geochim. Cocmochim. Acta* 43:247-258.

Nriagu, J.O., and S.S. Rao. 1987. Response of lake sediments to changes in trace metal emission from the smelters at Sudbury, Ontario. *Environ. Pollut.* 44:211-218.

Nyffeler, U.P., Y-H. Li, and P.H. Santschi. 1984. A kinetic approach to describe trace-element distribution between particles and solution in natural aquatic systems. *Geochim. Cocmochim. Acta* 48:1513-1522.

Ouellet, M. and H.G. Jones. 1983a. Historical changes in acid precipitation and heavy metals deposition originating from fossil fuel combustion in Eastern North America as revealed by lake sediment geochemistry. *Water Sci. Technol.* 15:115-130.

Ouellet, M. and H.G. Jones. 1983b. Paleolimnological evidence for long-range atmospheric transport of acidic pollutants and heavy metals into the Province of Quebec, Eastern Canada. *Can. J. Earth Sci.* 20:23- 26.

Peterson, M.L. and R. Carpenter. 1986. Arsenic distributions in porewaters and sediments of Puget Sound, Lake Washington, the Washington coast and Saanich Inlet, B.C. *Geochim. Cocmochim. Acta* 50:353-369.

Pierce, M.L. and C.B. Moore. 1982. Adsorption of arsenite and arsenate on amorphous iron hydroxide. *Water Res.* 16:1247-1253.

Santschi, P.H., U.P. Nyffeler, R.F. Anderson, S.L. Schiff, P. O'Hara, and R.H. Hesslein. 1986. Response of radioactive trace metals to acid-base titrations in controlled experimental ecosystems: evaluation of transport parameters for application to whole-lake radiotracer experiments. *Can J. Fish. Aquat. Sci.* 43: 60-77.

Schindler, P.W. 1981. Surface complexes at oxide-water interfaces. In *Adsorption of Inorganics at Solid-Liquid Interfaces*, eds. M.A. Anderson and A.J. Rubin, pp. 1-49. Ann Arbor, Michigan: Ann Arbor Sciences Publishers Inc.

Schindler, P.W., B. Fürst, R. Dick, and P.U. Wolf. 1976. Ligand properties of surface silanol groups. I. surface complex formation with $Fe^{3+}$, $Cu^{2+}$, $Cd^{2+}$, and $Pb^{2+}$. *J. Colloid Interf. Sci.* 55:469-475.

Schindler, D.W., R.H. Hesslein, R. Wagemann, and W.S. Broecker. 1980. Effects of acidification on mobilization of heavy metals and radionuclides from the sediments of a freshwater lake. *Can. J. Fish. Aquat. Sci.* 37:373-377.

Sigg, L., M. Sturm, and D. Kistler. 1987. Vertical transport of heavy metals by settling particles in Lake Zurich. *Limnol. Oceanogr.* 32:112-130.

Sigg, L., W. Stumm, and B. Zinder. 1984. Chemical processes at the particle-water interface; implications concerning the form of occurrence of solute and adsorbed species. In *Complexation of Trace Metals in Natural Waters*, eds. C.J.M. Kramer and J.C. Duinker, pp. 251-266. The Hague, Netherlands: Nyhoff/Junk.

Smith, R.M. and A.E. Martell. 1977. *Critical Stability Constants*. New York: Plenum Press.

Stumm, W., H. Hohl, and F. Dalang. 1976. Interaction of metal ions with hydrous oxide surfaces. *Croat. Chim. Acta* 48:491-504.

Stumm, W., C.P. Huang, and Jenkins, S.R. 1970. Specific chemical interaction affecting the stability of dispersed systems. *Croat. Chem. Acta* 42:223-245.

Stumm, W. and J.J. Morgan. 1981. *Aquatic Chemistry*. 2nd edition. New York: John Wiley and Sons.

Tessier, A., R. Carignan, B. Dubreuil, and F. Rapin. 1989. Partitioning of zinc between the water column and the oxic sediments in lakes. *Geochim. Cocmochim. Acta* 53:1511-1522.

Tessier, A., F. Rapin, and R. Carignan. 1985. Trace metals in oxic lake sediments: possible adsorption onto iron oxyhydroxides. *Geochim. Cocmochim. Acta* 49:183-194.

Tipping, E., D.W. Thompson, M. Ohnstad, and N.B. Hetherington. 1986. Effects of pH on the release of metals from naturally-occurring oxides of Mn and Fe. *Environ. Technol. Lett.* 7:109-114.

White, J.R. and C.T. Driscoll. 1987. Zinc cycling in an acidic Adirondack lake. *Environ. Sci. Technol.* 21:211-216.

# CHAPTER 9

# Partitioning of Organic Chemicals in Sediments: Estimation of Interstitial Concentrations Using Organism Body Burdens

Victor J. Bierman, Jr.

## Introduction

To assess the potential hazards of toxic chemicals in aquatic systems, it is necessary to determine spatial and temporal concentration distributions. For hydrophobic organic chemicals these concentrations are strongly influenced by sorption to particles (Karickhoff, 1984). Information on phase partitioning of these chemicals in sediments is essential for accurate process formulations of sediment-water interactions in whole system mass balance models (Schnoor et al., 1987) and for determining bioavailability to benthic organisms (Dickson et al., 1987).

Historically, reversible equilibrium models based on the Langmuir or the Freundlich isotherms have been used to describe experimental sorption data (Voice and Weber, 1983). However, the phenomenon of sorption-desorption hysteresis (Rao and Davidson, 1980; Di Toro and Horzempa, 1982), the potential association of hydrophobic organic chemicals with colloids and dissolved organic matter (Hassett and Anderson, 1979; Chiou et al., 1986; Carter and Suffet, 1982; Means and Wijayaratne, 1982), and the influence of particle concentration on observed partition coefficients (O'Connor and Connolly, 1980; Horzempa and Di Toro, 1983; Voice et al., 1983; Capel and Eisenreich, 1985; Swackhamer and Armstrong, 1987) have precluded a complete understanding of environmental phase partitioning. For sediments, this situation is further confounded because there is a paucity of direct measurements for contaminant concentrations in interstitial waters due to experimental difficulties in phase separation and acquisition of sufficient sample volumes for accurate analyses (Allan, 1986).

The purpose of this paper is to investigate phase partitioning of hydrophobic organic chemicals in sediments using existing field data for body burdens in benthic invertebrates to estimate dissolved phase chemical concentrations in sediment interstitial waters. Using equilibrium partitioning theory and data for animal body burdens and associated sediment contaminant concentrations, estimated sediment partition coefficients are computed. These partition coefficients are then compared with predictions of various phase partitioning models to determine the degree to which they can be used to discriminate among the models and/or to bound the values of various model parameters.

## Conceptual Framework for Phase Partitioning

The partitioning of hydrophobic organic chemicals among the relevant compartments in the sediments can be conceptualized as shown in Figure 1. The actual distribution volumes for these chemicals on sediment particles and in organisms are generally considered to be the weight fractions of organic carbon (Karickhoff, 1981) and tissue lipid (Konemann and van Leeuwen, 1980), respectively. Octanol is widely used in laboratory experiments as a convenient surrogate for these environmental distribution volumes.

At thermodynamic equilibrium the fugacity of a chemical must be the same in each phase distribution volume in the system (Mackay, 1979). For an aquatic system consisting of sediment particles, animals, and water this condition can be represented as:

$$F^{oc} = F^{l} = F^{w} \tag{1}$$

where the superscripts oc, l, and w refer to sediment organic carbon, lipid, and water, respectively. Following the development of Karkichoff (1984), fugacity can be related to chemical concentration in each phase, for example, for the aqueous phase:

$$F^{w} = \phi^{w} C^{w} \tag{2}$$

where $C^w$ is the aqueous phase concentration, and $\phi^w$ is the fugacity coefficient. The fugacity coefficient can be expressed as:

$$\phi^{w} = \gamma^{w} F_{o}^{w} \tag{3}$$

where $\gamma$ is an activity coefficient and $F_o$ is the reference state fugacity. In most environmental situations, concentrations are sufficiently dilute that fugacity coefficients are independent of chemical concentration.

Equilibrium partition coefficients are defined as the ratio of chemical

## Conceptual Framework for Phase Partitioning

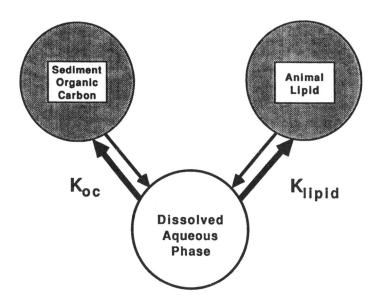

**Figure 1.** Conceptualization of phase partitioning for hydrophobic organic chemicals.

concentrations between a sorbed phase and the aqueous phase. For example, between sediment organic carbon and water:

$$K_{oc} = \frac{C^{oc}}{C^w} = \frac{\phi^w}{\phi^{oc}} = \beta \frac{\gamma^w}{\gamma^{oc}} \quad (4)$$

where $K_{oc}$ is the organic carbon-water partition coefficient, $\beta$ is a unit adjustment constant, and the reference state fugacities are the same in each phase. The $K_{oc}$ is related to the mass-based partition coefficient ($K_p$) by $K_{oc} = K_p/f_{oc}$, where $f_{oc}$ is the decimal fraction of particle organic carbon. In a similar manner, for sorption to tissue lipid and octanol:

$$K_l = \frac{C^l}{C^w} = \beta \frac{\gamma^w}{\gamma^l} \qquad (5)$$

and

$$K_{ow} = \frac{C^{oct}}{C^w} = \beta \frac{\gamma^w}{\gamma^{oct}} \qquad (6)$$

where the subscript oct refers to octanol and $K_l$ and $K_{ow}$ are lipid-water and octanol-water partition coefficients, respectively. The $K_l$ is related to the bioconcentration factor (BCF) by $K_l = BCF/f_l$, where $f_l$ is the decimal fraction of animal lipid.

If equilibrium phase partitioning is the operative mechanism for distribution of an organic chemical among the particle, animal, and interstitial water compartments in a sediment, then the aqueous phase concentration of the chemical is in joint equilibrium with the animal lipid and sediment organic carbon concentrations. In principle equations 4 and 5 can be used to determine partition coefficients for the two sorbed phases. The difficulty is that only the animal and particle concentrations ($C^l$, $C^{oc}$) of the chemical can be easily measured, not the aqueous phase concentration ($C^w$). However, if independent information could be found to characterize $K_l$ or $K_{oc}$ in the sediment, then an estimate could be made of the aqueous phase concentration.

For hydrophobic organic chemicals, differences in activity coefficients between water and organic solvents provide a basis for quantitative relationships among the above partition coefficients. Equations 4 and 5 can each be combined with equation 6 to yield, respectively:

$$K_{oc} = K^{ow} \left( \frac{\gamma^{oct}}{\gamma^{oc}} \right) \qquad (7)$$

and

$$K_l = K_{ow} \left( \frac{\gamma^{oct}}{\gamma^l} \right) \qquad (8)$$

Activity coefficients for hydrophobic organic chemicals are very low in lipid, organic carbon and octanol, compared with their values in water (Karickhoff, 1981; Mackay, 1982). Furthermore, most of the chemical-to-chemical variation in partition coefficients is due to variation in the activity coefficient for the aqueous phase (Karickhoff and Morris, 1987). Consequently, the activity coefficient ratios in equations 7 and 8 should be approximately unity, and predictable relationships should exist among $K_l$, $K_{oc}$, and $K_{ow}$. A large body of experimental evidence indicates that this is indeed the case (e.g., Lyman et al., 1982).

An approach for estimating sediment partition coefficients is now clear. For a particular chemical with known $K_{ow}$ an estimate can be made of $K_l$. Because $K_l = C^l/C^w$, this provides an estimate of $C^w$ if a direct measurement is available for $C^l$. In turn, because $K_{oc} = C^{oc}/C^w$, this provides an estimate of $K_{oc}$ if a direct measurement is available for the value of $C^{oc}$ associated with $C^l$. Note that to test phase partitioning models, $C^w$ must be determined independently using $K_l$ not $K_{oc}$.

This approach is based upon the following assumptions:

1. Equilibrium phase partitioning theory applies to the distribution of hydrophobic organic chemicals among particles, animals and interstitial water in aquatic sediments.

2. Such systems are actually in an equilibrium state.

3. Relationships between $K_l$ and $K_{ow}$ developed using pelagic species (primarily fish) are valid for benthic organisms.

4. The values of $C^l$ predicted from relationships between $K_l$ and $K_{ow}$ actually represent true dissolved aqueous concentrations.

Thomann and Connolly (1984) and Oliver and Niimi (1988) have shown that concentrations of hydrophobic organic chemicals in aquatic food chains depend on the trophic position of the animal. The ratio of bioaccumulation (uptake from food and water) to bioconcentration (uptake from water only) tends to increase with the number of trophic transfers. This suggests that chemical concentrations in animals near the base of the food chain are due primarily to simple phase partitioning between the exposure concentration and the lipid pool in the animal. Preliminary work by Bierman (1988) has shown that accumulation of organic chemicals in macrobenthic invertebrates is indeed consistent with simple bioconcentration. Other conditions that must be satisfied are no kinetic or steric hindrances to the uptake of the chemical and no significant degrees of degradation or metabolism.

It is nearly impossible to demonstrate that an environmental system is in a true equilibrium state. It is generally assumed that sediment systems are more likely to be in equilibrium than water column systems because the physical-chemical environment in the sediment is more temporally stable. This assumption notwithstanding, the above theory provides the basis for a presumptive test of sediment equilibrium. Combining equations 4 and 5 gives:

$$\frac{C^l}{C^{oc}} = \frac{\gamma^{oc}}{\gamma^l} \approx 1 \tag{9}$$

This implies that if equilibrium phase partitioning theory is the operative mechanism, and, if the system is in equilibrium, then the chemical will partition approximately equally between animal lipid and sediment organic carbon. This is the concept of thermodynamic bioaccumulation potential suggested by McFarland (1984) and Lake et al. (1987).

Relationships between $K_l$ and $K_{ow}$ developed using pelagic fish are probably valid for benthic organisms because the basic mechanism responsible for bioconcentration appears to be simple phase partitioning into the lipid pools in the animals (Mackay, 1982). The results of Landrum et al. (1985) support the assumption that the bioavailable fraction of hydrophobic organic chemicals is the freely dissolved aqueous form ($C^w$). It should be noted that the above approach for estimating $C^w$ can be no more accurate than the experimental values for $C^w$ used in developing relationships between $K_l$ and $K_{ow}$. This is an important point because $C^w$ is more difficult to measure than $C^l$ or $C^{oct}$, especially for chemicals with high $K_{ow}$ values because their aqueous phase solubilities are extremely small.

## Methods

Field data were obtained for animal and sediment concentrations of hydrophobic organic chemicals that were thought to be consistent with the assumptions in the above analysis. Sediment partition coefficients were then estimated using the approach previously described. Three different equilibrium phase partitioning models were selected to investigate relationships between these estimated partition coefficients and the $K_{ow}$ values of the chemicals.

### Field Data

The data used in this study consisted of polychlorinated biphenyls (PCBs) in oligochaete worms from Lake Ontario (Oliver and Niimi, 1988) and the Detroit River (Smith et al., 1985), and polyaromatic hydrocarbons (PAHs) in oligochaete worms and chironomid midges from Lake Erie and in the amphipod *Pontoporeia hoyi* from Lake Michigan (Eadie et al., 1983). The feeding habits of these animals have been reviewed by Adams (1987). Oligochaetes live within the surficial sediment and continuously ingest sediment particles. Chironomids are filter-feeders and sediment ingesters. Amphipods are found at the sediment-water interface and deposit-feed directly on detritus. Body burdens in oligochaetes and chironomids are probably the most representative of sediment exposure conditions. The amphipod data were used because they included PAH concentrations measured in the associated interstitial waters.

Chemical concentrations in animals and sediment, and fraction organic carbon, were reported in each of the above studies. Animal concentration data reported as dry weight were converted to wet weight by assuming that dry weight was 15% of wet weight (Oliver, 1984). Organism lipid contents were assigned as 1% of fresh weight for oligochaetes and chironomids and as 3% for amphipods (Oliver and Niimi, 1988). Data for PCBs were reported at the congener level and lumped at the homolog level for this study. All $K_{ow}$ values were obtained from the literature (Rapaport and Eisenreich, 1984; Hawker and Connell, 1988; Schnoor et al., 1987). The $K_{ow}$ values for used for PCB homologs were averages of $K_{ow}$ values for the individual congeners in each homolog group.

## Simulation of $K_{oc}$ Values

The general approach for estimating sediment $K_{oc}$ values was described above. The following relationship was used to estimate bioconcentration factors as a function of $K_{ow}$ (Mackay, 1982):

$$\log K_b = \log K_{ow} - 1.32 \qquad (10)$$

or

$$K_b = 0.048 \, K_{ow} \qquad (11)$$

where $K_b$ (= BCF) is the ratio of the concentration in the animal to the concentration in the water. This relationship was used because it is based on a large number of different chemicals (n = 36) that span a range of $K_{ow}$ values from $10^2$ to $10^6$. It also includes most of the same chemicals used in the relationship developed by Veith et al. (1979).

A shortcoming of the relationship in equations. 10 and 11 is that it was not based on lipid-normalized animal concentration data. There are few lipid-normalized relationships in the literature, and they generally do not include large numbers of different chemicals or animals. For example, Konemann and van Leeuwen (1980) proposed a lipid-normalized relationship that was based on uptake of six chlorobenzenes by guppies. However, McFarland and Clarke (1986) compared this relationship to the above relationship by Mackay and found only slight differences in terms of predicting thermodynamic bioaccumulation potential.

## Phase Partitioning Models

Three equilibrium phase partitioning models were included in this study: the conventional two phase model, a three phase model, and a particle interaction model proposed by Di Toro (1985). The first two models are based on the assumptions of complete sorption-desorption reversibility and invariance of the partition coefficients with particle concentration. Equations 4-6 are examples of the conventional two phase model. The three phase model attempts to account for the influence of non separable particles (e.g., colloids) and dissolved organic matter that may confound the interpretation of experimental sorption data. The relationship between the two and three phase models can be expressed as (Gschwend and Wu, 1985):

$$K_{oc}^{apparent} = \frac{K_{oc}^{true}}{1 + K_{oc}^{nsp} DOC} \quad (12)$$

where $K_{oc}^{true}$ is the partition coefficient for separable particles, $K_{oc}^{nsp}$ is the partition coefficient for non separable particles plus dissolved organic matter (the "third phase"), DOC is the concentration of the third phase expressed in units of organic carbon, and $K_0^{apparent}$ is the apparent partition coefficient that would be observed if the third phase were not explicitly taken into account. In the limit as DOC approaches zero, $K_{oc}^{apparent}$ approaches $K_{oc}^{true}$, and the three phase model reduces to the two phase model.

The Di Toro model attempts to describe non singular isotherm behavior and the reported particle concentration effect on partition coefficients. It postulates an additional desorption reaction resulting from physical particle-particle interactions. The approach involves separation of sorption data into resistant and reversible partition coefficients and then relating the reversible component partition coefficient to changes in the concentration of adsorbing particles. This model can be expressed as (Di Toro, 1985):

$$\Pi = \frac{f_{oc} K_{oc}^x}{1 + m f_{oc} K_{oc}^x/\nu} \quad (13)$$

where $\Pi$ is the mass-based partition coefficient, m is particle concentration, $\nu$ is a constant, and the superscript x refers to reversible component partitioning. In the limit as m becomes large, $\Pi$ becomes small and is completely a function of the particle concentration. In the limit as m becomes small, $\Pi$ approaches the classical value of $f_{oc}K_{oc}^x$, and the particle interaction model approaches the conventional two phase model.

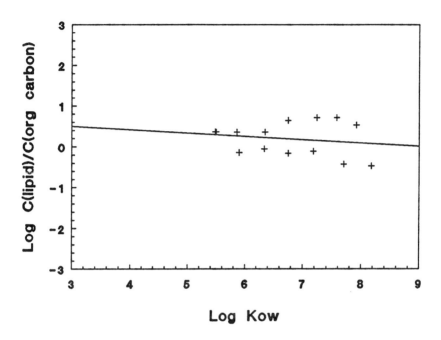

**Figure 2.** Relationship between the ratio $C^l/C^{oc}$ and $K_{ow}$ for PCB homologs in oligochaetes from Lake Ontario and the Detroit River. Data from Oliver and Niimi (1988) and Smith *et al.* (1985), respectively.

## Results

### Phase Distributions Between Animal Lipid and Sediment Organic Carbon

As discussed above, approximately equal partitioning of a hydrophobic organic chemical between animal lipid and sediment organic carbon ($C^l/C^{oc} \approx 1$) constitutes a presumptive test for equilibrium conditions in the sediment. The actual value of this ratio can only be established experimentally. McFarland and Clarke (1986) suggested a value between 1.7 and 1.9, based on existing empirical relationships among $K_l$, $K_{oc}$ and $K_{ow}$. Preliminary experiments by Lake *et al.* (1987) were consistent with a value of approximately 2.

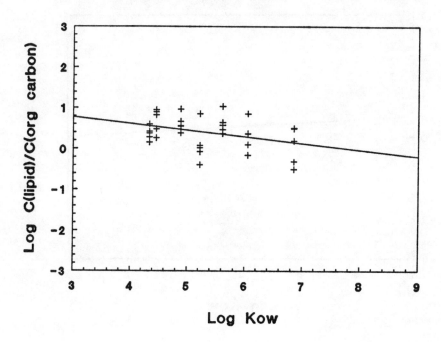

**Figure 3.** Relationship between the ratio $C^l/C^{oc}$ and $K_{ow}$ for PAHs in Lake Erie oligochaete worms and chironomid midges. Data from Eadie *et al.* (1983).

Figures 2-4 contain values for the ratio $C^l/C^{oc}$ as a function of log $K_{ow}$ for all data used in this study. Average values (± 1 std dev) were 2.23 (1.73), 3.50 (2.78), and 3.45 (2.41) for the PCB homologs, and the PAHs in Lakes Erie and Michigan, respectively. The slopes of the regressions of log $(C^l/C^{oc})$ versus log $K_{ow}$ were not significant at $\rho < 0.05$ for PCB homologs and PAHs in Lake Michigan. This slope was significant at $\rho < 0.05$ for PAHs in Lake Erie but was not significant at $\rho < 0.03$. All three of these average ratios were within a standard deviation of the expected range of 1 to 2. Although arithmetic statistics were used to average these data, it may have been more appropriate to log transform the data because the underlying distributions are probably lognormal.

It is not clear if the higher values of $C^l/C^{oc}$ for PAHs as compared to PCBs have any physical significance. Although some chemical-to-

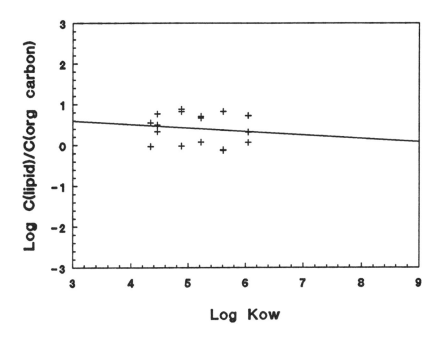

**Figure 4.** Relationship between the ratio $C^l/C^{oc}$ and $K_{ow}$ for PAHs in Lake Michigan *Pontoporeia hoyi*. Data from Eadie et al. (1985).

chemical variation is to be expected, there are many potential sources of error in the individual determinations of $C^l$ and $C^{oc}$, and these errors can be magnified when taking their ratio. Furthermore, this ratio is very sensitive to biological variability because it is inversely proportional to $f_l$, the fraction lipid. Lipid content is not measured on every sample, and a constant $f_l$ must be assigned for all animals of a particular feeding type. For the benthic invertebrates in this study, an absolute error in $f_l$ of only 1% corresponds to an error factor of 1.3 to 2 in the ratio $C^l/C^{oc}$ because the assigned values of $f_l$ were 1% (oligochaetes and chironomids) and 3% (*P. hoyi*).

## Relationships Between Estimated $K_{oc}$ and $K_{ow}$

Figures 5-7 illustrate the relationships between estimated sediment $K_{oc}$

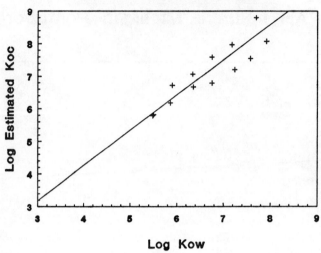

**Figure 5.** Relationship between estimated $K_{oc}^{true}$ for PCB homologs (Lake Ontario and Detroit River) and $K_{ow}$.

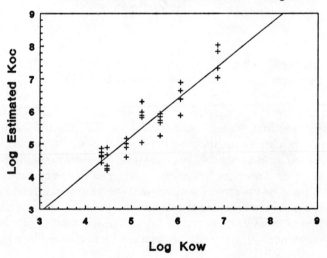

**Figure 6.** Relationship between estimated $K_{oc}^{true}$ for PAHs (Lake Erie) and $K_{ow}$.

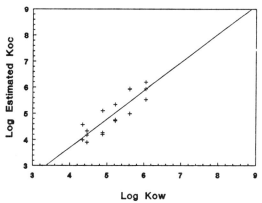

Figure 7. Relationship between estimated $K_{oc}^{true}$ for PAHs (Lake Michigan) and $K_{ow}$.

values and $K_{ow}$. Table 1 contains a summary of statistical results for linear regressions of log estimated $K_{oc}$ versus log $K_{ow}$. Between 77 and 88 percent of the variability in log estimated $K_{oc}$ is explained by log $K_{ow}$. All of the slopes of the regressions lines were approximately unity and none of the intercepts were significantly different from zero ($p < 0.05$).

These results are consistent with reported relationships between $K_{oc}$

Table 1

Summary of Statistical Results for Linear Regressions of Log Estimated $K_{oc}$ on Log $K_{ow}$

| Data Group | Intercept (Std Err) | Slope | $r^2$ | n |
|---|---|---|---|---|
| PCB Homologs (Lake Ontario and Detroit River) | 0.054 (0.880)[1] | 1.080 (0.129) | 0.85 | 14 |
| PAHs (Lake Erie) | 0.571 (0.404)[1] | 1.160 (0.075) | 0.88 | 35 |
| PAHs in (Lake Michigan) | 0.650 (0.787)[1] | 1.086 (0.152) | 0.77 | 17 |

[1]Not significant ($p < 0.05$).

and $K_{ow}$ for hydrophobic organic chemicals in more dilute suspensions (Karickhoff, 1984). These suspensions generally represent a range of solids concentrations from $10^1$ to $10^4$ mg/L. For a typical surficial sediment with a solids density of 2.5 gm/cm$^3$ and a porosity of 0.90, the bulk solids concentration is 2.5 x $10^5$ mg/L.

## Comparisons of Phase Partitioning Models

The general lack of a particle concentration effect on sediment partition coefficients can be seen in the results of Figures 5-7. In physical terms, these results imply that most of the variability in sediment partition coefficients can be explained by organic carbon normalization and variability in $K_{ow}$. If particle concentration had a significant effect on the partition coefficients, then the slopes of the regression lines would be much less than unity and, in fact, would approach zero at solids concentrations in the sediment. This is because the particle interaction model predicts that at sufficiently high solids concentrations, partitioning is completely controlled by solids and is independent of $K_{ow}$.

The discrepancy between estimated sediment partition coefficients and predictions of the particle interaction model can be illustrated for the particular case of hexachlorobiphenyl. Figure 8 shows the predicted relationship between log $K_{oc}$ ($K_{oc} = \Pi/f_{oc}$) and log m for the particle interaction model (Equation 13) using parameters for hexachlorobiphenyl (Di Toro, 1985) and $f_{oc} = 0.01$. Values for $K_{oc}$ are predicted to decrease by approximately an order of magnitude for each order of magnitude increase in solids concentration. There is a difference of approximately five orders of magnitude between the estimated $K_{oc}$ value for hexachlorobiphenyl in this study (at log $K_{ow} = 7.28$ in Figure 5) and the prediction of the particle interaction model (at m = 2.5x$10^5$ in Figure 8).

The two and three phase partitioning models can be compared by relating $K_{oc}^{apparent}$ and $K_{oc}^{true}$ using Equation 12. The $K_{oc}$ values in Figures 5-7 are estimates of $K_{oc}^{true}$ because they were determined using estimates of truly dissolved aqueous concentrations. For each value of $K_{oc}^{true}$, a value of $K_{oc}^{apparent}$ can be determined that corresponds to specified values for DOC and the partition coefficient for the third phase ($K_{oc}^{nsp}$). The $K_{oc}^{apparent}$ is the partition coefficient that would be observed if the chemical fraction bound to the third phase was not separated but instead was operationally included as part of the "dissolved aqueous concentration." The $K_{oc}^{apparent}$ would then be an underestimate of the true partition coefficient because the true dissolved aqueous concentration would be overestimated.

Figures 9-11 show the relationships between $K_{oc}^{apparent}$ and $K_{ow}$ for a range of third phase concentrations from 1 to 20 mg/L as organic carbon. A range of 5 to 20 mg/L is representative of concentrations found in sediment interstitial waters (Landrum *et al.*, 1987; Brownawell and

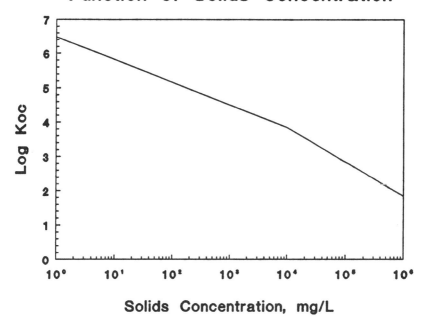

**Figure 8.** Predicted relationship between $K_{oc}$ ($f_{oc}$ = 0.01) for hexachlorobiphenyl and solids concentration using the particle interaction model (DiToro 1985).

Farrington, 1986). Results for 1 mg/L are presented for illustrative purposes. It was assumed that $K_{oc}^{true} = K_{oc}^{nsp}$. The straight lines correspond to the regressions of log $K_{oc}$ versus log $K_{ow}$ from Figures 5-7.

The influence of a third phase can be substantial, depending on its concentration and the $K_{ow}$ of the particular chemical. At log $K_{ow}$ = 8 there is a difference of almost three orders of magnitude between true and apparent partition coefficients for a third phase concentration of 1 mg/L; however, at log $K_{ow}$ = 6 this difference decreases to less than one order of magnitude (Figure 9). For log $K_{ow}$ between 5 and 6, the influence of the third phase begins to diminish, and, below log $K_{ow}$ = 5, the true and apparent partition coefficients converge to within the uncertainty of the data (Figures 10 and 11). For log $K_{ow}$ between 6 and 8 the log of the apparent partition coefficient is approximately inversely proportional to the third phase concentration (Figure 9). Below log $K_{ow}$

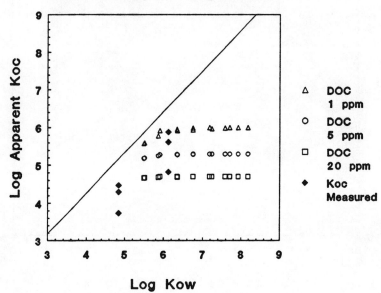

**Figure 9.** Relationship between estimated $K_{oc}^{apparent}$ for PCB homologs (Lake Ontario and Detroit River) and $K_{ow}$ for third phase concentrations of 1-20 mg/L as organic carbon.

= 6 the apparent partition coefficient becomes less sensitive to the third phase concentration and eventually converges to the true partition coefficient (Figures 10 and 11).

Figure 9 also contains results for measured apparent partition coefficients for Aroclor 1254 (log $K_{ow}$ = 6.12) and Aroclor 1242 (log $K_{ow}$ = 4.84) in the interstital waters of Lake Michigan sediments (Eadie et al., 1983). Although these values are not directly comparable to the $K_{oc}$ values determined using the oligochaete data from Lake Ontario and the Detroit River, they are reasonably consistent with predictions of the three phase model for the range of DOC concentrations (5 - 15 mg/L) reported to exist in Lake Michigan sediments (Landrum et al., 1987).

Figure 11 contains results for apparent partition coefficients that were measured in association with the PAHs in Lake Michigan *Pontoporeia hoyi*.

## Relationship of Apparent Koc to Kow
### PAH in Lake Erie Worms and Midges

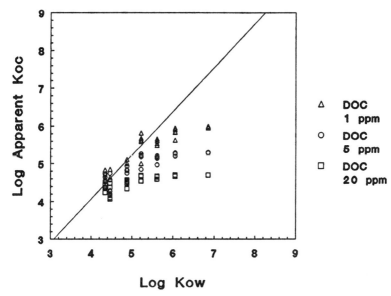

Figure 10. Relationship between estimated $K_{oc}^{apparent}$ for PAHs (Lake Erie) and $K_{ow}$ for third phase concentrations of 1-20 mg/L as organic carbon.

These measured partition coefficients agree with the trend of the three phase model predictions in that they diverge from the true partition coefficients at higher log $K_{ow}$ and converge at lower log $K_{ow}$. However, they do not agree in terms of absolute magnitude, especially at higher log $K_{ow}$. In fact, it is not clear from these data that the measured apparent partition coefficients bear any relationship to log $K_{ow}$. Unfortunately, benzo(ghi)perylene (log $K_{ow}$ = 6.85) was not detected in *P. hoyi*, thus limiting the range over which comparisons can be made.

## Discussion

The particle interaction model has been successfully applied to experimental sorption data for a large number of hydrophobic organic chemicals over a range of suspended solids concentrations from $10^1$ to $10^4$

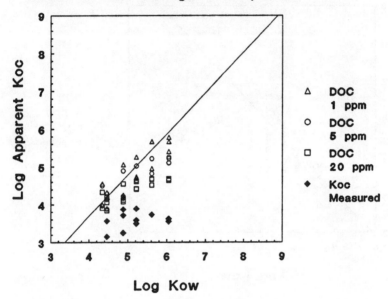

**Figure 11.** Relationship between estimated $K_{oc}^{apparent}$ for PAHs (Lake Michigan) and $K_{ow}$ for third phase concentrations of 1-20 mg/L as organic carbon.

mg/L (Di Toro, 1985). A question is whether the predictions of this model can be extrapolated from solids in agitated suspensions to stationary solids characteristic of bedded sediments. Results from long-term sediment diffusion experiments suggest that due to the absence of particle motion, sediment partition coefficients should correspond to values measured in low particle concentration suspended sediment experiments (Di Toro *et al.*, 1985). In this limit, partition coefficients are not controlled by particle concentrations, and they approach values that are consistent with the conventional two phase model.

The results in the present study indicate that sediment partition coefficients are not influenced by particle concentrations. Under the assumptions in this analysis, partitioning of hydrophobic organic chemicals in sediments can be completely described in terms of organic carbon normalization and the $K_{ow}$ of the chemical. Furthermore, because $K_{oc}$ and $K_{ow}$ were found to be very nearly equal, this implies that sorption of these chemicals to sediment organic carbon is essentially the same as partitioning into octanol.

A major obstacle to understanding phase partitioning of hydrophobic organic chemicals in sediments is separation and characterization of the third phase. Under the assumptions in this analysis, a three phase model was consistent with the partitioning of PCBs in the sediments (Figure 9); however, this model did not adequately describe the partitioning of PAHs (Figure 11). An important assumption in this model was that the partition coefficients for separable particles and DOC were equal ($K_{oc}^{true}$ = $K_{oc}^{nsp}$). Landrum et al. (1984) showed that there could be considerable variation in $K_{oc}^{nsp}$ for the same chemical among different natural sources of DOC. In a more detailed study, Landrum et al. (1987) found that log $K_{ow}$ accounted for only 50% of the observed variability in the partitioning of PAHs to different natural sources of DOC from sediment interstitial waters.

These findings greatly complicate the potential application of three phase models to sediment partitioning because model coefficients for the third phase appear to be a function of geographic location and/or season. This behavior is a possible explanation for the conflicting results obtained above for the partitioning of PCBs and PAHs. An unresolved question is whether DOC constitutes an adequate surrogate for the third phase or even whether organic carbon is the most appropriate normalization for the third phase.

An important ancillary aspect of this study is that it included a test of the concept of thermodynamic bioaccumulation potential. The results in Figures 2-4 indicate that uptake of PCBs and PAHs depends only on sediment organic carbon and animal lipid, not on properties of the individual chemicals. For sediment systems and organic chemicals that satisfy the assumptions in this analysis, it appears that organic carbon normalized sediment concentrations can be used to predict upper bounds for organism body burdens.

## Acknowledgments

The assistance of Rob Bankoske and Levia Stein in compiling and reducing the data in this study is gratefully acknowledged.

## References

Adams, W.J. 1987. Bioavailability of neutral lipophilic organic chemicals contained on sediments: a review. In: *Fate and Effects of Sediment-Bound Chemicals in Aquatic Systems*. K.L. Dickson, A.W. Maki and W.A. Brungs (eds.), Society of Environmental Toxicology and Chemistry Special Publication Series,

Pergamon Press, New York, pp. 219-244.

Allan, R.J. 1986. The role of particulate matter in the fate of contaminants in aquatic systems. Scientific Series No. 142, National Water Research Institute, Canada Centre for Inland Waters, Burlington, Ontario.

Bierman, V.J., Jr. 1988. Bioaccumulation of organic chemicals in Great Lakes benthic food chains. In: *Proceedings of the Workshop on Aquatic Food Chain Modelling*. Y. Hamdy and G. Johnson (eds.), Ontario Ministry of the Environment, Toronto, Canada, pp. 82-119.

Brownawell, B.J. and J.W. Farrington. 1986. Biogeochemistry of PCBs in interstitial waters of a coastal marine sediment. *Geochim. et Cosmochim. Act.*, 50:157-169.

Capel, P.D. and S.J. Eisenreich. 1985. PCBs in Lake Superior, 1978-1980. *J. Great Lakes Res.*, 11(4):447-461.

Carter, C.W. and I.H. Suffet. 1982. Binding of DDT to dissolved humic materials. *Environ. Sci. Technol.*, 16(11):735-740.

Chiou, C.T., R.L. Malcolm, T.I. Brinton, and D.E. Kile. 1986. Water solubility enhancement of some organic pollutants and pesticides by dissolved humic and fulvic acids. *Environ. Sci. Technol.*, 20(5):502-508.

Dickson, K.L., A.W. Maki and W.A. Brungs. 1987. *Fate and Effects of Sediment-Bound Chemicals in Aquatic Systems*. Society of Environmental Toxicology and Chemistry Special Publication Series, Pergamon Press, New York.

Di Toro, D.M. and L.M. Horzempa. 1982. Reversible and resistant components of PCB adsorption-desorption: isotherms. *Environ. Sci. Technol.*, 16(9):594-602.

Di Toro, D.M. 1985. A particle interaction model of reversible organic chemical sorption. *Chemosphere.*, 14(10):1503-1538.

Di Toro, D.M., J.S. Jeris, and D. Ciarcia. 1985. Diffusion and partitioning of hexachlorobiphenyl in sediments. *Environ. Sci. Technol.*, 19(12):1169-1176.

Eadie, B.J., J.A. Robbins, P.F. Landrum, C.P. Rice, M.S. Simmons, M.J. McCormick, S.J. Eisenreich, G.L. Bell, R.L. Pickett, K. Johansen, R. Rossman, N. Hawley and T. Voice. 1983. The cycling of toxic organics in the Great Lakes: A 3-year status report. National Oceanic and Atmospheric Administration, Great Lakes Environmental Research Laboratory, Ann Arbor, Michigan. NOAA Technical Memorandum ERL GLERL-45.

Gschwend, P.M. and S. Wu. 1985. On the constancy of sediment-water partition coefficients of hydrophobic organic pollutants. *Environ. Sci. Technol.*,19(1):90-96.

Hassett, J.P. and M.A. Anderson. 1979. Association of hydrophobic organic compounds with dissolved organic matter in aquatic systems. *Environ. Sci. Technol.*, 13(12):1526-1529.

Hawker, D.W. and D.W. Connell. 1988. Octanol-water partition coefficients of polychlorinated biphenyl congeners. *Environ. Sci. Technol.*, 22(4):382-387.

Horzempa, L.M. and D.M. Di Toro. 1983. PCB partitioning in sediment-water systems: the effect of sediment concentration. *J. Environ. Qual.*, 12(3):373-380.

Karickhoff, S.W. 1981. Semi-empirical estimation of sorption of hydrophobic pollutants on natural sediments and soils. *Chemosphere*, 10(8):833-846.

Karickhoff, S.W. 1984. Organic pollutant sorption in aquatic systems. *J. Hydraulic Engineering*, 110(6):707-735.

Karickhoff, S.W. and K.R. Morris. 1987. Pollutant sorption: relationship to bioavailability. In: *Fate and Effects of Sediment-Bound Chemicals in Aquatic Systems*, K.L. Dickson, A.W. Maki and W.A. Brungs (eds.), Society of Environmental Toxicology and Chemistry Special Publication Series, Pergamon Press, New York, pp. 75-82.

Konemann H. and K. van Leeuwen. 1980. Toxicokinetics in fish: accumulation and elimination of six chlorobenzenes by guppies. *Chemosphere*, 9:3-19.

Lake, J.L., N. Rubinstein, and S. Pavignano. 1987. Predicting bioaccumulation: development of a simple partitioning model for use as a screening tool for regulating ocean disposal of wastes. In: *Fate and Effects of Sediment-Bound Chemicals in Aquatic Systems*, K.L. Dickson, A.W. Maki and W.A. Brungs (eds.), Society of Environmental Toxicology and Chemistry Special Publication Series, Pergamon Press, New York, pp. 151-166.

Landrum, P.F., S.R. Nihart, B.J. Eadie, and W.S. Gardner. 1984. Reverse-phase separation method for determining pollutant binding to aldrich humic acid and dissolved organic carbon of natural waters. *Environ. Sci. Technol.*, 18(3):187-192.

Landrum, P.F., M.D. Reinhold, S.R. Nihart, and B.J. Eadie. 1985. Predicting the availability of organic xenobiotics to *Pontoporeia hoyi* in the presence of humic and fulvic materials and natural dissolved organic matter. *Environ. Toxicol. Chem.*, 4:459-467.

Landrum, P.F., S.R. Nihart, B.J. Eadie, and L.R. Herche. 1987. Reduction in bioavailability of organic contaminants to the amphipod *Pontoporeia hoyi* by dissolved organic matter of sediment interstitial waters. *Environ. Toxicol. Chem.*, 6:11-20.

Lyman, W.J., W.F. Reehl, and D.H. Rosenblatt. 1982. *Handbook of Chemical Property Estimation Methods*, McGraw-Hill, New York.

Mackay, D. 1979. Finding fugacity feasible. *Environ. Sci. Technol.*, 13(10):1218-1223.

Mackay, D. 1982. Correlation of bioconcentration factors. *Environ. Sci. Technol.*, 16(5):274-278.

McFarland, V.A. 1984. Activity-based evaluation of potential bioaccumulation from sediments. In: *Dredging and Dredged Material Disposal*, Vol. 1, Proceedings of the Conference Dredging '84. R.L. Montgomery and J.W. Leach (eds.), American Society of Civil Engineers, New York, pp. 461-466.

McFarland, V.A. and J.U. Clarke. 1986. Testing bioavailability of polychlorinated biphenyls from sediments using a two-level approach. U.S. Army Corps of Engineers Committee on Water Quality, *Sixth Seminar Proceedings*, New Orleans, Louisiana, pp. 220-229.

Means, J.C. and R. Wijayaratne. 1982. Role of natural colloids in the transport of hydrophobic pollutants. *Science*, 215:968-970.

O'Connor, D.J. and J.P. Connolly. 1980. The effect of concentration of adsorbing solids on the partition coefficient. *Water Res.*, 14:1517-1523.

Oliver, B.G. 1984. Uptake of chlorinated organics from anthropogenically contaminated sediments by oligochaete worms. *Can. J. Fish. Aquat. Sci.*, 41:878-883.

Oliver, B.G. and A.J. Niimi. 1988. Trophodynamic analysis of polychlorinated biphenyl cogeners and other chlorinated hydrocarbons in the Lake Ontario ecosystem. *Environ. Sci. Technol.*, 22(4):388-397.

Rao, P.S.C. and J.M. Davidson. 1980. Estimation of pesticide retention and transformation parameters required in nonpoint source pollution models. In: *Environmental Impact of Nonpoint Source Pollution*, M.R. Overcash and J.M. Davidson (eds.), Ann Arbor Science Publishers, Ann Arbor, MI, pp. 23-67.

Rapaport, R.A. and S.J. Eisenreich. 1984. Chromatographic determination of octanol-water partition coefficients ($K_{ow}$) for 58 polychlorinated biphenyl congeners. *Environ. Sci. Technol.*, 18(3):163-170.

Schnoor, J.L., C. Sato, D. McKechnie, and D. Sahoo. 1987. Processes, coefficients, and models for simulating toxic organics and heavy metals in surface waters. U.S. Environmental Protection Agency, Environmental Research Laboratory, Athens, GA, EPA/600/3-87/015.

Smith, V.E., J.M. Spurr and J.C. Filkins. 1985. Organochlorine contaminants of wintering ducks foraging on Detroit River sediments. *J. Great Lakes Res.*, 11(3):231-246.

Swackhamer, D.L. and D.E. Armstrong. 1987. Distribution and characterization of PCBs in Lake Michigan water. *J. Great Lakes Res.*, 13(1):24-36.

Thomann, R.V. and J.P. Connolly. 1984. Model of PCB in the Lake Michigan lake trout food chain. *Environ. Sci. Technol.*, 18(2):65-71.

Veith, G.D., D.L. DeFoe, and B.V. Bergstedt. 1979. Measuring and estimating the bioconcentration factor of chemicals in fish. *J. Fish. Res. Board Ban.*, 36:1040-1048.

Voice, T.C. and W.J. Weber, Jr. 1983. Sorption of hydrophobic compounds by sediments, soils and suspended solids - I. Theory and background. *Water Res.*, 17(10):1433-1441.

Voice, T.C., C.P. Rice, and W.J. Weber, Jr. 1983. Effect of solids concentration on the sorptive partitioning of hydrophobic pollutants in aquatic systems. *Environ. Sci. Technol.*, 17(9):513-518.

# CHAPTER 10

# Predicting Metals Partitioning During Resuspension Events

Joseph V. DePinto
Thomas L. Theis
Thomas C. Young
Susan Thompson Leach

## Introduction

In spite of major efforts to reduce the impacts of toxic chemicals on human health and aquatic communities, continued problems of toxicity to aquatic organisms and bioaccumulation in aquatic food chains persist. One of the factors contributing to the persistance of degraded conditions in many systems is the presence of contaminated bottom sediments. Bottom sediments in many aquatic ecosystems have become contaminated through a gradual process that involves the deposition of external inputs by subjecting these allochthonous contaminants to a variety of physical, chemical, and biological water column and sediment processes. Although the ultimate fate of most of these "in-place pollutants" is to be inactivated by diagenesis and burial in bottom sediments, this is a very slow succession, which allows for shorter time scale impacts to occur.

Among the potential pathways for expression of water column effects long after the initial introduction of toxics to the system is the physical resuspension of contaminated bottom sediments. In addition to its importance for the general assessment of biological exposure and effects, the overall process of contaminant resuspension must be understood in order to make accurate site-specific decisions regarding regulatory and remediation concerns. Among these are: quantitative assessment of the relative significance of "in-place pollutants" as a downstream source of contamination (e.g., an additional source to be considered in performing toxic waste load allocations); the development and prioritization of site remediation programs; and a potentially significant internal feedback process in system-wide toxic mass balance modeling efforts. Consideration of factors that govern contaminant resuspension has also been proposed for the development of sediment quality criteria (Kreis, 1988).

Proper assessment of the impact of bottom sediment resuspension events on an aquatic ecosystem requires a quantitative understanding of four aspects of the overall problem:
- prediction of the resuspension flux of particulate matter as a function of controlling environmental conditions;
- prediction of the advective and dispersive transport of resuspended sediments in the water column;
- prediction of the contaminant phase redistribution during a resuspension event;
- prediction of the biological exposure and effects of contaminants associated with a resuspension event.

Our understanding of the first two aspects of the problem are presented in several papers in this book (Lick, 1991; Bedford, 1991; Bonner et al., 1991; Ziegler et al., 1991). Assessing biological effects of contaminated sediments has been the subject of numerous studies; however, generalizing these efforts is restricted by our ability to define the "active" fraction of the suite of contaminants associated with any given sediment system. This is the rationale for the chemical phase/speciation aspect of the above framework, which is the subject of this manuscript.

There are two basic types of approaches that might be taken in computing the expected contaminant exposure regime during bottom sediment resuspension: those based on an assumption of equilibrium and those that employ a kinetic description of the distribution process. In this paper we will attempt to present a discussion of the factors to consider in selecting an approach as well as the development and preliminary application of a time-dependent model for simulating the kinetic partitioning of heavy metals during sediment resuspension.

# Equilibrium Metals Partitioning

In analyzing metals phase distribution during a resuspension event, the simplest approach is to assume that the total resuspended metal rapidly achieves an equilibrium partitioning between the two phases so that an equilibrium model can be used to calculate the distribution for each set of controlling conditions. A great deal of research has been conducted on the sorption of metals to solid surfaces in aquatic systems. While it is beyond the scope of this paper to develop the theory of metal sorption completely, it is pertinent to review the factors that control the equilibrium partitioning of metals in aqueous systems. These factors are discussed in some detail in a number of reviews on metals sorption in aquatic systems (e.g., Forstner and Wittman, 1981; Jenne and Zachara, 1984; Theis, 1985; Davis and Hayes, 1986; Schindler and Stumm, 1987; Elder, 1988; Honeyman and Santschi, 1988; and Allen, 1991).

Models describing the equilibrium partitioning between heavy metals and sediments in natural aquatic systems can range from simple empirical partition coefficients (derived from the slope of site-specific linear adsorption isotherms) to theoretically based chemical equilibrium models, such as the one proposed by Davis, James, and Leckie (Davis et al., 1978; James et al., 1978), that combine surface complexation theory with electric double-layer theory. More recently, these models have been extended to "triple layer models" to interpret more accurately the effects of ionic strength on ion adsorption (Hayes and Leckie, 1987). These more complex equilibrium adsorption models have been shown to have a sound theoretical basis (Westall and Hohl, 1980; Morel, 1981) and have shown very good agreement with experimental observations in well-characterized laboratory metal-ligand-sorbent systems (e.g., Davis and Leckie, 1978; Theis and Richter, 1980; Benjamin and Leckie, 1981; James et al., 1981). However, successful application of these models to natural aquatic systems has been hampered by the need for data on the numerous intrinsic system parameters that can potentially influence metal distribution calculations.

Despite the lag between the development of theoretical metal adsorption models and their practical application to natural systems, there has developed general agreement on many of the characteristic features of these adsorption reactions. Metal adsorption to sediments is considered to be analogous to the formation of soluble complexes, with the only difference being that the ligand in the reaction is a surface site (Stumm et al., 1980; Benjamin and Leckie, 1981). Therefore, the extent to which metals become operationally associated with natural aquatic sediments (i.e., the partitioning process) depends on many of the same factors that affect soluble complex formation.

One thing is clear from all this work: the application of relatively simple empirical expressions (exchange equilibrium equations, adsorption isotherms of the Freundlich or Langmuir type, or simple partition coefficients) leads to a parameterization that is <u>conditional</u> in that it is limited to the specific system from which it was derived. Variability of environmental conditions (pH, p$\varepsilon$; ionic strength; adsorbate solubility and molecular size; degree of metal complexation and type of complexes formed; concentration of competing sorbates; sorbent physical and chemical characteristics; and sorption site density, selectivity, and location) can significantly alter metal-sediment partitioning (Vuceta and Morgan, 1978; Benjamin and Leckie, 1982; Oakley et al., 1981; Benjamin and Leckie, 1982; Lion et al., 1982; Rygwelski, 1984; McIlroy et al., 1986; Theis, 1985, Tessier et al., 1991, among many others).

In an effort to determine the extent to which a relatively simple equilibrium partitioning model could be employed to calculate metal partitioning during sediment resuspension, a combination field and laboratory experimental investigation was undertaken using the Trenton Channel of the Detroit River as the study site (DePinto et al., 1989). In this study sediment resuspension experiments were conducted in the field by applying a calibrated shear stress to intact sediment cores with an oscillating grid. The design and calibration of the experimental apparatus is presented by Tsai and Lick (1986), and the results and interpretation of the metals partitioning data are presented in Theis et al. (1988a). These experiments, along with concurrent laboratory equilibrium partitioning and sediment sequential extraction experiments (DePinto et al., 1989), produced several valuable observations about the process. Metals partitioning behavior during sediment resuspension events was shown to be clearly influenced by pH, shear stress, cationic metal studied, resuspended solids concentration, time of resuspension, and, to a lesser extent, particle size distribution and site-specific sediment characteristics. An illustration of some of these effects is presented in Figure 1, which illustrates the large variability of measured partition coefficients for field measurements made between 10 and 30 minutes after the initiation of resuspension experiments (experiments were 30 minutes in duration and it took resuspended solids 10 minutes to come to quasi-steady-state).

Comparison of field shaker experimental results with laboratory adsorption studies revealed differences that led to, among others, the hypothesis that assessment of metals flux between solid and solution phase during resuspension events required a time-variable approach. This realization led to the development and preliminary application of the kinetic adsorption-desorption model presented below.

## Development of ME-SORB

### Model Framework and Assumptions

The main concern associated with using the equilibrium partitioning approach for calculating release of sediment-bound metals during a resuspension event is related to the time it takes to reach desorption equilibrium. There is evidence in the literature suggesting that equilibrium is not fully attained when contaminated bottom sediments are resuspended into the water column (Young et al., 1987; Coates and Elzerman, 1986; Isaacson and Frink, 1984; Theis et al., 1988b). If there truly are kinetic limitations for metals release from resuspended sediments, then using the equilibrium approach would overestimate the net contribution of bioavailable metals to the water column during a resuspension event. Over the course of many such events, the error could propagate into a significant discrepancy between prediction of

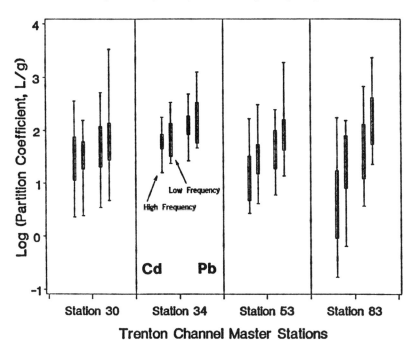

**Figure 1.** Field resuspension data on Cd and Pb partitioning in suspended sediments of Trenton Channel shaker experiments. Plot shows variability in measured partition coefficients among metals, stations, and grid oscillation frequency.

long-term system response to the presence of contaminated sediments and the actual observed response. In an effort to develop a kinetic approach for describing metal phase distribution during sediment resuspension, a preliminary dynamic model was developed and calibrated to field resuspension data from the Trenton Channel of the Detroit River.

The model developed herein (ME-SORB) is intended to be used in formulating a mechanistic description of metal ion exchange between the aqueous and solid phases during a bottom sediment resuspenion event. It is a dynamic model that takes into account the relative rates of the transport and reaction processes involved. Other time-variable models of this nature have been developed and applied to well-characterized, metal-sorbent laboratory systems (e.g., Hachiya et al., 1984; Hayes and Leckie, 1986; Theis et al., 1988b). The application of a kinetic model such as ME-SORB to field resuspension data from a natural system is the unique aspect of this study.

The model examines five transport and adsorption processes. Selection of these steps as important parameters is based on the characteristics of the adsorbent and knowledge of typical adsorption kinetics. Since the adsorbent is a river sediment, it is composed of a combination of silt, sand, and natural organic material. This material is known to be porous, thus the model must include intraparticle diffusion and adsorption as well as the surface phenomena. As model development continues, it may become apparent that the effects of one or more of these steps may be insignificant when compared to the others.

The steps include:

1. Bulk transport - transfer of material from liquid bulk solution to the liquid surface film surrounding the particle. In a well mixed suspension such as a resuspension event, this step is normally rapid and not rate limiting.
2. Film transport - transfer of the material through the surface film (a layer of quiescent liquid near the particle surface) between the bulk solution and the adsorbent particle surface.
3. Surface adsorption - adsorption of material on to the surface of the adsorbent particle.
4. Pore diffusion - transport of the adsorbate radially between the particle surface and the center of the particle through the intra-particle pore spaces.
5. Pore adsorption - adsorption of diffused material on to the walls of the pores.

Each of these steps can occur in either the forward or reverse direction depending on the sorbate gradients at a particular point in the idealized sorbate-sorbent system.

The sorption reaction is treated as a pseudo first-order reversible reaction based on the following equation:

$$[Me^{+z}] + \{S^{-z+1}\} \underset{k_r}{\overset{k_f}{\rightleftharpoons}} \{SMe\} + [H^+] \tag{1}$$

where
$[Me^{+z}]$ = sorbate metal concentration $(M/L^3)$
$\{S^{-z+1}\}$ = sorption site density $(M/L^2)$
$\{SMe\}$ = adsorbed metal density $(M/L^2)$
$[H^+]$ = hydrogen ion concentration $(M/L^3)$
$k_f$ = forward reaction rate $(L^3/M\text{-}t)$
$k_r$ = reverse reaction rate $(L^3/M\text{-}t)$.

As stated above, the adsorbent is assumed to be a porous media. In addition, all adsorption sites are assumed to be equivalent (both internal and external) and to represent average particle characteristics; therefore, the reaction rate constants are assumed to be the same in the pores as on the particle surface. The site density, however, may vary between internal and external surfaces. Hence, any difference between internal and external adsorption is controlled by the differences in sorbate concentration gradients due to pore diffusion and in site density.

Additionally, all particle sorbent physical and chemical parameters are assumed to represent average properties of the suspension. Temperature and pH are assumed to be constant for a given event, and rate coefficients are independent of solution phase concentrations.

## ME-SORB Mathematical Formulation

ME-SORB was formulated by incorporating flux terms for each of the five transport and adsorption steps listed above into differential mass balance equations for each of five state variables in the model. The state variables are the metal concentrations in each of five compartments of the system: bulk solution, surface film layer, pore solution, outer surface sorbed metal, and pore surface sorbed metal.

Both the bulk and film diffusion steps are based on Fick's First Law. For the bulk solution:

$$J_B = \frac{V_B}{A_B} \frac{dC_B}{dt} = k_B (C_{Bo} - C_B) \tag{2}$$

where  $J_B$ = flux from bulk solution (M/L²-t)
  $V_B$ = bulk solution volume (L³)
  $A_B$ = interfacial area (L²)
  $k_B$ = bulk mass transfer coefficient (L/t)
  $C_{Bo}$ = initial bulk solution concentration (M/L³)
  $C_B$ = bulk solution concentration (M/L³).

Since the model is intended to simulate metal partitioning during resuspension events, it was assumed that mixing was sufficient throughout the bulk solution to create a uniform concentration. This assumption allowed us to eliminate bulk diffusion from the calculation.

Similarly, for film solution transfer to the particle surface:

$$J_F = \frac{V_F}{A_F}\frac{dC_F}{dt} = k_M(C_B - C_F) \quad (3)$$

where  $J_F$ = flux from film solution (M/L²-t)
 $V_F$ = film solution volume (L³)
 $A_F$ = film interfacial area (L²)
 $k_M$ = film mass transfer coefficient (L/t)
 $C_B$ = film solution concentration at outer boundary (M/L³)
 $C_F$ = solution concentration at particle surface (M/L³).

The formulation of the equation for the adsorption process is based on the previously discussed equation for adsorption. The reaction mechanism in equation (1) can be expressed as a rate equation:

$$V_F\frac{d[Me^{+2}]}{dt} = A_S(-k_f[Me^{+2}]\{S^{-z+1}\} + k_r\{MeS\}[H^+]) \quad (4)$$

Finally, the diffusion of metal ions into the pore spaces must be accounted for. This is based on the equation for a solute diffusing into a sphere (Crank, 1975):

$$\frac{dC_P}{dt} = \frac{D}{r^2}\frac{d}{dr}(r^2\frac{dC_P}{dr}) \quad (5)$$

where  $C_P(r,t)$ = sorbate metal concentration in pore (M/L³)
 $D$ = longitudinal diffusion coefficient in pore (L²/t)
 $r$ = distance along radial pore from center of particle (L).

Combining these equations, a system of differential equations is developed that forms the basis of the computer model, ME-SORB. The differential equations are as follows:

$$V_B\frac{dC_B(t)}{dt} = A_S[k_M(C_F(t) - C_B(t))] \quad (6)$$

$$V_F\frac{dC_F(t)}{dt} = A_S[k_M(C_B(t) - C_F(t)) - k_fC_F(t)S_S + k_rC_{AS}(t)[H^+]] \quad (7)$$

$$A_S \frac{dC_{AS}(t)}{dt} = A_S [k_f C_F(t) S_S - k_r C_{AS}(t)[H^+]] \quad (8)$$

$$\frac{dC_P(r,t)}{dt} = \frac{D}{r^2} \frac{d}{dr}\left(r^2 \frac{dC_P(r,t)}{dr}\right) + \frac{A_P}{V_P}[-k_f C_P(r,t) S_P + k_r C_{AP}(r,t)[H^+]] \quad (9)$$

$$A_P \frac{dC_{AP}(r,t)}{dt} = A_P [k_f C_P(r,t) S_P - k_r C_{AP}(r,t)[H^+]] \quad (10)$$

where  $A_P$ = surface area of pores (L²)
$C_{AP}(r,t)$ = density of metal adsorbed to pore surface (M/L²)
$C_{AS}(r,t)$ = density of metal adsorbed to particle surface (M/L²)
$S_S(t)$ = unbound site density at particle surface (sites/L²)
$S_{ST}$ = total site density at particle surface (sites/L²)
$S_P(r,t)$ = unbound site density at pore surface (sites/L²)
$S_{PT}$ = total site density at pore surface (sites/L²)
R = particle radius (L)
$V_P$ = pore volume (L³)

and  $S_S(t) = S_{ST} - C_{AS}(t)$
$S_P(r,t) = S_{PT} - C_{AP}(r,t)$.

The initial conditions for solving this set of equations include establishing the initial metal concentration in the bulk solution, setting the initial metal activity within the particle pores to zero (for adsorption), and performing a mass balance around the film layer of the particle at time = 0. These initial conditions are expressed in equations (11-13), respectively:

$$C_B(0) = C_{Bo} \quad (11)$$

$$C_P(r,0) = C_{AP}(r,0) = 0 \quad (12)$$

The boundary conditions for this set of equations include setting the film metal concentration equal to the pore concentration at the particle

$$V_F C_F(0) + A_S C_{AS}(0) + \frac{3}{R^3}\int_0^R (C_p(r,0) + (\frac{A_P}{V_P})C_{AP}(r,0))r^2\,dr = Const \quad (13)$$

surface, setting the flux of metal at the center of the particle equal to zero, and conserving mass flux at the particle surface pore entrance. These conditions are established by equations (14-16), respectively:

$$C_p(R,t) = C_F(t) \quad (14)$$

$$\frac{dC_p(0,t)}{dt} = \frac{dC_{AP}(0,t)}{dt} = 0 \quad (15)$$

$$\frac{3}{R^3}\frac{d}{dt}\int_0^R [C_p(r,t) + (\frac{A_P}{V_P})C_{AP}(r,t)]r^2\,dr = \frac{A_F}{V_F}(K_M[C_B - C_F(t)]) \quad (16)$$

Since the system of equations is parabolic, it was be solved using the Crank-Nicholson technique. This is an implicit finite difference technique that has been proven to be unconditionally stable for parabolic partial differential equations. Also, the truncation error is relatively small (Burden and Faires, 1978). The tridiagonal linear matrix resulting from the Crank-Nicholson discretization was solved using a Crout Reduction technique.

## Model Calibration

Two field resuspension experiments were used for the calibration of ME-SORB: experiments 18 and 22 from the Trenton Channel field shaker studies discussed earlier (Theis et al. 1988a). Experiment 18 was conducted at station 30 using a shaker mixing frequency of 12.5 strokes/second (corresponds to 5.0 dynes/cm$^2$ bottom shear stress). Experiment 22 was conducted at station 34 with a lower resuspension frequency (7.14 strokes/s or 2.5 dynes/cm$^2$ shear stress). As a result the steady-state resuspended solids in the two experiments were 3.16 and 0.56 g/L for experiments 18 and 22, respectively. Both experiments maintained a relatively constant overlying water pH value of 7.6, and bulk solution metal data followed a definite trend that could be modeled.

Table 1. Measured model input data used for calibration of ME-SORB to Trenton Channel sediment resuspension experiments.

| Characteristic | Station 30 | Station 34 |
|---|---|---|
| Sediment Water Content (%) | 57.64 | 55.31 |
| Total Organic Carbon (%) | 7.4 | 10.0 |
| Cation Exchange Capacity (meq/g) | 1.04 | 1.67 |
| Ave. Particle Diameter (µm) | 1.0 | 1.0 |
| Particle Porosity | 0.55 | 0.53 |
| Pore Area-Hg intrusion (m$^2$/g) | 29.01 | 18.66 |
| Ave. Pore Diameter (µm) | 0.076 | 0.114 |
| Apparent Particle Density (g/ml) | 2.27 | 2.53 |
| Total Metal Content (mg/kg) | | |
| Cadmium | 24.7 | 27.8 |
| Chromium | 89.33 | 213.0 |
| Cobalt | 26.1 | 16.6 |
| Copper | 111.3 | 62.2 |
| Lead | 177.3 | 248 |
| Nickel | 64.6 | 62.2 |
| Zinc | 562.7 | 3920. |

Use of these two experiments for calibration permitted investigation of the influence of shear stress and sediment properties on metal adsorption kinetics.

A number of input parameters must be specified for application of ME-SORB to a given metal-sediment system. Many of these parameters that could be measured *a priori* were specified as shown in Table 1 and held constant during the calibration. *A priori* specification of these parameters left three calibration parameters: reactive site density, $S_{ST}$ (sites/nm$^2$), surface film mass transfer rate, $K_M$ (cm/min), and forward reaction rate, $k_f$ (L/mole-min). Although each of these was permitted to vary in effecting a calibration to the above data, their calibration values were bounded by the range of previous experimental work and their relative values were required to be consistent with system experimental conditions and known metal partitioning behavior. For example,

Table 2. Coefficients used for calibration of ME-SORB to experiments 18 and 22 of the Trenton Channel field resuspension study.

| Metal | Film Mass Transfer $K_M$ (cm/min) | Forward Reaction Rate $K_f$ (L/mol-min) | Site Density $S_{ST}$ (sites/nm$^2$) |
|---|---|---|---|
| Exp18(Sta30) | | | |
| Cd | 1.0 | $3.0 \times 10^4$ | 0.4 |
| Co | 1.0 | $2.1 \times 10^5$ | 0.1 |
| Cr | 1.0 | $1.0 \times 10^5$ | 0.5 |
| Ni | 1.0 | $4.0 \times 10^4$ | 0.4 |
| Pb | 1.0 | $3.4 \times 10^4$ | 0.4 |
| Exp22(Sta34) | | | |
| Cd | $5.0 \times 10^{-2}$ | $1.0 \times 10^4$ | 1.0 |
| Co | $5.0 \times 10^{-2}$ | $2.0 \times 10^5$ | 0.15 |
| Cr | $5.0 \times 10^{-2}$ | $1.0 \times 10^5$ | 0.2 |
| Ni | $5.0 \times 10^{-2}$ | $4.0 \times 10^4$ | 0.5 |
| Pb | $5.0 \times 10^{-2}$ | $3.4 \times 10^4$ | 0.5 |

experiment 18 had a much higher mixing rate than experiment 22; therefore, the film transfer coefficient for experiment 18 was constrained to be higher than that for experiment 22. Also, previous experimental work suggested a rather large range of forward reaction rate between $9.0 \times 10^2$ and $1.0 \times 10^9$, with less mobile metals like Pb exhibiting the largest adsorption rates (Hachiya et al., 1984; Kaul, 1985). Finally, reactive site density is very difficult to predict for natural sediments; however, a range of values between 3 and 10 sites/nm$^2$ have been determined for pure mineral surfaces by Benjamin and Leckie (1981).

The calibration was performed by selecting reasonable values for each of the parameters from the literature and then adjusting these values within limits defined by experimental conditions to simulate both data sets satisfactorily. The results of the calibration are presented in Figures 2-6 for Cd, Co, Cr, Ni, and Pb, respectively. Calibration coefficients for each experimental system are presented in Table 2. Since a higher shear stress was applied in experiment 18, the film transfer rate for that experiment was higher than for experiment 22. The major effect of this parameter was on the slope of the initial portion of the adsorption curve.

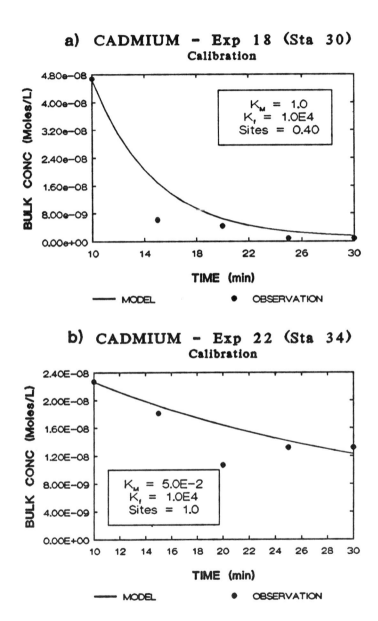

**Figure 2.** ME-SORB calibration plots for cadmium in experiments 18 (a) and 22 (b).

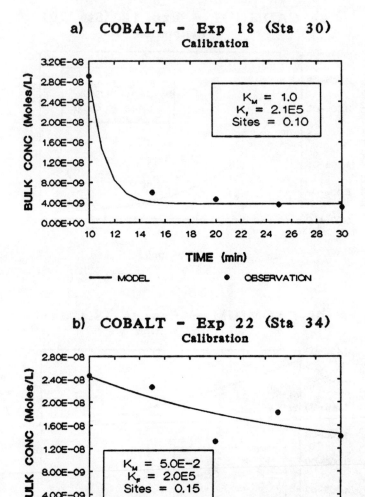

**Figure 3.** ME-SORB calibration plots for cobalt in experiments 18(a) and 22(b).

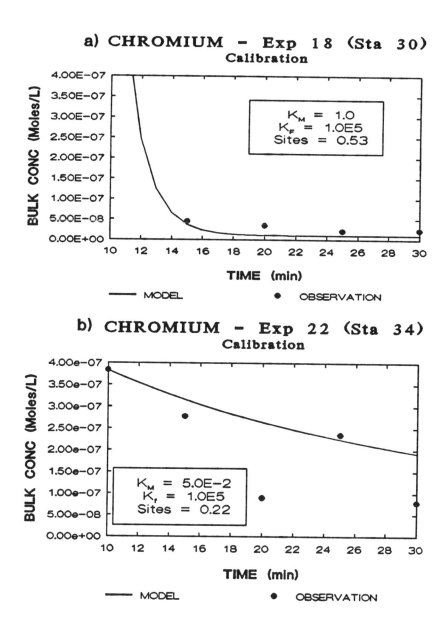

**Figure 4.** ME-SORB calibration plots for chromium in experiments 18(a) and 22(b).

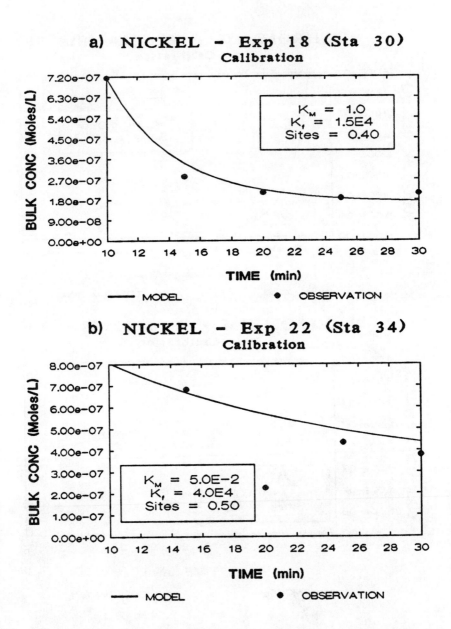

Figure 5. ME-SORB calibration plots for nickel in experiments 18(a) and 22(b).

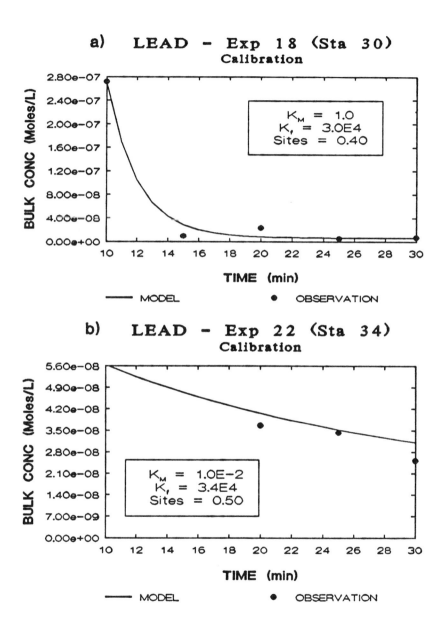

Figure 6. ME-SORB calibration plots for lead in experiments 18(a) and 22(b).

In experiment 18, for which film transfer was not limiting because of the high shear stress, the forward reaction rate had a significant effect on the rate at which equilibrium was approached. As shown in Figures 2-6, all metals bulk solution values approached equilibrium much more rapidly at the higher shear stess. In experiment 22, on the other hand, film transfer appeared to be much more rate-limiting, and the forward raction rate had little effect on the slope of the adsorption curves.

Taking both experiments into account, the forward rate ion rate seemed to be more dependent on adsorbate type than on adsorbent properties, although the sediment characteristics controlling reaction rate may be similar for the two stations. Additional experiments with a wider range of sediment types would be necessary to confirm this observation.

The main effect of the site density was on the equilibrium concentration, thus controlling the latter part of the adsorption curve. A higher site density yielded a lower bulk metal concentration as the systems approached equilibrium. The calibration values for site density suggested that station 34 (experiment 22) had a slightly higher reactive site density, a result consistent with most of the equilibrium experimental observations as well as the sediment characterization work (e.g., cation exchange capacity). Calibration values from both experiments were lower than those determined by Benjamin and Leckie (1981) for pure hydrous oxide surfaces; however, this result is not unexpected given the heterogeneous nature of sediments and the heavy contamination of those in the Trenton Channel study.

## Model Application

Since ME-SORB is based on a reversible reaction at the particle surface, it is instructive to use the calibrated model to investigate the conditions under which desorption might occur during resuspension of Trenton Channel sediments. This was done by using the calibration input parameters for experiment 18 with the only exception being that the initial bulk concentration was set to zero. Experiment 18 was chosen for the desorption runs, because the applied shear stress was more representative of a significant resuspension event in the river.

Since, under these conditions, the model suggests that adsorption/desorption kinetics are reaction controlled, we investigated the effect of pH on desorption of metals during resuspension. The model simulation of lead and nickel desorption under a variety of pH conditions is presented in Figure 7. For both metals, desorption rate increases as pH is decreased, although nickel is more mobile than lead in that it desorbs to a greater extent at a given pH value. This observation can be understood by recognizing that the reverse (desorption) reaction rate, $K_r$, is calculated from the equilibrium constant for the sorption reaction and

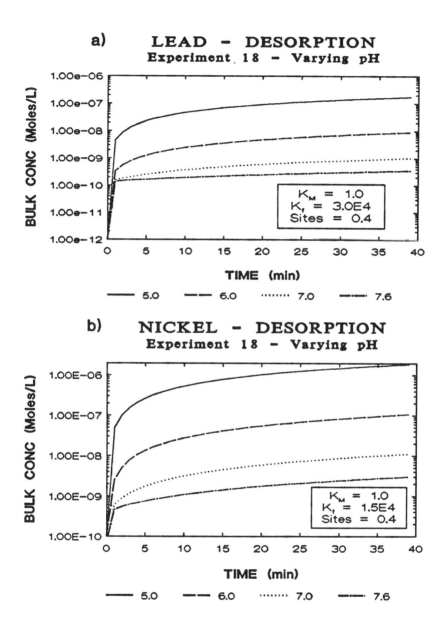

**Figure 7.** ME-SORB prediction of bulk solution concentration versus time for Pb (a) and Ni (b) desorption from Trenton Channel sediments as a function of solution pH.

the calibrated forward reaction rate according to the following relationship:

$$K_{eq} = \frac{K_f}{K_r} = \frac{\{SMe\}[H^+]}{\{S\}[Me]} \tag{17}$$

where $K_{eq}$ is the conditional equilibrium constant (dimensionless) for the reaction. Since the equilibrium constant as defined above is independent of [H⁺], increasing the hydrogen ion concentration requires a corresponding increase in the reverse rate. Therefore, a system like experiment 18, which is reaction rate limited, shows a positive desorption response to decreasing pH. However, given the value of $K_{eq}$, pH must be less than about 7 to get measurable metal desorption.

The above expression also explains why the observed partition coefficient for these cationic metals -- defined as {SMe}/{S}[Me] -- decreases with decreasing pH, causing an increase in the equilibrium bulk metal concentration.

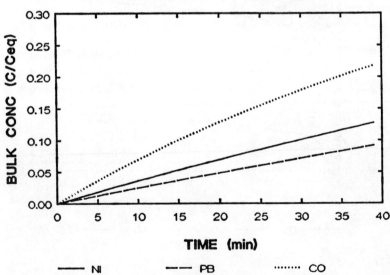

Figure 8. Comparison of approach to equilibrium at pH = 5.0 for Ni, Pb, and Co.

When conditions favor adsorption, as in the calibration data sets, the time-to-equilibrium (time to approach within 5% of the equilibrium bulk concentration) was in the range of minutes to hours (see Figures 2-6). Desorption, however, tends to have a much longer time-to-equilibrium. Shown in Figure 8 is a temporal plot of Ni, Pb, and Co desorption at pH = 5.0 with the ordinate expressed as a fraction of the bulk equilibrium concentration attained at the specified time. It shows that only about 10-15% of the bulk equilibrium metal concentration has been achieved after 40 minutes of desorption. Extrapolated out, these model applications suggest that desorption time-to-equilibrium in the pH range of 5.0-7.5 is in the range of several hours to days, even when film diffusion is not rate-limiting.

Slow desorption kinetics, as has been observed by Young *et al.* (1987) for Raisin River sediments, may be explained using the ME-SORB framework as a pore diffusion limitation. As shown in Figure 7, most desorption curves exhibited a initial, rapid release of metal to bulk solution followed by a much slower release rate for the remainder of its approach to equilibrium. Figure 9 depicts an expanded version of the lead desorption at pH = 7.6 shown in Figure 7. It better illustrates the two-stage sorbate release which has been observed by others (Young *et al.*, 1987; Karickhoff, 1980; Karickhoff and Morris, 1985; Coates and Elzerman, 1986). ME-SORB supports the hypothesis that the fast release stage is an initial desorption of metal from the particle surface, especially in systems such as experiment 18 for which film diffusion is large. The slow second stage is the result of metal release from the pores of the sediment particles, which involves desorption from the pore sites followed by diffusion at a relatively slow rate to the film layer at the surface of the particle. Here the metal may become readsorbed at the surface on sites left unoccupied by the more rapid, first-stage release to bulk solution. The relative amount of first-stage versus second-stage desorption for a given system is dependent on the relative amount of metal adsorbed to the particle surface versus in its porous structure.

## Conclusion

Preliminary application of ME-SORB has shown that under typical solution chemistry conditions in the Trenton Channel (in situ pH around 7.5) there is not a strong tendency for metal release to bulk solution during resuspension. However, when pH falls below about 7.5 there is a driving force for desorption, although the release rate is slow enough t warrant use of a kinetic description to accurately predict release during the course of a typical resuspension event. Using an equilibrium model would lead to over estimation of metals release to a dissolved phase.

In addition to pH, other factors that govern the rate and extent of metals release during sediment resuspension include: other aspects of bulk solution chemistry (e.g., presence of metal-binding ligands), the degree of sediment contamination, physical and chemical characteristics of the sediments, and the intensity of the resuspension event. Each of these factors will affect the relationship between mass transfer, surface reaction kinetics, and diffusion processes necessary to predict metal phase transfer. This study has demonstrated that additional research is needed on metals desorption kinetics as affected by particle concentration, particle size distribution dynamics, and bulk solution chemistry on metals desorption kinetics. Also, using laboratory data for parameterizing metals sorption kinetic models for site-specific application to natural systems remains a challenge.

## Acknowledgment

This research was conducted in partial fulfillment of U.S. Environmental Protection Agency Cooperative Agreement No. CR812570 granted to Clarkson University. The authors would like to acknowledge all the staff at the Large Lakes Research Station in Grosse Ile, MI for their active

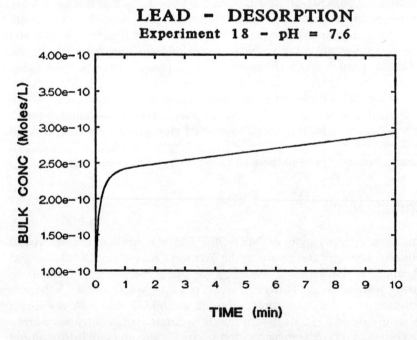

**Figure 9.** ME-SORB simulation of initial phase of Pb desorption at pH = 7.6, showing two-stage release kinetics.

participation and tremendous spirit of cooperation on this project. In particular, we would like to thank John Filkins and David Dolan for serving as project officers, Russell Kreis for serving as the Trenton Channel in-place pollutant project coordinator, and William L. Richardson, the LLRS director.

# References

Allen, H.E. 1991. Partitioning of toxic metals in natural water-sediment systems. Chapter 7 in: *Transport and Transformation of Contaminants Near the Sediment-Water Interface*. Springer-Verlag Publishers, New York.

Bedford, K.W. 1991. *In situ* measurement of entrainment, resuspension and related processes at the sediment-water interface. Chapter 5 in: *Transport and Transformation of Contaminants Near the Sediment-Water Interface*. Springer-Verlag Publishers, New York.

Benjamin, M.M., and J.O. Leckie. 1981. Multiple site adsorption of Cd, Cu, Zn, and Pb on amorphous iron oxyhydroxide. *J. Coll. Int. Sci.* 79:209-221.

Benjamin, M.M. and J.O. Leckie. 1982. Effects of complexation by Cl, $SO_4$, and $S_2O_3$, on adsorption of Cd on oxide surfaces. *Environ. Sci. Technol.* 16:162-170.

Benjamin, M.M. K.F. Hayes, and J.O. Leckie. 1982. *Removal of toxic metals from power generation waste streams by adsorption and coprecipitation*. J. Water Pollut. Control Fed. 54:1472-1481.

Bonner, J.S., A.N. Ernest, R.L. Autenrieth, and S.L. Ducharme. 1991. Parameterizing models for contaminated sediment transport. Chapter 14 in: *Transport and Transformation of Contaminants Near the Sediment-Water Interface*. Springer-Verlag Publishers, New York.

Burden, R.L. and J.D. Faires. 1978. *Numerical Analysis*. Prindle, Weber, Schmidt, Boston.

Coates, J.T. and A.W. Elzerman. 1986. Desorption kinetics for selected PCB congeners from river sediments. J. Contam. Hydrol. 1:191-210.

Crank, J. 1975. *The Mathematics of Diffusion*. Oxford University Press, London.

Davis, J.A. and K.F. Hayes. 1986. Geochemical process at mineral surfaces: An overview. In: *Geochemical Processes at Mineral Surfaces*. J.A. Davis and K.F. Hayes (eds). ACS Symposium Series 323. American Chemical Society, Washington DC.

Davis, J.A. and J.O. Leckie. 1980. Surface ionization and complexation at the oxide/water interface: 3. Adsorption of anions. *J. Coll. Int. Sci.* 64:32-43.

Davis, J.A., R.O. James, and J.O. Leckie. 1978. Surface ionization and complexation at the oxide/water interface: 1. Computation of electric double layer properties in simple electrolytes. *J. Colloid Int. Sci.* 63:480-499.

Davis, J.A. and J.O. Leckie. 1978. Effects of adsorbed complexing ligands on trace metal uptake by hydrous oxides. *Environ. Sci. Technol.* 12:1309-1315.

DePinto, J.V., T.L. Theis, T.C. Young, D. Vanetti, M. Waltman, and S. Leach. 1989. Exposure and biological effects of in-place pollutants: sediment exposure potential and particle-contaminant interactions. Environ. Engr. Rept. Ser. No. 90-3, Civil and Environ. Engr. Dept. Clarkson Univ., Potsdam NY.

Elder, J.F. 1988. Metal biogeochemistry in surface-water systems -- A review of principles and concepts. U.S. Geological Survey Circular 1013, USGS, Denver, CO.

Forstner, U. and G.T.W. Wittman. 1981. *Metal Pollutants in the Aquatic Environment.* Springer-Verlag Publishers, New York.

Hachiya, K., M. Sasaki, T. Ikeda, N. Mikami, and T. Yasunaga. 1984. Static and kinetic studies of adsorption-desorption of metal ions on a $\gamma$-$Al_2O_3$ surface. 2. Kinetic study by means of pressure-jump technique. *J. Phys. Chem.* 88:27-31.

Hayes, K.F. and J.O. Leckie. 1987. Modeling ionic strength effects on cation adsorption at hydrous oxide/solution interfaces. *J. Coll. Int. Sci.* 115:564-572.

Hayes, K.F. and J.O. Leckie. 1986. Mechanism of lead ion adsorption at the goethite-water interface. in: *Geochemical processes at mineral surfaces.* eds. J.A. Davis and K.F. Hayes. American Chemical Society, Washington DC.

Honeyman, B.D. and P.H. Santschi. 1988. Metals in aquatic systems. *Environ. Sci. Technol.* 22:862-871.

Isaacson, P.J. and C.R. Frink. 1984. Nonreversible sorption of phenolic compounds by sediment fractions: the role of sediment organic matter. *Environ. Sci. Technol.* 18(1):43-48.

James, R.O., J.A. Davis, and J.O. Leckie. 1978. Surface ionization and complexation at the oxide/water interface: 2. Surface properties of amorphous iron oxyhdroxide and adsorption of metal ions. *J. Colloid Int. Sci.* 63:331.

James, R.O., P.J. Stiglich, and T.W. Healy. 1981. The $TiO_2$/aqueous electrolyte system - applications of colloid models and model colloids. In: *Adsorption from Aqueous Solutions*, P.H. Tewari (ed), Plenum Press, New York, pp. 19-40.

Jenne, E.A. and J.M. Zachara. 1984. Factors influencing the sorption of metals. In: *Fate and Effects of sediment-bound chemicals in aquatic systems*. K.L. Dickson, A.W. Maki, and W.A. Brungs (eds), Pergamon Press, New York, p. 83-98.

Karickhoff, S.W. 1980. Sorption kinetics of hydrophobic pollutants in natural sediments. In: *Contaminants and Sediments, Vol. 2: Analysis, Chemistry, Biology*, R.A. Baker (ed), Ann Arbor Science, Ann Arbor, MI, pp.193-206.

Karickhoff, S.W. and K.R. Morris. 1985. Sorption dynamics of hydrophobic pollutants in sediment suspensions. *Environ. Sci. Technol.* 4:469-479.

Kaul, L.W. 1985. Mini-column study of transient sorption of $\alpha$-FeOOH. M.S. Thesis, University of Notre Dame.

Kreis, R.G. (ed). 1988. Integrated study of exposure and biological effects of in-place pollutants in the upper connecting channels: Interim results. Final report to the Upper Connecting Channels Study activities workgroup, U.S. EPA, ERL-Duluth, Large Lakes Research Station, Grosse Ile, MI.

Lick, W. 1991. The flocculation, deposition, and resuspension of fine-grained sediments. Chapter 4 in: *Transport and Transformation of Contaminants Near the Sediment-Water Interface*. Springer-Verlag Publishers, New York.

Lion, L.W., R.S. Altmann, and J.O. Leckie. 1982. Trace metal adsorption characteristics of estuarine particulate matter: evaluation of contributions of Fe/Mn oxide and organic surface coatings. *Environ. Sci. Technol.* 16:660-666.

McIlroy, L., J.V. DePinto, T.C. Young, and S.C. Martin. 1986. Partitioning of heavy metals to suspended solids of the Flint River, Michigan. *Environ. Toxicol. Chem.* 5:609-623.

Morel, F.M.M., J.C. Westall, and J.G. Yeasted. 1981. Adsorption models, a mathematical framework of general equilibrium calculations. In. *Adsorption of Inorganics at Solid-Liquid Interfaces*. Anderson, M.A. and A.J. Rubin, eds. Ann Arbor Sci., Ann Arbor MI.

Oakley, S.M., P.O. Nelson, and K.J. Williamson. 1981. Model of true metal partitioning in marine sediments. *Environ. Sci. Technol.* 15:474-480.

Rygwelski, K.R. 1984. Partitioning of toxic trace metals between solid and liquid phases in the Great Lakes. In: *Toxic Contaminants in the Great Lakes*, J.O. Nriagu and M.S. Simons, eds. John Wiley & Sons, New York, pp. 321-333.

Schindler, P.W. and W. Stumm. 1987. The surface chemistry of oxides, hydroxides, and oxide minerals. In: *Aquatic Surface Chemistry: Chemical Processes at the Particle-Water Interface*, W. Stumm, ed. John Wiley & Sons, Inc., New York.

Stumm, W., R. Kummert, and L. Sigg. 1980. A ligand exchange model for the adsorption of inorganic and organic ligands at hydrous oxide interfaces. *Croatica Chem.* Acta 53:291-312.

Tessier, A., R. Carignan, and N. Belzile. 1991. Reactions of trace elements near the sediment-water interface. Chapter 8 in: *Transport and Transformation of Contaminants Near the Sediment-Water Interface*. Springer-Verlag Publishers, New York.

Theis, T.L. and R.O. Richter. 1980. Adsorption reactions of nickel species at oxide surfaces. In: *Particulates in Water*, M.C. Kavanaugh and J.O. Leckie, eds. Advances in Chemistry Series #189, ACS, Washington, DC, pp. 78-96.

Theis, T.L. 1985. Equilibrium and kinetic descriptions of sorption reactions on hydrous oxide surfaces. In: Report No. 600/X-85/122, U.S. EPA, Office of Exploratory Research, Washington, DC.

Theis, T.L., T.C. Young, and J.V. DePinto. 1988a. Factors affecting metal partitioning during resuspension of sediments from the Detroit River. *J. Great Lakes Res.* 14:216-226.

Theis, T.L., R. Iyer, and L.W. Kaul. 1988b. Kinetic studies of cadmium adsorption on goethite. *Environ. Sci. Technol.* 18:43-48.

Tsai, C.H., S. Iacobellis, and W. Lick. 1987. Flocculation of fine-grained lake sediments due to a uniform shear stress. *Jour. Great Lakes Res.* 13(2):135-146.

Tsai, C.H. and W. Lick. 1986. A portable device for measuring sediment resuspension. *J. Great Lakes Res.* 12(4): 314-321.

Vuceta, J., and J.J. Morgan. 1978. Chemical modeling of trace metals in fresh waters: role of complexation and adsorption. *Environ. Sci. Technol.* 12:1302-1309.

Westall, J. and H. Hohl. 1980. A comparison of elctrostatic models for the oxide/solution interface. *Adv. Coll. Interface Sci.* 12:265-294.

Young, T.C., J.V. DePinto, and T.W. Kipp. 1987. Adsorption and desorption of Zn, Cu, and Cr by sediments from the Raisin River (Michigan). *J. Great Lakes Res.* 13(3):353-366.

Ziegler, C.K., W. Lick, and J. Lick. 1991. The transport of fine-grained sediments in the Trenton Channel of the Detroit River. Chapter 12 in: *Transport and Transformation of Contaminants Near the Sediment-Water Interface.* Springer-Verlag Publishers, New York.

# CHAPTER 11

# Biological and Chemical Mechanisms of Manganese Reduction in Aquatic and Sediment Systems

Charles R. Myers
Kenneth H. Nealson

## Introduction

Microbial redox reactions are important mechanisms for mobilizing metals and organic compounds in natural anaerobic aquatic environments. Important microbial reactions include those involving the oxidation of organic matter coupled to the reduction of nitrate ($NO_3^-$), ferric iron [Fe(III)], manganese oxides [containing Mn(IV) and Mn(III)], or sulfate ($SO_4^{2-}$), and the conversion of organic matter to carbon dioxide ($CO_2$) and methane ($CH_4$) (Henrichs and Reeburgh, 1987).

Many studies on anaerobic sediments have focused on the microbial reduction of nitrate and sulfate and the production of methane (Henrichs and Reeburgh, 1987), and a great deal is known of the microbial mechanisms involved in these processes. In contrast, while oxidation and reduction of metals (e.g., manganese) have been recognized as microbially catalyzed reactions since the beginning of this century (Beijerinck, 1913), remarkably little is known about the mechanisms involved (Nealson et al., 1988a). In the environment, microbes are major catalysts of the cycling of Mn (Ehrlich, 1987; Ghiorse, 1988; Nealson et al., 1988a,b) and Fe (Lovley, 1987), and these metals may represent principal electron acceptors for organic matter oxidation in sedimentary environments where they are enriched (Lovley, 1987; Lovley and Phillips, 1988). Whether their reduction is mediated directly via bacterial electron transport-linked respiratory processes or indirectly by the excretion of reducing chemicals, Mn and Fe oxides could serve as external sinks for reducing equivalents, allowing for increased bacterial metabolism and/or growth.

Manganese is cycled in the environment between the oxidized and reduced forms. The oxidized forms, Mn(III) and Mn(IV), are favored by conditions of high pH and high Eh, while the reduced form, Mn(II), is favored by low pH and low Eh (Stumm and Morgan, 1981). Thus, Mn cycling may occur at redox interfaces or in response to changes in redox conditions. Iron and manganese reduction would likely be important in the zone between the regions of oxygen removal and sulfate reduction in marine sediments; in freshwater sediments, which are relatively low in both $NO_3^-$ and $SO_4^{2-}$, metal reduction would occur between the regions of oxygen depletion and $CO_2$ reduction (methanogenesis). In support of this, pore water profiles of anaerobic sediments always show a zone of manganese reduction consistent with the redox potential of reduction of manganese oxides (Froelich et al., 1979; Nealson, unpublished), usually contiguous with the zone where nitrate disappears and just above the zone of iron reduction. This chapter will focus on the mechanisms of bacterial reduction of Mn oxides.

## Manganese Reduction Mechanisms

The mechanisms of bacterial reduction of Mn oxides can be separated into direct and indirect processes (Table 1). Several studies (Burdige and Nealson, 1986; Dubinina, 1979a,b; Ghiorse, 1988; Myers and Nealson, 1988b; Stone, 1987a,b; Stone and Morgan, 1984a,b; Stone and Morgan, 1987) have shown that a variety of organic and inorganic chemicals, many of which are potential bacterial excretion products, are able to reduce Mn oxides indirectly (Table 1). Thus, the sediment profiles of Mn reduction could well result from the nature of the chemical environment rather than direct microbial reduction.

Direct reduction of Mn by bacteria has been postulated for many years (see reviews by: Ehrlich, 1987; Ghiorse, 1988), but in only a few cases have studies with pure cultures yielded insights into either the mechanisms involved or the importance of such organisms in nature. Until recently (Ehrlich, 1987; Lovley and Phillips, 1988; Myers and Nealson, 1988a), no reports of microbially mediated direct manganese reduction coupled to the growth of an organism using Mn oxide as a terminal electron acceptor had been presented. The remainder of this chapter will discuss recent findings on direct microbial reduction of Mn oxides and compare and contrast these with indirect mechanisms of reduction.

Table 1

Possible Modes of Mn(IV) Reduction by Bacteria

---

I.  Indirect Reduction (Production of Chemical Reductants)

    1.    Inorganic reductants

        a. hydrogen peroxide ($H_2O_2$)
        b. sulfide ($S^{2-}$)
        c. ferrous iron ($Fe^{2+}$)

    2.    Organic reductants

        a. organic acids
        b. organic thiols
        c. quinones, phenols, catechols

II. Direct Reduction

    1.    Enzymes that catalyze Mn(IV) reduction

        a. Mn(IV) reductase systems

---

## Direct Manganese Reduction

We define direct reduction of Mn as those cell-mediated processes that result in the reduction of Mn oxides as a result of respiratory electron transport-linked Mn reductase systems. While an early study by Hochster and Quastel (1951) demonstrated that Mn oxides could substitute for $O_2$ as an anaerobic electron acceptor in certain biological redox reactions, several studies have shown that $O_2$ (an energetically more favorable electron acceptor than $MnO_2$) does not inhibit microbial Mn reduction (Ehrlich, 1987; Ghiorse, 1988; Trimble and Ehrlich, 1968, 1970; Troshanov, 1968). With the exception of two recent reports (Lovley and Phillips, 1988; Myers and Nealson, 1988a), only a single report of obligately anaerobic microbially mediated Mn reduction exists (Burdige and Nealson, 1985), and this study was done with mixed cultures. Key characteristics, recently stated by Ehrlich (1987), that one would expect to be associated with obligately anaerobic direct Mn-reducing bacteria are:

1.    $O_2$ should inhibit the reduction of Mn oxides

2.    cell contact with the insoluble Mn oxides should be required for Mn reduction

3. the conditions necessary for Mn reduction should be consistent with an enzymatic process

4. the bacteria should couple anaerobic growth to the reduction of Mn oxides

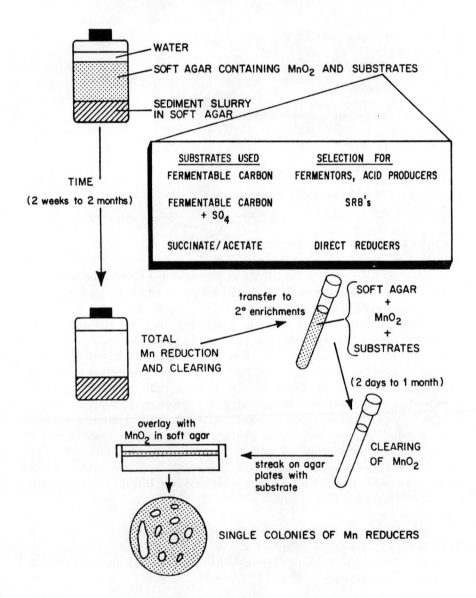

Figure 1. Enrichment Protocol Used to Isolate MR-1. See Myers and Nealson (1988a,b) for complete procedural details.

Recently, two different bacteria have been isolated that meet these criteria (Lovley and Phillips, 1988; Myers and Nealson, 1988a). The approach used to isolate both organisms was similar (i.e., to enrich for organisms that would grow anaerobically on nonfermentable carbon sources using Mn or Fe oxides as terminal electron acceptors).

The first of these organisms, *Alteromonas putrefaciens* strain MR-1, was isolated (according to the enrichment protocol outlined in Figure 1) from the anaerobic sediments of Oneida Lake, New York (Myers and Nealson, 1988a,b) as a Mn(IV)-reducer. Mn reduction by MR-1 is proportional to cell density and is optimal at 30° to 35°C and pH 6 to 7 (Myers and Nealson, 1988a); reduction of manganese by MR-1 was, however, also detectable at 7° and 15°C, and at pHs as high as 7.8. The temperature of the sediment-water interface in Oneida Lake (from which MR-1 was isolated) would typically be 4°C during the winter months and approximately 20°C during the summer months (Dean *et al.*, 1981; Mills *et al.*, 1978), and the pH of these sediments ranges from 7.5 to 8.2 (Dean *et al.*, 1981; Mills *et al.*, 1978; Nealson, unpublished). Molecular oxygen ($O_2$) inhibits Mn reduction by this strain, as do several metabolic poisons, including: antimycin A, azide, 2-heptyl-hydroxyquinolone-N-oxide (HQNO), dicumarol, dinitrophenol, carbonyl cyanide *m*-chlorophenyl hydrazone (CCCP), and formaldehyde (Myers and Nealson, 1988a); Mn reduction by this bacterium is not inhibited by cyanide.

In the absence of other electron acceptors, MR-1 can couple its growth to the reduction of $MnO_2$ (Figure 2). MR-1 does not grow in the same medium if $MNO_2$, or an alternate electron acceptor, is lacking (Figure 2). The observed growth of MR-1 in the presence of $MnO_2$ as the sole electron acceptor gives molar growth yield values of 9 to 45 grams of cell dry weight produced per mole $MnO_2$ reduced (Myers and Nealson, 1988a). These values are similar to growth yield values for *Paracoccus denitrificans*, which range from 16 to 39 grams of cells per mole nitrate or oxygen reduced (Stouthamer, 1977).

MR-1 is a nonfermentative, obligately respiratory bacterium (Myers and Nealson, 1988a); like other nonfermentative bacteria under anaerobic conditions, it must obtain its energy from respiratory reactions in which terminal electron acceptors other than oxygen are used. In addition to Mn(IV), MR-1 can couple its growth to the reduction of a large variety of other electron acceptors, including: oxygen ($O_2$), nitrate, nitrite, Fe(III), trimethylamine N-oxide (TMAO), sulfite, thiosulfate, tetrathionate, fumarate, and glycine; it is unable to use sulfate, carbon dioxide, or molybdate as electron acceptors (Myers and Nealson, 1988a). Other recent studies (Arnold *et al.*, 1986; Semple and Westlake, 1987) report the ability of several other strains of *Alteromonas putrefaciens* to use sulfite, thiosulfate, and Fe(III) as terminal electron acceptors. The remarkable respiratory versatility of *A. putrefaciens* could provide a distinct advantage

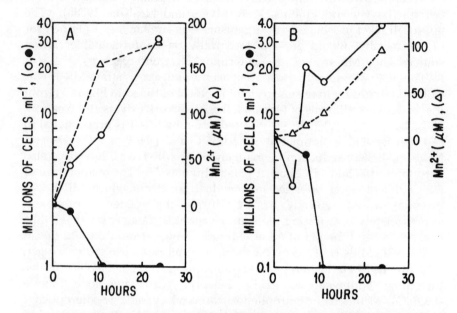

Figure 2. Anaerobic growth of MR-1 with MnO2 as the sole terminal electron acceptor. Growth in the presence of $MnO_2$ (2 mM) was assessed by increases in cell number (o); the particulate $MnO_2$ in each sample was reduced with a small amount of sodium dithionite ($Na_2S_2O_4$) just prior to plating the cell counts. Increases in microbial growth paralleled reduction of manganese (Δ), whereas no growth was noted in the absence of a terminal electron acceptor (●); the cell counts shown at baseline on the x-axis represents <103 cells ml -1. The experiment in A was conducted in a defined mediumm and that in B was conducted in a Lake Oneida water-based medium, with succinate as the carbon source in both. Growth of MR-1 coupled to the reduction of manganese was also demonstrated with lactate as the carbon source. See Myers and Nealson (1988a) for complete procedural details.

in environments where the concentration of certain electron acceptors may change with time and sediment depth.

To further understand the direct reduction of Mn oxides by MR-1, as well as the use of other anaerobic electron acceptors, we have generated electron transport mutants of MR-1. Nine classes of electron transport mutants have been isolated (Table 2). Class I mutants have lost the ability to use any of the tested compounds anaerobically, suggesting that some component(s) are common to all anaerobic electron transport mechanisms; these pleiotropic mutations may also be in genes required for the expression of "anaerobic" genes, analogous for the *fnr* and *ant* genes of *Escherichia coli* (Jones and Gunsalus, 1987; Yerkes et al., 1984).

Class VII mutants have only lost the ability to reduce Fe(III), class VIII mutants the ability to reduce Mn(IV), and class IX mutants the ability to reduce nitrate (Table 2). This suggests that there are components unique to the reduction of each of these three terminal electron acceptors. That we are able to isolate mutants (class VII) which could not reduce Fe(III) but could reduce Mn(IV), suggests that Mn reduction by this bacterium is not mediated indirectly via Fe(III) reduction (see below).

MR-1 can use hydrogen ($H_2$), formate, lactate, and pyruvate as electron donors for the reduction of Mn(IV) and Fe(III) (Lovley et al., 1989). Another strain of A. putrefaciens, NCMB 1735, oxidizes the amino acids serine and cysteine to $CO_2$ under anaerobic conditions with TMAO as the electron acceptor (Ringo et al., 1984); we have not yet tested MR-1 for the ability to use these amino acids as electron donors for the reduction for Mn(IV) and Fe(III).

The other bacterium isolated to date which directly reduces Mn(IV) oxides was isolated as an Fe(III)-reducer from the sediments of the Potomac River, Maryland (Lovley and Phillips, 1988). This bacterium, GS-15, can couple its growth to the reduction of Fe(III), Mn(IV), or nitrate. GS-15 optimally reduces Fe(III) at 30° to 35°C and at pH 6.7 to 7; no Fe(III) reduction is detected at temperatures of 10°C or less (Lovley and Phillips, 1988). Acetate, ethanol, butyrate, and propionate can serve as electron donors for Fe(III) reduction by GS-15, whereas $H_2$ or other organic compounds cannot (Lovley and Phillips, 1988). GS-15 has not yet been classified taxonomically, but it is clearly different from MR-1 in that it is an obligate anaerobe, whereas MR-1 is facultative.

The isolation and characterization of these organisms, MR-1 and GS-15, in pure culture clearly indicates that the process of dissimilation of Fe(III) and Mn(IV) provides a mode for survival and growth for some organisms. With regard to mechanisms of direct metal reduction by these bacteria, little is known other than electron transport-linked respiratory processes are implicated for MR-1 on the basis of inhibition of Mn reduction by electron transport inhibitors (see above). It is, however, clear that organisms are now in hand which should provide material for dissection of Mn and Fe reductases and the associated electron-transfer systems.

## Indirect Reduction: Interactions with Iron and Sulfur

As shown in Table 1 and briefly discussed above, there are a variety of chemicals which act as indirect reductants of Mn oxides. For example, there have been several studies demonstrating Mn reduction by various organic compounds (Stone, 1987a,b; Stone and Morgan, 1984a,b; Stone and Morgan, 1987). These reactions are strongly affected by pH; nearly all organic acids tested catalyzed Mn reduction at pH $\leq 5$, while only

Table 2

Electron Acceptor Mutants of MR-1[a]

| Class | No. of Mutants | Growth on/reduction of[b]: | | | | | |
|---|---|---|---|---|---|---|---|
| | | Fe(III) | Mn(IV) | $NO_3$ | $S_2O_3^{2-}/S_4O_6^{2-}$ | TMAO[c] | $O_2$ |
| I | 7 | 0 | 0 | 0 | 0 | 0 | + |
| II | 1 | 0 | 0 | 0 | 0 | + | + |
| III | 4 | 0 | 0 | 0 | + | + | + |
| IV | 9 | 0 | 0 | + | + | + | + |
| V | 1 | 0 | 0 | + | + | 0 | + |
| VI | 2 | 0 | + | 0 | + | + | + |
| VII | 2 | 0 | + | + | + | + | + |
| VIII | 7 | + | 0 | + | + | + | + |
| IX | 2 | + | + | 0 | + | + | + |

[a] Mutants unable to use certain electron acceptors for anaerobic growth were isolated as follows: mid-log phase aerobic cells were subjected to mutagenesis with N-methyl-N'-nitrosoguanidine (Carlton and Brown, 1981), and surviving cells were plated and grown aerobically. Colonies were screened for inability to use particular electron acceptors for anaerobic growth (we did not screen for mutants unable to use glycine, fumarate, sulfite, or nitrite).
[b] + = positive for growth/reduction; 0 = negative for growth/reduction.
[c] TMAO, trimethylamine N-oxide.

pyruvate and oxalate were active in Mn reduction at pH 7.2 (Stone and Morgan, 1984b). Therefore, any organism that excretes organic acids and lowers the pH could presumably use amorphous Mn oxides as an external electron sink. However, some crystalline forms of $MnO_2$ (e.g., electrolytically produced commercial $MnO_2$) do not significantly react with organic acids, even under acidic conditions (H. Ehrlich, personal communication).

Evidence has also been presented that Mn reduction can occur as a function of hydrogen peroxide production by bacteria (Dubinina, 1979a,b; Ghiorse, 1988). Reduction of Mn oxides by peroxide will, however, be restricted to environments with $O_2$ and moderately low pH; there are no known mechanisms for the generation of $H_2O_2$ in the absence of $O_2$, and, at high pH, $H_2O_2$ acts as an oxidant of Mn(II).

Hydrogen sulfide ($H_2S$) is also a known potent reductant of Mn oxides (Burdige and Nealson, 1986). The reaction occurs very rapidly and proceeds according to:

$$MnO_2 + HS^- + 3H^+ \rightarrow Mn^{2+} + 2H_2O + S°$$

Therefore, any bacteria that generate free sulfide as an endproduct of dissimilatory reduction of sulfur compounds (e.g., $SO_4^{2-}$, $S_2O_3^{2-}$) will be potential Mn-reducing bacteria. This has been demonstrated for two isolates of *Desulfovibrio*, a sulfate-reducing bacterium; growth of these isolates was coupled to sulfide production, and sulfide production was coupled to Mn(IV) reduction (Burdige and Nealson, 1986). In addition to sulfate-reducing bacteria, many bacteria are able to dismutate reductively thiosulfate ($S_2O_3^{2-}$) to sulfite ($SO_3^{2-}$) and sulfide ($S^{2-}$) (Barrett and Clark, 1987). This can result in rapid reduction of Mn(IV), as shown for strain MR-1 for which the rate of Mn reduction is markedly increased in the presence of thiosulfate (Figure 3a). While the reduction of Mn is a strictly inorganic chemical reaction, the generation of the sulfide is a bacterial process.

In contrast to certain other electron acceptors (e.g., nitrate, fumarate) that partially inhibited Mn reduction by MR-1 (Myers and Nealson, 1988b), we repeatedly noted that the rate of Mn reduction by MR-1 is slightly enhanced by the presence of Fe(III) (Figure 3b). MR-1 can reduce Fe(III) to Fe(II) (Myers and Nealson, 1988b), and Fe(II) can act as a potent chemical reductant of Mn(IV) under both acidic (Postma, 1985) and neutral pH conditions (Figure 4). The reaction should proceed according to:

$$2Fe^{2+} + MnO_2 + H_2O \rightarrow Fe_2O_3 + Mn^{2+} + 2H^+$$

The predicted stoichiometry of this reaction is that 2 moles of Fe(II) will be oxidized to Fe(III) for every mole of Mn(IV) reduced to Mn(II); the values of moles Fe(II) oxidized/moles Mn(IV) reduced that we observed in repeat experiments ranged from 1.87 to 2.23, in good agreement with the predicted ratio of 2.0. Irrespective of the mechanism, however, reduction of Fe(III) by MR-1 will result in the production of Fe(II) which may then reduce additional Mn(IV); this may explain the enhanced reduction of Mn(IV) in the presence of Fe(III) (Figure 3b). This predicts that any proficient iron-reducing bacterium is potentially a good Mn-reducing bacterium if the conditions are appropriate. Since almost all sediments that are rich in Mn oxides also contain Fe, one should not consider the reduction of one in isolation from the other.

Given the chemical interactions between Mn oxides and Fe(II) or sulfide, the presence of Mn oxides in the environment could allow for substantial reduction of either iron or sulfate (or thiosulfate) without the reduced products ($Fe^{2+}$ or $S^{2-}$) appearing. That Fe(II) does not accumulate until the reduction of Mn oxides is complete is shown for strain MR-1 (Figure 5). Also, the addition of $MnO_2$ to cultures of MR-1 that have accumulated Fe(II) results in the "immediate" appearance of Mn(II) and the concurrent disappearance of Fe(II) (Myers and Nealson, 1988b). The important implication of this in terms of an environmental perspective is that while one process (e.g. Mn reduction) may be observed by measurement, the driving force and hence the bacteria responsible for the

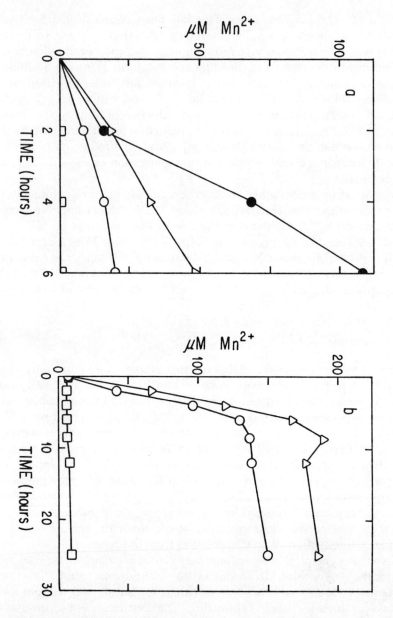

Figure 3. Manganese reduction by MR-1 in the presence of thiosulfate or ferric iron. (a) Mn reduction by MR-1 in the presence of 2mM (●), 0.5 mM (Δ), or no thiosufate (o); no reduction of Mn by thiosufate occured in the absence of cells. (□) (b) Mn reduction by MR-1 in the presence of 1 mM Fe(III) (Δ), or no Fe(III) (o); no reduction of $MnO_2$ by Fe(III) occured in the absence of cells.(□) Note that the scale of both axes is different in panels a and b. The initial $MnO_2$ concentration in both a and b was 0.215 mM. See Myers and Nealson (1988b) for complete procedural details.

process may be quite different (e.g., iron or sulfate reduction). Also, if one is monitoring the reduction of iron or sulfur by the accumulation of Fe(II) and sulfide, respectively, then these processes might be scored as falsely low in the presence of Mn oxides.

For all of the indirect processes of Mn reduction, the mechanism of reduction is linked to the physiology of given processes in bacterial cells (e.g., sulfide production, Fe(II) production, organic acid excretion, etc.). While Mn reduction is not mediated directly by the bacteria per se, the generation of the chemical reductants is linked to bacterial metabolism. Such processes must be kept in mind as they may predominate in certain environments and may even obviate the need for direct reduction by respiratory processes as described above.

## Summary of Manganese Reduction Mechanisms

Manganese reduction mechanisms involve both direct and indirect processes. Certain microbes (e.g., MR-1) can reduce Mn oxides by more than one mechanism (Figure 6); direct reduction is one mechanism by which MR-1 mediates Mn reduction. In this mechanism (top of Figure 6), an electron donor is oxidized, with the electrons being transferred to a Mn reductase system, for which $MnO_2$ serves as the terminal electron acceptor. The process is a form of anaerobic respiration, analogous to the use of nitrate as a terminal electron acceptor. In this diagram, the Mn reductase system is designated with the letter "X" as its mechanism is not yet understood; this diagram is an oversimplification as there are no doubt many components that are involved in producing the Mn reductase activity. The bacterium GS-15 is also apparently able to mediate directly the reduction of Mn oxides and therefore also could be included as an example of direct reduction.

Indirect reduction via the activities of iron(III) reductase and thiosulfate reductase are the other mechanisms by which MR-1 mediates Mn reduction (Figure 6). In the iron reductase-mediated process (middle of Figure 6), an electron donor is oxidized, with the electrons being transferred to an Fe reductase system, for which Fe(III) serves as the electron acceptor (the Fe reductase system is designated "X" as its components are not yet known; this designation does not imply that this reductase system is the same as the Mn reductase components). The important point concerning Mn reduction, however, is that Fe(II), generated by the Fe reductase activity, is a potent chemical reductant of Mn oxides. Similarly, sulfide, another chemical reductant of manganese, is generated by the thiosulfate reductase system (bottom of Figure 6), in which thiosulfite serves as the electron acceptor. Certainly, other bacteria that can generate ferrous iron or sulfide by these mechanisms may also indirectly mediate the reduction of Mn oxides. An interesting possibility concerning the iron reductase system is that the Fe(III) generated by the

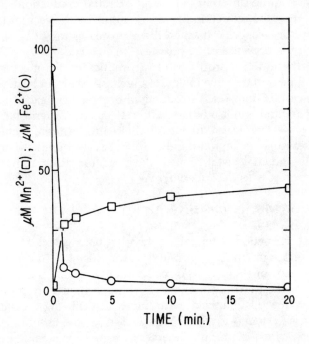

Figure 4. Reduction of $MnO_2$ by Fe(II). The experiment was conducted under anaerobic conditions at 24°C at pH 7.4. The medium was made 92μm with respect to Fe(II) and the zero time sample was taken. $MnO_2$ was then added to a concentration of 260μm, and samples were taken at timed intervals and analyzed for Mn(II) (□) and Fe(II) (o). Repeat experiments with different concentrations of $MnO_2$ and Fe(II) resulted in analogous results. See Myers and Nealson (1988b) for complete details.

chemical reduction of Mn oxides by Fe(II) might be "reutilized" by the bacteria as an electron acceptor; if this is the case, then a small amount of iron may serve, in a cyclical fashion, to reduce large amounts of Mn oxides.

The mechanisms of microbially mediated Mn reduction, which MR-1 does not possess, are shown in Figure 7; these modes are all indirect in nature. The generation of the reductant sulfide via dissimilatory sulfate reduction (top of Figure 7) is one such mechanism and is probably quite important in those environments in which sulfate reduction is a dominant process. The production of various organic reductants (e.g., organic acids) (middle of Figure 7) represents another mechanism for the indirect reduction of Mn oxides. The last known microbial mechanism, the production of hydrogen peroxide (bottom of Figure 7), may be important

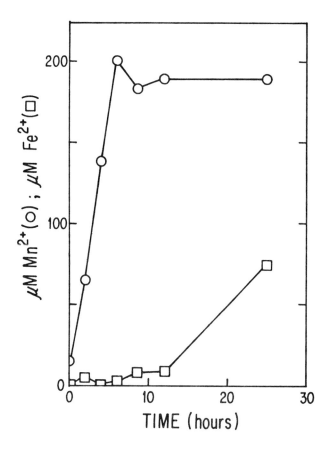

Figure 5. Lack of accumulation of Fe(II) in the presence of $MnO_2$. MR-1 was incubated in the presence of 0.215 mM $MnO_2$ and 2 mM Fe(III). Levels of Mn(II) (o) rose rapidly and indicated that $MnO_2$ reduction was essentially complete by 6 hours. Fe(II) (□) levels remained essentially at zero until Mn reduction was complete. Repeat experiments using different concentrations of Fe(III) and/or $MnO_2$ gave analogous results. See Myers and Nealson (1988b) for complete details.

in oxygenated zones; it is unique in this respect as all other mechanisms discussed would be confined to anoxic zones. While these six mechanisms (Figures 6, 7) represent the extent of knowledge of microbial mechanisms of Mn reduction to date, we would not necessarily be surprised if additional mechanisms will be uncovered in the future.

At this point in time, it is difficult to estimate the relative contributions of these direct and indirect processes in various environments. We are,

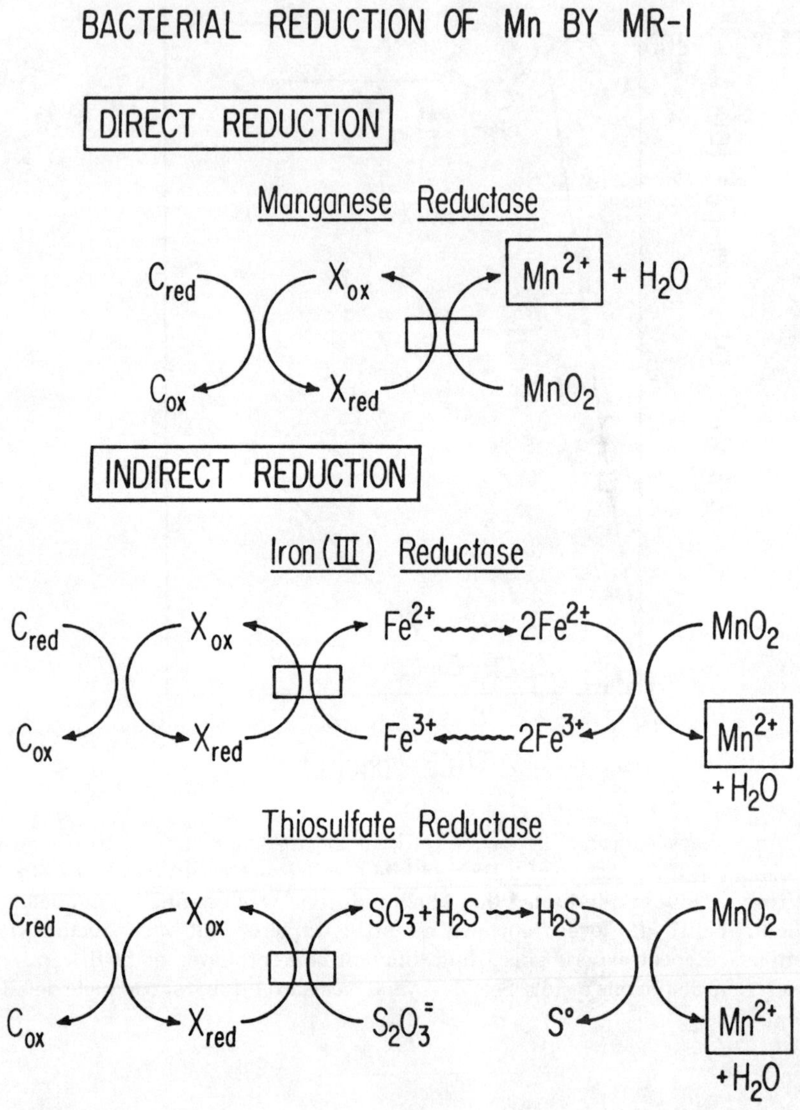

## BACTERIAL REDUCTION OF Mn BY MR-1

### DIRECT REDUCTION

#### Manganese Reductase

$C_{red} \rightarrow X_{ox} \rightarrow Mn^{2+} + H_2O$
$C_{ox} \leftarrow X_{red} \leftarrow MnO_2$

### INDIRECT REDUCTION

#### Iron (III) Reductase

$C_{red} \rightarrow X_{ox} \rightarrow Fe^{2+} \rightsquigarrow 2Fe^{2+} \rightarrow MnO_2$
$C_{ox} \leftarrow X_{red} \leftarrow Fe^{3+} \leftarrow 2Fe^{3+} \leftarrow Mn^{2+} + H_2O$

#### Thiosulfate Reductase

$C_{red} \rightarrow X_{ox} \rightarrow SO_3 + H_2S \rightarrow H_2S \rightarrow MnO_2$
$C_{ox} \leftarrow X_{red} \leftarrow S_2O_3^= \quad S° \leftarrow Mn^{2+} + H_2O$

Figure 6. Mechanisms of Manganese Reduction by MR-1.

however, confident that microbes play major roles in the reductive dissolution of Mn oxides in the environment, both directly via respiratory processes and indirectly via the excretion of metabolic endproducts that act as chemical reductants. Changing environmental conditions to those

## OTHER MODES OF INDIRECT REDUCTION:

### Dissimilatory Sulfate Reduction

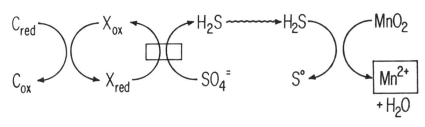

### Fermentation

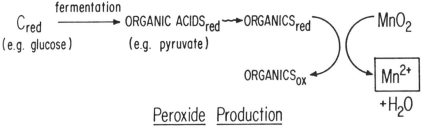

### Peroxide Production

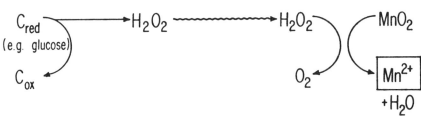

Figure 7. Mechanisms of Mn reduction by other bacteria.

that increase the microbial dissolution of Mn and Fe oxides could result in the mobilization of toxic compounds adsorbed to these oxides. We therefore feel that further research in this area is needed so that we may better understand the mechanisms involved and how these processes may be affected by environmental conditions.

# Acknowledgments

We wish to acknowledge the following colleagues who provided us with preprints, reprints, and discussion, and who allowed us to cite unpublished results: W. Davison, W.C. Ghiorse, J.V. Klump, D.R. Lovley, C.C. Remsen, R.A. Rosson, and A.T. Stone. The views presented in this chapter, however, are solely the interpretations and perspectives of the authors (as this was not intended to be a complete treatise on the subject, some contributions may not have received proper citation for which we apologize). In addition we thank B. Wimpee and L. Yewer for technical assistance and B. Wimpee for the drawings. This work was suppported in part by NSF grant NAGW-1047 and NASA grant 8609778 to K.H.N. C.R.M. was supported in part by NIEHS toxicology training grant ES07043.

# References

Arnold, R.G., DiChristina, T.J., and Hoffmann, M.R. 1986. Inhibitor studies of dissimilative Fe(III) reduction of *Pseudomonas sp.* strain 200 (*"Pseudomonas ferrireductans"*). *Appl. Environ. Microbiol.*, 52:281-289.

Barrett, E.L. and Clark, M.A. 1987. Tetrathionate reduction and production of hydrogen sulfide from thiosulfate. *Microbiol. Rev.*, 51:192-205.

Beijerinck, M.W. 1913. Oxydation des manganbikarbonates durch bakterein und schimmelpilze. *Folia Microbiol.* (Delft), 2:123-134.

Burdige, D.J. and Nealson, K.H. 1985. Microbial manganese reduction by enrichment cultures from coastal marine sediments. *Appl. Environ. Microbiol.*, 50:491-497.

Burdige, D.J. and Nealson, K.H. 1986. Chemical and microbiological studies of sulfide-mediated manganese reduction. *Geomicrobiol. J.*, 4:361-387.

Carlton, B.C. and Brown, B.J. 1981. Gene mutation. In: *Manual of Methods for General Bacteriology*, ed. P. Gerhardt et al., pp. 222-242, Washington, DC, American Society for Microbiology.

Dean, W.E., Moore, W.S., and Nealson, K.H. 1981. Manganese cycles and the origin of manganese nodules, Oneida Lake, New York, U.S.A. *Chem. Geol.*, 34:53-64.

Dubinina, G.A. 1979a. Mechanism of the oxidation of divalent iron and manganese by iron bacteria growing at neutral pH of the medium. *Microbiology U.S.S.R.*, 47:471-478.

Dubinina, G.A. 1979b. Functional role of bivalent iron and manganese oxidation in *Leptothrix pseudoochraceae*. *Microbiology U.S.S.R.*, 47:631-636.

Ehrlich, H.L. 1987. Manganese oxide reduction as a form of anaerobic respiration. *Geomicrobiol. J.*, 5:423-431.

Froelich, P.N., Klinkhammer, G.P., Bender, M.L., Luedtke, N.A., Heath, G.R., Cullen, D., Dauphin, P., Hammond, D., Hartman, B. and Maynard, V. 1979. Early oxidation of organic matter in pelagic sediments of the eastern equatorial Atlantic: suboxic diagnosis. *Ceochim. Cosmochim. Acta*, 43:1075-1090.

Ghiorse, W.C. 1988. Microbial reduction of manganese and iron. In: *Biology of Anaerobic Microorganisms*, ed. A.J. Zehnder, pp. 305-331, New York, John Wiley and Sons.

Henrichs, S.M. and Reeburgh, W.S. 1987. Anaerobic mineralization of marine sediment organic matter: rates and the role of anaerobic processes in the oceanic carbon economy. *Geomicrobiol. J.*, 5:191-237.

Hochster, R.M. and Quastel, J.H. 1951. Manganese dioxide as a terminal hydrogen acceptor in the study of respiratory systems. *Arch. Biochem. Biophys.*, 36.132-146.

Jones, H.M. and Gunsalus, R.P. 1987. Regulation of *Escherichia coli* fumarate reductase (*frdABCD*) operon expression by respiratory electron acceptors and the *fnr* gene product. *J. Bacteriol.*, 169:3340-3349.

Lovley, D.R. 1987. Organic matter mineralization with the reduction of ferric iron: a review. *Geomicrobiol. J.*, 5:375-399.

Lovley, D.R. and Phillips, E.J.P. 1988. Novel mode of microbial energy metabolism: organic carbon oxidation coupled to dissimilatory reduction of iron or manganese. *Appl. Environ. Microbiol.*, 54:1472-1480.

Lovley, D.R., Phillips, E.J.P., and Lonergan, D.J. 1989. Hydrogen and formate oxidation coupled to dissimilatory reduction of iron or manganese by *Alteromonas putrefaciens*. *Appl. Environ. Microbiol.*, 55:700-706.

Mills, E.L., Forney, J.L., Clady, M.D., and Schaffner, W.R. 1978. In: *Lakes of New York State*, Volume II, ed. J.A. Bloomfield, pp. 367-451, New York, Academic Press.

Myers, C.R. and Nealson, K.H. 1988a. Bacterial manganese reduction and growth with manganese oxide as the sole electron acceptor. *Science*, 240:1319-1321.

Myers, C.R. and Nealson, K.H. 1988b. Microbial reduction of manganese oxides: interactions with iron and sulfur. *Geochim. Cosmochim. Acta*, 52:2727-2732.

Nealson, K.H., Rosson, R.A., and Myers, C.R. 1988a. Mechanisms of oxidation and reduction of manganese. In: *Metal Ions and Bacteria*. ed. T.J. Beveridge and R.J. Doyle, pp. 383-411, New York, John Wiley and Sons.

Nealson, K.H., Tebo, B.M., and Rosson, R.A. 1988b. Occurrence and mechanisms of microbial oxidation of manganese. *Adv. Appl. Microbiol.*, 33:279-318.

Postma, D. 1985. Concentration of Mn and separation from Fe in sediments - I. Kinetics and stoichiometry of the reaction between birnessite and dissolved Fe(II) at 10°C. *Geochim. Cosmochim. Acta*, 49:1023-1033.

Ringo, E., Stenberg, E., and Strom, A.R. 1984. Amino acid and lactate catabolism in trimethylamine oxide respiration of *Alteromonas putrefaciens* NCMB 1735. *Appl. Environ. Microbiol.*, 47:1084-1089.

Semple, K.M. and Westlake, D.W.S. 1987. Characterization of iron-reducing *Alteromonas putrefaciens* strains from oil fluid fields. *Can. J. Microbiol.*, 33:366-371.

Stone, A.T. 1987a. Microbial metabolites and the reductive dissolution of manganese oxides: oxalate and pyruvate. *Geochim. Cosmochim. Acta*, 51:919-925.

Stone, A.T. 1987b. Reductive dissolution of manganese (III/IV) oxides by substituted phenols. *Environ. Sci. Technol.*, 21:979-988.

Stone, A.T. and Morgan, J.J. 1984a. Reduction and dissolution of manganese (III) and manganese (IV) oxides by organics: 1. Reaction with hydroquinone. *Environ. Sci. Technol.*, 18:450-456.

Stone, A.T. and Morgan, J.J. 1984b. Reduction and dissolution of manganese (III) and manganese (IV) oxides by organics: 2. Survey of reactivity of organics. *Environ. Sci. Technol.*, 18:617-624.

Stone, A.T. and Morgan, J.J. 1987. Reductive dissolution of metal oxides. In: Aquatic Surface Chemistry, ed. W. Stumm, pp. 221-254, New York, John Wiley and Sons.

Stouthamer, A.H. 1977. Energetic aspects of the growth of micro-organisms. In: *Microbial Energetics*, ed. B.A. Haddock and W.A. Hamilton, pp. 285-315, New York, Cambridge Univ. Press.

Stumm, W. and Morgan, J.J. 1981. *Aquatic Chemistry*, 2nd edition, New York, Wiley-Interscience.

Trimble, R.B. and Ehrlich, H.L. 1968. Bacteriology of manganese nodules. III. Reduction of $MnO_2$ by two strains of nodule bacteria. *Appl. Microbiol.*, 16:695-702.

Trimble, R.B. and Ehrlich, H.L. 1970. Bacteriology of manganese nodules. IV. Induction of an $MnO_2$-reductase system in a marine *Bacillus*. *Appl. Microbiol.*, 19:966-972.

Troshanov, E.P. 1968. Iron- and manganese-reducing microorganisms in ore-containing lakes of Kareliam isthmus. *Microbiology U.S.S.R.*, 37:786-790.

Yerkes, J.H., Casson, L.P., Honkanen, A.K., and Walker, G.C. 1984. Anaerobiosis induces expression of *ant*, a new *Escherichia coli* locus with a role in anaerobic electron transport. *J. Bacteriol.*, 158:180-186.

# CHAPTER 12

# The Transport of Fine-Grained Sediments in the Trenton Channel of the Detroit River

Carl Kirk Ziegler
Wilbert Lick
James Lick

## Introduction

The Detroit River connects Lake St. Clair to Lake Erie. At its lower end just before it enters Lake Erie, a significant fraction of the flow (21% of the total) is diverted by an island, Grosse Ile, into the Trenton Channel (see Figure 1). The western shore of this channel is heavily industrialized and numerous wastewater drains from factories flow into the channel and its tributaries. In addition, groundwater seepages from solid waste dumpsites along the channel contaminate the river water. Hazardous pollutants are thus released into the channel at various point sources along its western shore. These contaminants tend to become adsorbed to fine-grained, cohesive sediments suspended in the river water. Subsequent deposition of the sediments concentrates pollutants in the sediment bed. Wind waves generated during storms can then resuspend the polluted sediments. The contaminants associated with the resuspended sediments may then be released into the water in relatively high concentrations and also transported to another location. For these reasons, an analysis of the sediment transport processes is an important and necessary part of any water quality study concerning the channel.

This paper presents the results of a numerical analysis of the deposition, resuspension, and transport of the fine-grained, cohesive sediments in the Trenton Channel. The equations of motion governing

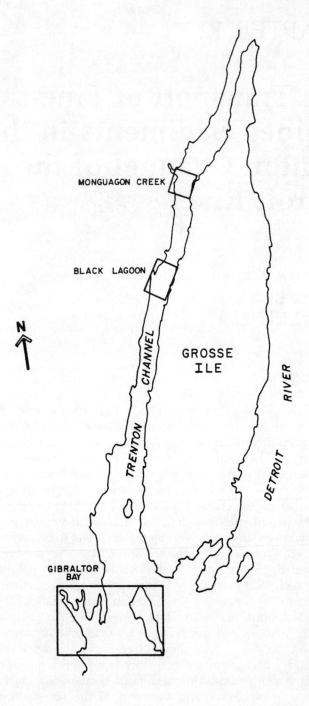

Figure 1. Trenton Channel of the Detroit River with study areas outlined

Sediment Transport in Trenton Channel

the hydrodynamics and sediment mass transport in the model are presented first. Following this is a discussion of the dynamics at the sediment-water interface, which have been incorporated into the numerical analysis. Wind waves generated during storms will resuspend sediment by increasing the bottom shear stress. Expressions for these stresses are given in the next section. Results of the numerical calculations are then presented. The deposition and resuspension characteristics of three different areas of the Trenton Channel (Monguagon Creek, Black Lagoon, and Gibraltar Bay) were studied and compared. A summary of the pertinent results concludes the paper.

## Governing Equations

The Trenton Channel is shallow, with depths of less than 8 meters, and the flow is generally well mixed in the vertical direction. Under these conditions, a valid approximation to the three-dimensional equations of motion for a viscous, incompressible fluid are the two-dimensional, vertically integrated hydrodynamic equations, which can be written as

$$\frac{\partial \eta}{\partial t} + \frac{\partial U}{\partial x} + \frac{\partial V}{\partial y} = 0 \tag{1}$$

$$\frac{\partial U}{\partial t} + gh\frac{\partial \eta}{\partial x} = \tau_x^w - \tau_x^b + A_H\left(\frac{\partial^2 U}{\partial x^2} + \frac{\partial^2 U}{\partial y^2}\right) - \frac{\partial}{\partial x}\left(\frac{U^2}{h}\right) - \frac{\partial}{\partial y}\left(\frac{UV}{h}\right) \tag{2}$$

$$\frac{\partial V}{\partial t} + gh\frac{\partial \eta}{\partial y} = \tau_y^w - \tau_y^b + A_H\left(\frac{\partial^2 V}{\partial x^2} + \frac{\partial^2 V}{\partial y^2}\right) - \frac{\partial}{\partial x}\left(\frac{UV}{h}\right) - \frac{\partial}{\partial y}\left(\frac{V^2}{h}\right) \tag{3}$$

where $h = h_o + \eta$ is the total water depth, $h_o(x,y)$ is the equilibrium water depth, and $\eta(x,y)$ is the surface displacement with respect to $h_o$. The vertically integrated velocities, $U$ and $V$, are defined as $U = \int_{-h_o}^{\eta} u\,dc$ and $V = \int_{-h_o}^{\eta} v\,dc$ where $u$ and $v$ are the velocities in the $x$ and $y$ directions, respectively, and $z$ is the vertical coordinate that is positive upwards. In the above equations, $\tau^w$ is the wind stress, $\tau^b$ is the bottom stress which has components in the x and y directions of $\tau_x^b = c_f\, u|u|$ and $\tau_y^b = c_f\, v|v|$ where $c_f$ is a friction factor, $A_H$ is the horizontal eddy viscosity, and g is the acceleration due to gravity.

A vertically integrated transport equation for the suspended sediment concentration, C, can be expressed as

$$\frac{\partial(hC)}{\partial t} + \frac{\partial(UC)}{\partial x} + \frac{\partial(VC)}{\partial y} = hD_H\left(\frac{\partial^2 C}{\partial x^2} + \frac{\partial^2 C}{\partial y^2}\right) + q_s \qquad (4)$$

where $D_H$ is the horizontal eddy diffusivity, and $q_s$ is the net flux at the sediment-water interface. The turbulent processes that affect the fluid momentum and the sediment mass transport are assumed to be similar for the flows considered here. The numerical values of $A_H$ and $D_H$ are thus the same in any particular calculation.

A quantitative analysis of sediment transport must consider the fact that sediments consist of a mixture of particles with widely varying sizes. This distribution can be taken into account by separating the sediments into components, with each component described by its own average quantities. The above equation is then valid for each component. In this case, the source term, $q_s$, must also include changes in component concentrations due to aggregation and disaggregation.

## Sediment Dynamics

An accurate description of the sediment flux at the sediment-water interface is necessary in order to predict accurately the transport of fine-grained sediments. The net flux at the sediment-water interface, $q_s$, is the difference between the resuspension rate, E, and the deposition rate, D, so that $q_s = E-D$.

The deposition rate for particle of a single size can be expressed as $D = \beta C$ where $\beta$ is an effective average settling speed. For a mixture of different particle sizes, it is assumed that $D = \sum_m \beta_m C_m$ where each component (particle size) of the mixture has its own average settling speed, $\beta_m$, and concentration, $C_m$.

The aggregation and disaggregation of fine-grained, cohesive sediments have a major impact on the effective sizes and densities of the resulting flocs and therefore the settling speeds of the flocs. Recent experimental and theoretical studies (Tsai et al., 1987; Burban et al., 1988; Lick and Lick, 1988) on the flocculation of fine-grained lake sediments have made it possible to describe the flocculation process quantitatively.

However, the numerical calculations required to include an analysis of the aggregation and disaggregation of sediments are quite time-consuming at the present time and, because of this, have not been included in the present calculations. Work is presently being done to increase the efficiency of the numerical calculations, and an accurate description of the flocculation process will be included in future modeling.

For the present, an approximate method (Lick and Lick, 1988) for determining the quasi-steady-state floc diameter has proven to be very useful. Experimental data indicate that the following relationship is valid as a first approximation

$$C \, d_m^2 \, G = \alpha_o \tag{5}$$

where $C$ is the sediment concentration (gm/cm³), $d_m$ is the median floc diameter (cm), $G$ is the fluid shear stress (dynes/cm²), and $\alpha_o$ is an experimentally determined constant. For the freshwater sediments studied here, $\alpha_o = 1 \times 10^{-8}$ gm²/cm³-s². In the water column, the shear stress varies from a maximum of $\tau_b$ at $z = -h_o$ to zero at $z = \eta$. The flows considered here are generally turbulent and analytical expressions for $G(z)$ are difficult to use. As a first-approximation, it will be assumed that $G$ varies linearly with depth, and an average value of $0.5 \, \tau_b$ will be used. Thus, the median floc diameter, $d_m$, can be determined at any point in the solution domain once $C$ and $G$ are calculated.

The settling speed is, of course, dependent on floc size, but it is also dependent on the effective density of the floc. As described previously, this effective density is determined by the suspended sediment concentration and the fluid shear. Suspended sediment concentrations in the Trenton Channel are on the order of 10 mg/L, so experimental results obtained at this concentration will be used here. For this concentration, a valid first approximation that neglects the effect of fluid shear is

$$\beta = 95 \, d_m^{2.4} \tag{6}$$

where $\beta$ has units of cm/s and $d_m$ is in cm. The maximum floc diameter that has been observed in the experiments is approximately 200 μm. This limit on the floc size has been used in the present calculations yielding a maximum settling speed of 79 μm/s.

Experimental work on the resuspension of fine-grained, cohesive sediments (Lee et al., 1981; MacIntyre et al., 1986; Lick and Kang, 1987;

Tsai and Lick, 1988) has shown that only a finite amount of these sediments can be resuspended for any particular shear stress applied to the sediment bed. The reason for this phenomenon is that only the finer sediment particles can be resuspended at low stresses. The coarser particles remain on the sediment bed and armor it. Compaction of the bed also contributes to limiting the amount of fine-grained sediment that can be resuspended because cohesion between particles increases with depth in the sediment bed. In contrast, noncohesive, uniform-size sediments are resuspended at a constant rate, and they will be entrained until the sediment source is exhausted.

The experimental work cited above has shown that the total amount of sediment resuspended, $\varepsilon$, primarily depends upon the turbulent stress at the sediment-water interface and the water content, i.e., time after deposition, of the deposited sediments. An expression for $\varepsilon$ that expresses this functional dependence is (Ziegler and Lick, 1988)

$$\varepsilon = \frac{a_o}{t_d^2}\left(\tau - \tau_o\right)^2, \qquad \text{for } \tau \geq \tau_o$$
$$= 0, \qquad \text{for } \tau < \tau_o \tag{7}$$

where $\varepsilon$ is the net amount of resuspended sediment per unit surface area in gm/cm$^2$, $a_o$ is a constant, $t_d$ is the time after deposition in days, $\tau$ is the bottom shear stress (dynes/cm$^2$) due to waves and currents, and $\tau_o$ is an effective critical shear stress. Both $a_o$ and $\tau_o$ are dependent on the sediment properties at a particular site and must be determined experimentally. For the sediments in the Trenton Channel, a portable resuspension device called a shaker (Tsai and Lick, 1986) has been used in field studies to determine that $a_o = 0.0068$ and $\tau_o = 1.2$ dynes/cm$^2$.

The total amount of resuspended sediment, given by Eq. (7), is resuspended over a time period of approximately one hour. A valid approximation for numerical computations is to assume that the resuspension rate, E, is equal to its initial value. The resuspension rate is then held constant until all of the available sediment, i.e., an amount equal to $\varepsilon$, has been resuspended. The resuspension rate is then set to zero until additional sediment is deposited and becomes available for resuspension.

## Bottom Stresses Due to Currents and Waves

Currents and waves are the causes of the resuspension of fine-grained sediments in the Trenton Channel. Waves generated by storms can propagate into depositional areas along the shores of the river, increase the bottom shear stress, and resuspend the sediments deposited during calm weather. Contaminants that are associated with these resuspended sediments may then be released into the water column and transported to other sections of the river.

The bottom shear stress due to both currents and waves must be determined in order to calculate the net resuspension during a storm event. The bottom stress due to currents alone is given by

$$\tau^B = c_f (u_c^2 + v_c^2) \tag{8}$$

where $u_c$ and $v_c$ are the current velocities in the x and y directions, respectively, and $c_f$ is a friction coefficient. Since the sediment bed considered here consists of fine-grained sediments, the value of $c_f$ used in all calculations was 0.002 (Sternberg, 1972).

Wave-current interaction in the bottom boundary layer is an active area of research being studied by several investigators (e.g., Grant and Madsen, 1979; Glenn and Grant, 1987; Davies et al., 1988). The wave-current models discussed in these papers take into account turbulence and the nonlinear interactions between waves and currents. These models are generally one dimensional in the vertical direction, complex to apply, and have not been experimentally verified. Due to these factors, none of these wave-current interaction models have been used here. In the present study, the wave and current velocities have been assumed not to interact. This simplifying assumption is valid since the interaction appears to cause only minor differences in the calculated bottom shear stresses (Davies et al., 1988). The total bottom shear stress due to waves and currents is therefore

$$\tau^B = c_f \left[ (u_c + u_w)^2 + (v_c + v_w)^2 \right] \tag{9}$$

where $u_w$ and $v_w$ are the x and y direction components, respectively, of the maximum horizontal velocity generated by periodic wind waves at the sediment-water interface.

Wave parameters must be determined in order to calculate the maximum horizontal velocity due to waves, $Q_w$. Shallow water Sverdup-Munk-Bretschneider (SMB) theory has been used to calculate the significant wave height, $H_s$, and period, $T_s$, of wind waves. The factors that determine these parameters are the wind velocity, fetch, and average water depth along the fetch. Once $H_s$ and $T_s$ are calculated, the maximum horizontal velocity can be found using

$$Q_w = \frac{\pi H_s}{T_s \sinh(2\pi h_o/L_d)} \qquad (10)$$

where $L_d$ is the wave length and is given by

$$L_d = \frac{gT_s^2}{2\pi} \tanh(2\pi h_o/L_d). \qquad (11)$$

The wind waves are assumed to propagate in the direction of the wind so that the components of the maximum horizontal velocity, $u_w$ and $v_w$, can be readily determined.

The bottom shear stress oscillates between a minimum, $\tau_{min}^B$, and a maximum, $\tau_{max}^B$, during one wave period. In these calculations, $\tau_{max}^B$ has been used to determine the total amount of resuspended sediment, $\varepsilon$. The wave-current environment can affect the resuspension rate depending on what fraction of the wave period the critical shear stress, $\tau_o$, has been exceeded. The resuspension rate, E, has been assumed to be unaffected if $\tau_{min}^B > \tau_o$, which means that the sediment is resuspended over a period of one hour. However, if $\tau_{min}^B < \tau_o < \tau_{max}^B$, then E has been assumed to be reduced to $\delta E$ where

$$\delta = \frac{\tau_{max}^B - \tau_o}{\tau_{max}^B - \tau_{min}^B}. \qquad (12)$$

## Results of Numerical Calculations

Sediment transport calculations were made for three different areas of the Trenton Channel. These were Monguagon Creek, Black Lagoon, and Gibraltar Bay (Figure 1). These sites were selected because deposits of highly contaminated sediments were found in these areas.

Monguagon Creek is a shallow creek, about 1 m deep, that flows into the Trenton Channel on its western shore. The average flow rate is about 0.7 m³/s (Rathbun, 1985) with a velocity of approximately 5 cm/s. In the past, the creek has served as a drainage ditch for industrial wastewater and, as a consequence, the sediment bed of the creek is highly polluted. Sediments transported into the Trenton Channel from the creek are generally deposited in a narrow strip along the western shore. Because of this, a 450 m long section of the Trenton Channel has been

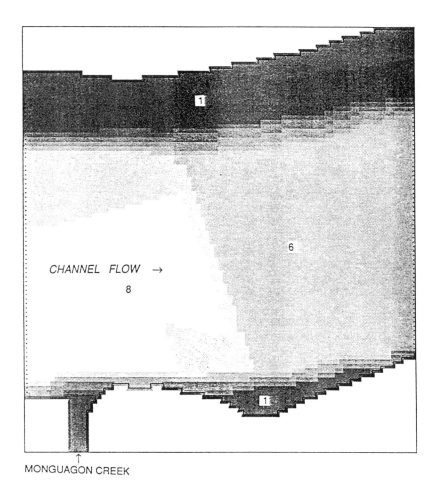

Figure 2 Monguagon Creek computational grid. Bathymetry displayed by a varying gray scale from 1 m (darkest) to 8 m (lightest).

studied including a portion of Monguagon Creek and the shore area downstream of the creek mouth (see Figure 2).

Black Lagoon is a shallow area along the western shore of the channel with depths of less than 1 m. It is approximately 2.5 km downstream of Monguagon Creek. Suspended sediments from the Trenton Channel are transported into the slack water of the lagoon and deposited there. A 900 m length of the channel, starting about 450 m north (upstream) of Black Lagoon, was studied (Figure 3).

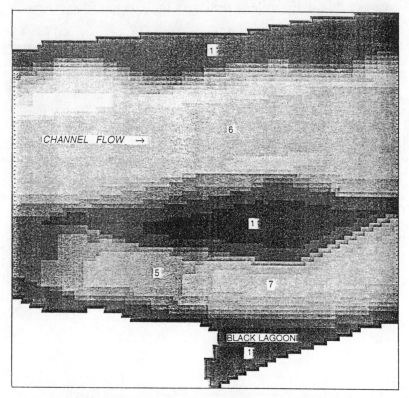

Figure 3. Black Lagoon computational grid. Bathymetry displayed by a varying gray scale from 1 m (darkest) to 7 m (lightest).

A small bay exists at the mouth of the Trenton Channel and to the west of Celeron Island. For some strange reason, this bay has been denoted as Gibraltar Bay in this study but it should not be confused with the Gibraltar Bay at the south end of Grosse Ile (see Figure 1). Gibraltar Bay will hereafter refer only to the study area. This bay is shallow, less than 1 m deep, and is similar to Black Lagoon but much larger. The area is directly exposed to Lake Erie so that it will have a greatly different wave environment than Monguagon Creek or Black Lagoon, which are much more sheltered from wind waves. Approximately 3 km of the channel between Celeron Island and the western shore has been modeled (see Figure 4).

# Sediment Transport in Trenton Channel

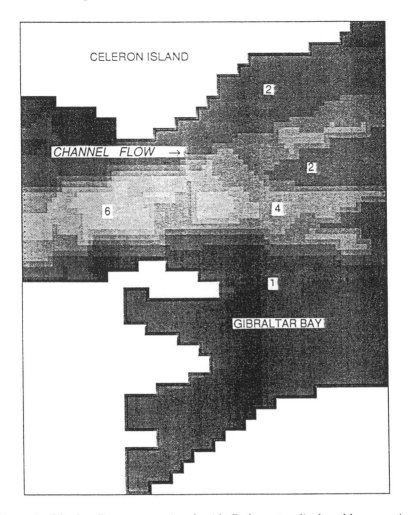

Figure 4. Gibraltar Bay computational grid. Bathymetry displayed by a varying gray scale from 1 m (darkest) to 6 m (lightest).

In these studies, two types of calculations were made. First, a steady-state calculation determined depositional sites and depositional rates as affected by steady currents but in the absence of wave action. High (summer) and low (winter) flow rates in the Trenton Channel were considered. The second type of calculation examined the effect of wind waves. In this latter calculation, a model storm was assumed which had a wind velocity of 15 m/s (29 knots) blowing from the direction of maximum fetch for each area. The structure of the sediment bed was determined by the previous steady-state calculation for that area. From the steady-state calculations, the amount deposited in one day was determined. The sediment bed was then assumed to be formed by steady deposition, i.e., the bed consisted of layers with this amount of sediment in each layer and with each successive layer differing by one day in the time after deposition. The resuspension potential of each layer was then calculated from Eq. (7). The duration of the storm was 2 hours by which time all available sediment had been resuspended by the wave action. A calm period then followed the end of the storm so that transport and redeposition of the resuspended sediments could occur.

In all of these calculations, the hydrodynamic equations, Eqs. (1)-(3), were solved first using a previously developed numerical model (Ziegler and Lick, 1986). The sediment transport equation, Eq (4), along with the experimental information on floc size and settling speed, summarized by Eqs. (5) and (6), were then used to predict sediment transport and deposition. The SMB theory and experimental results on resuspension, Eq. (7), were then used to predict resuspension. It should be noted that all of the sediment transport calculations were carried out using parameters based entirely on experimental results. A posteriori fine-tuning of the numerical model was not necessary and was not done.

## Monguagon Creek

The flow in the Detroit River is fairly constant with the mean flow rate ranging from a winter low of 4810 m$^3$/s to a high in summer of 5950 m$^3$/s. The Trenton Channel transports 21% of the total river flow for a flow rate range of 1010 m$^3$/s to 1250 m$^3$/s (Derecki, 1984). Mean flow in Monguagon Creek was assumed to vary seasonally from 0.65 m$^s$/s (low) to 0.8 m$^3$/s (high), which is consistent with river flow rate variations. The channel inlet velocity ranged from 44 cm/s (winter) to 54 cm/s (summer), and the velocity was assumed to be constant across the channel in all of the calculations. The velocity in Monguagon Creek was varied from a low of 4.3 cm/s to a high of 5.3 cm/s, and it was assumed to be constant across the width of the creek. The value of the eddy viscosity, $A_H$, was set at 7500 cm$^2$/s. The computational grid for this area used square elements, 5 m on a side.

The steady-state velocity field for the Monguagon Creek area with a Trenton Channel inlet flow rate of 1250 m$^3$/s (high) is shown in Figure 5. For clarity only every other velocity vector has been plotted. The flow is fairly uniform with appreciable velocity gradients only occurring near the shore. Due to its very low flow rate, Monguagon Creek has a negligible effect on the currents in the channel. The velocity field for the low flow rate is similar to Figure 5 except for lower values.

Contaminated sediments that are in suspension, either due to an upstream source or resuspension from the sediment bed, will eventually be deposited somewhere in the Trenton Channel or Lake Erie.

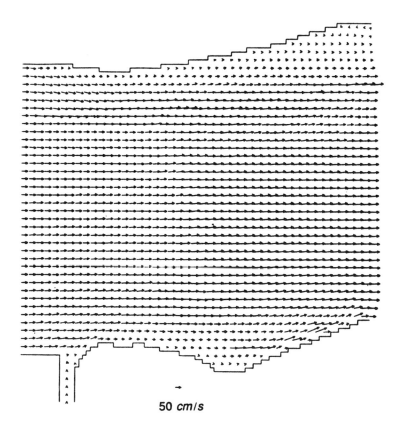

Figure 5. Steady-state current velocities for Monguagon Creek, high flow.

Identifying these deposition sites and quantifying the deposition rates in these areas is important and necessary for pollution management.

Suspended sediment concentration measurements were made in the Trenton Channel during May, 1987. The concentrations were distributed from a high of 8.9 mg/L near the western shore to a minimum of 5.4 mg/L in the middle of the channel (U.S. Environmental Protection Agency, 1987). This concentration distribution was used at the upstream inlet of the computational grid in all of the steady-state deposition calculations presented in this study. Suspended sediment concentration measurements at the mouth of Monguagon Creek were made over a 6 month period in 1986 (Kreis, 1988). The mean value of ten measurements during this time was 15 mg/L. This value was assumed to be constant across the width of the creek, and it was used in all of the Monguagon Creek calculations.

The hydrodynamic calculations discussed above have been used to

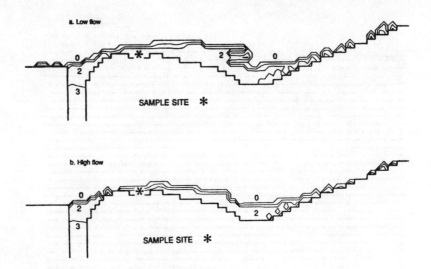

Figure 6. Steady-state deposition rates (gm/cm² -year) for Monguagon Creek area, 1 gm/cm²-year (~2 cm/year) contour interval. a) low flow rate; b) high flow rate.

determine the steady-state deposition sites in the Monguagon Creek area for both winter and summer flow rates. These sites are shown in Figure 6. The deposition rate, D, is expressed in gm/cm²-year, and it can be converted to a growth rate, T (cm/year), using the formula

$$T = \frac{D}{(1 - \gamma) \rho_s} \qquad (14)$$

where $\gamma$ is the volume fraction of water in the deposited sediment and $\rho_s$ is the density of the solid sediment. For the fine-grained sediments examined here, $\rho_s = 2.65$ g/cm³. If it is assumed that the volume fraction of water in the deposited sediments after compaction is reduced to 80%, i.e., $\gamma=0.8$, then the deposition rates expressed here can be multiplied by 2 to get an approximate growth rate in cm/year.

Depositional areas downstream of the mouth of Monguagon Creek extend about 15 m to 20 m into the channel from the western shore. The current velocities outside of the depositional sites are high enough so that any freshly deposited sediment is immediately resuspended, and the net sediment flux is therefore zero. Lower velocities in the channel are thus the reason for the wider depositional strip during low flow than that produced under high flow conditions. The maximum deposition rate along the western shore is approximately 2 gm/cm$^2$ -year with a rapid decrease in the rate at the edge of the deposition site. Fine-grained sediment samples have been found during field studies at the location denoted on Figure 6. This sample site is located in a predicted deposition area, for both flow rates, which qualitatively confirms the accuracy of the numerical model.

The deposition rate in Monguagon Creek is greater than 3 gm/cm$^2$-year due to the higher sediment concentration. Numerical experiments have shown that a major fraction of the suspended sediments in Monguagon Creek is deposited in the creek. The Monguagon Creek plume extends less than 20 m from the western shore into the channel due to the low flow rate of the creek. Thus, nearly all of the sediments transported from the creek to the Trenton Channel are confined to the narrow depositional strip along the western shore.

Sediments deposited in the areas shown in Figure 6 will remain there until they are resuspended due to an increase in the local bottom shear stress. An increased current velocity or the wind waves from a storm can both cause the resuspension of these sediments. The Trenton Channel current is fairly constant with flow variations generally occurring over a seasonal time period. The depositional area during winter (low) flow is larger than that during summer (high) flow. Some sediments deposited during the winter will thus be resuspended and transported downstream during the summer. This seasonal resuspension due to a higher flow is a slow process, though. In contrast, storms occur over a much shorter period of time, generally on the order of hours. Wind waves generated by storms can thus resuspend large quantities of sediment quickly. High concentrations of contaminants could then be released into the water column from the resuspended sediments during a storm.

Results of the deposition rate studies discussed previously were used to determine the structure of the sediment bed in the wave resuspension calculations. The calculated steady-state deposition rate at a particular grid point was used to determine the thickness, $T_{day}$ (gm/cm$^2$), of a sediment layer deposited in one day. The sediment bed at any particular point was then assumed to be composed of a series of layers of thickness $T_{day}$. The top layer was assumed to be one day old and the age of each layer increased by one day with increasing depth in the sediment bed. This structure was assumed to only exist in the depositional areas shown in Figure 6. Outside of the deposition areas, it was assumed that the sediment bed had been scoured of fine-grained sediments and the bed was composed of either sand or gravel in those

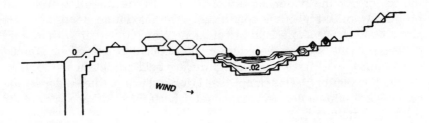

Figure 7. Sediment resuspension (gm/cm$^2$-year) in Monguagon Creek area due to wind waves, 0.01 gm/cm$^2$- (~0.02 cm) contour interval, high flow. Wind velocity of 15 m/s from 30° North.

locations. In order to isolate the effects of wave action in these calculations, it was assumed that the suspended sediment concentration was initially zero at all points in the solution domain, and it remained at zero at the upstream inlet during the entire calculation.

The maximum fetch for the Monguagon Creek area is in an upstream direction with the wind blowing from 30° north. The length of this fetch is 3 km with an average depth of 7.6 m. The narrow width of the fetch, only 400 m, reduces the effective fetch for this portion of the river to 960 m. Applying SMB theory then yields a significant wave height of 38 cm with a significant wave period of 2.2 s. The wind was allowed to blow for 2 hours, at which time all of the available fine-grained sediment had been resuspended.

The resuspension areas for the Monguagon Creek area resulting from this storm at high flow are shown in Figure 7. Erosional areas in this figure are denoted by negative depth contours, and the zero contour line represents the initial position of the sediment bed. Most of the resuspension has occurred in the shallow near-shore area approximately 250 m downstream of the mouth of Monguagon Creek. The sediment bed has been eroded to a maximum depth of approximately 0.036 gm/cm$^2$ (≈0.07 cm), which represents 5 days of deposited sediment. The average age of the resuspended layers in the sediment bed was 3 days for both flow rates. More sediment was resuspended during low flow, 430 kg, than for high flow, 310 kg, because of the larger depositional area produced by the low flow hydrodynamics. All of the resuspended sediment was transported downstream by the currents after it was entrained into the overlying water column.

## Black Lagoon

The channel inlet flow velocity for Black Lagoon varied from 45 cm/s (winter) to 56 cm/s (summer), nearly identical to that in the Monguagon Creek calculations. A 10 m square element was used for the Black Lagoon grid. The eddy viscosity, $A_H$, was increased by a factor of two from that used in Monguagon Creek to 15,000 cm$^2$/s. This increase kept the ratio between $A_H$ and the grid size constant between the two areas. This procedure ensured that the effects of viscosity were the same in the different regions.

Similar to the Monguagon Creek area, the channel flow in the Black Lagoon region is generally uniform except for the currents in the lagoon itself, illustrated in Figure 8 for high flow. Water from the Trenton Channel enters the shallow lagoon and accelerates in the upper part of the lagoon as the water depth decreases. A weak vortex is generated in the lower part of the lagoon with velocities of approximately 1 cm/s. No velocity vector has been plotted at grid points where the current velocity

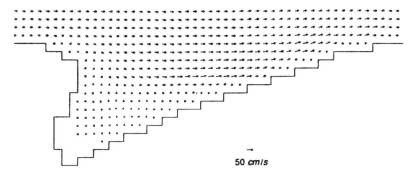

Figure 8. Steady-state current velocities for Black Lagoon, high flow.

is less than 0.1 cm/s. The low flow velocity field closely resembles that for high flow but with lower velocities.

The Black Lagoon depositional areas for the summer flow are presented in Figure 9. The heaviest deposition occurs in the upstream end of Black Lagoon where the channel flow enters the lagoon. Higher velocities in the central portion of the lagoon create a resuspension area where net deposition does not occur. Sediment is deposited in the deeper water along the edge of the lagoon during low winter flow because of lower current velocities. The weak vortex in the lagoon transports only a minor amount of the suspended sediment to the rear of the lagoon causing the low deposition rate in that area. A mud bar has

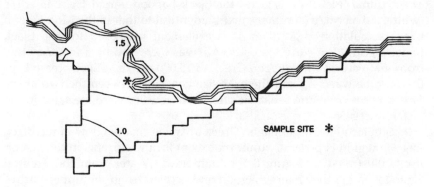

Figure 9. Steady-state deposition rates (gm/cm$^2$) in Black Lagoon, 0.5 gm/cm$^2$-year (~1cm/year) contour interval, high flow.

thus been created in the lagoon due to these factors. This bar has been observed in the predicted location during field studies in Black Lagoon. In addition, fine-grained sediments have been collected at the sample site shown on Figure 9. Both of these observations help to confirm qualitatively the predictions of the numerical analysis.

Maximum deposition rates in Black Lagoon are approximately 0.5 gm/cm$^2$-year less than those found along the western shore downstream

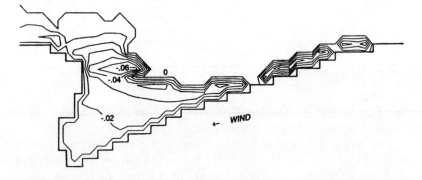

Figure 10. Sediment resuspension (gm/cm$^2$) in Black Lagoon due to wind waves, 0.01 gm/cm2 (~0.02 cm) contour interval, high flow. Wind velocity of 15 m/s from 190° North.

of Monguagon Creek. The reason for this decrease is that deposition occurs upstream of the lagoon due to low velocities along the western shore of the Trenton Channel. The suspended sediment load transported into Black Lagoon is thus reduced to a concentration of approximately 7 mg/L from a high of 8.9 mg/L at the upstream inlet.

Black Lagoon's maximum fetch is in a downstream direction, and the wind would be blowing from 190° north. This long fetch, 10.9 km, is very narrow, 600 m, so that the effective fetch becomes 1950 m. The average depth over this fetch is 4.6 m, which produces wind waves with $H_s$ = 48 cm and $T_s$ = 2.4 s. The assumed duration of the model storm was 2 hours followed by 3 hours of no wind. The calm period allowed resuspended sediment in the lagoon time to be transported and redeposited on the sediment bed.

The structure of the sediment bed in Black Lagoon at the end of the numerical experiment is illustrated in Figure 10. The mean age of the resuspended layers was 4 days for winter flow and 5 days during summer conditions. Maximum erosion of about 0.066 gm/cm$^2$ ($\approx$0.13 cm), or 8 days of deposited sediment, occurs along the mud bar in the central section of the lagoon. The reasons for the maximum occurring in this area are that the highest bottom shear stress due to waves and currents, 24 dyne/cm$^2$, occurs at this location, and there is also a large quantity of sediment available for resuspension. The sediment bed layers decrease in thickness in the back portion of the lagoon due to the low deposition rate there, reducing the amount of fine-grained sediment that can be resuspended. Only a small quantity of sediment, less than 0.02 gm/cm$^2$ ($\approx$0.04 cm), is eroded in that area because the sediment bed becomes armored at a shallow depth due to the thin sediment layers. Similar to the Monguagon Creek calculations, more sediment was resuspended during the winter (low) flow than the summer (high) flow with 6400 kg and 4900 kg eroded, respectively. Nearly all of the resuspended sediment was transported downstream. A negligible amount of sediment was transported by currents to the back of the lagoon and deposited there.

## Gibraltar Bay

The flow rate in the channel between Celeron Island and the western shore accounts for 15% of the total river flow with a range of 722 m$^3$/s to 893 m$^3$/s (Derecki, 1984). The upstream inlet velocity for the Gibraltar Bay channel ranged from 21 cm/s to 26 cm/s. The decrease in the current velocity from that upstream in the Trenton Channel is due to widening of the channel and a lower flow rate. Square grid elements of 45 m size were used in this area. The eddy viscosity, $A_H$, was increased to 67,500 cm$^2$/s, which is in proportion to the size increase of the grid element from that in Black Lagoon.

The Gibraltar Bay area velocity field for high flow is presented in Figure 11, where every other velocity vector has been plotted for clarity. Velocity vectors have been omitted from this plot where the velocity is less than 0.1 cm/s. The channel currents are not as uniform here as they were in the Monguagon Creek and Black Lagoon regions. The bottom topography in the Gibraltar Bay channel is more variable causing acceleration and deceleration of the flow as the depth changes. Velocities in Gibraltar Bay are generally less than 5 cm/s. The low flow velocity field is not appreciably different from that for high flow.

Steady-state deposition sites in Gibraltar Bay are shown in Figure 12. Upstream deposition claims a large portion of the suspended sediment before it is transported into the bay. The reduction in

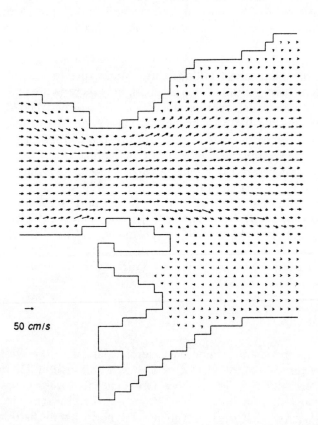

Figure 11. Steady-state current velocities for Gibraltar Bay, high flow.

suspended sediment load, from 8.9 mg/L at the upstream inlet to less than 5 mg/L at the entrance of the bay, causes the maximum deposition rates in Gibraltar Bay to be about 0.5 gm/cm$^2$-year less than that in Black Lagoon. Sediments carried into the bay from the Trenton Channel are not transported very far before deposition occurs due to the weak currents in Gibraltar Bay. The deposition rate decreases by a factor of two over a distance of about 250 m at the entrance of the bay. A large portion of Gibraltar Bay thus has a very low deposition rate, less than 0.5 gm/cm$^2$-year. The complex bathymetry and variable currents in the Gibraltar Bay channel cause the depositional sites in the channel to be very patchy and irregular. Field studies may be affected by this variability since the composition of sediment bed samples taken within a few meters of one another at a site can change from fine-grained sediments to sand or gravel.

Gibraltar Bay is exposed to the western basin of Lake Erie from a southeastern direction. The maximum fetch across Lake Erie is 57 km

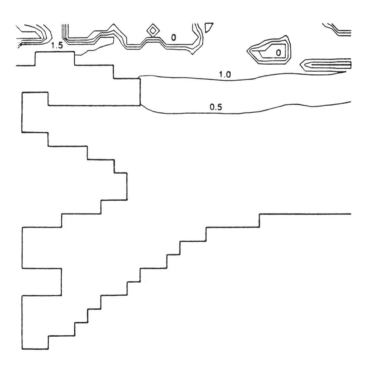

Figure 12. Steady-state deposition rates (gm/cm$^2$-year) for Gibraltar Bay, 0.5 gm/cm$^2$-year (~ 1 cm/year) contour interval, high flow.

with an average depth of 8 m. The wind blows from 160° north along this fetch. Waves generated across Lake Erie have a height of 134 cm and a period of 4.6 s when they encounter the shallow waters at the mouth of the Detroit River. Bottom friction in these shallower depths will cause the wind waves to decay (U.S. Army Coastal Engineering Research Center, 1977). When the waves reach the downstream edge of the computational grid, the wave height has been reduced to 101 cm. These waves will become unstable and break when they enter the shallow water of Gibraltar Bay, which is less than 1 m deep. Bottom shear stresses generated by breaking waves are poorly understood at the present time, and the stresses due to breaking waves have been neglected here. To account for breaking, wave heights in the bay were limited to the maximum stable wave height of $H_{s,max} = 0.78 \, h_o$ (U.S. Army Coastal Engineering Research Center, 1977). This limit on the wave height has been used in the present calculations causing the wind waves to decay to a height of 24 cm in the shallowest portions of Gibraltar Bay. The wave period was assumed to remain constant as the wave height decays.

The Gibraltar Bay storm calculations, 2 hours of wind followed by a 10 hour calm, produced areas of both resuspension and net deposition as shown in Figure 13. The highest resuspension occurs at the entrance of the bay where bottom shear stresses due to waves and currents are highest, with a maximum of 45 dynes/cm$^2$. The sediment bed was eroded to a maximum depth of approximately 0.10 gm/cm$^2$ ($\approx$0.20 cm) in this region, resuspending 14 days of deposited sediment for both flow rates. An average of 21 days of deposited sediment was resuspended by the storm in Gibraltar Bay. The mean age of the eroded layers is greater than that at the maximum depth because of the high bottom shear stresses in the bay, ranging from 5 to 20 dynes/cm$^2$. Also, the low deposition rate in Gibraltar Bay created very thin sediment layers. A relatively high number of these thin layers were resuspended over a large area in the bay, and the mean value was thus increased. By the end of the storm, a total of 510,000 kg had been resuspended from Gibraltar Bay for winter flow and 489,000 kg during summer flow. This dramatic increase in eroded sediment, about 3 orders of magnitude greater than the Monguagon Creek area and a 2 order of magnitude increase from that in Black Lagoon, was primarily due to the much greater area of Gibraltar Bay that is exposed to wave action.

An interesting result of the storm calculations was the creation of an area of net deposition in Gibraltar Bay (see Figure 13). Redeposition was negligible in both the Monguagon Creek and Black Lagoon storm calculations with the resuspended sediments being transported downstream. Of the total amount of sediment resuspended during the Gibraltar Bay storm, 3000 kg (winter) and 2900 kg (summer) of this material was redeposited over approximately 1 km$^2$ of the bay. This net

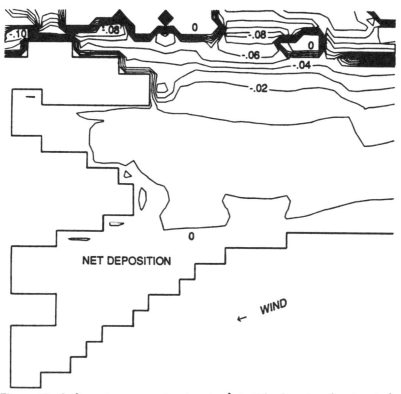

Figure 13. Sediment resuspension (gm/cm$^2$) in Gibraltar Bay due to wind waves, 0.01 gm/cm$^2$ (~0.02 cm) contour interval, high flow. Wind velocity of 15 m/s from 160° North.

deposition accounted for less than 1% of the total quantity of resuspended sediment, the remainder of which was transported downstream toward Lake Erie or redeposited in erosional areas in the bay where net deposition did not occur. The source of the freshly deposited sediment was the eastern section of Gibraltar Bay where the deposition rate is less than 0.5 gm/cm$^2$-year (Figure 12). Sediment is resuspended in this region and then transported further into the bay by the low velocity currents, less than 5 cm/s. Once the storm has ended, the bottom shear stress is reduced below the critical value, and the suspended sediments are redeposited in the area indicated in Figure 13. This calculation illustrates a mechanism by which sediments can be transported from the entrance of Gibraltar Bay to areas near the western shore.

The effect of storm duration on sediment resuspension and deposition in Gibraltar Bay was also investigated. The above storm calculations were repeated except that the wind was allowed to blow for 6 hours followed by a 6 hour calm period. The increased duration of the storm caused the net deposition of sediment in Gibraltar Bay to increase. Resuspended sediment was kept in suspension longer during the extended storm because freshly deposited sediment was quickly resuspended by the wave action during the storm. Larger quantities of resuspended sediments could then be transported by the currents to the back portion of the bay where redeposition occurred after the storm ended. Even though net deposition increased, the total amount of sediment resuspended by wave action was unchanged because the sediment bed had been armored by the end of the first 2 hours of the storm.

## Summary and Concluding Remarks

Recent experimental results concerning the deposition and resuspension of fine-grained, cohesive sediments have been incorporated into a numerical analysis of the transport of fine-grained sediments in the Trenton Channel. Of primary importance was the inclusion of an approximate method for determining the effective diameter of suspended flocs. Using the calculated floc diameter, the settling speed of the flocs could be determined from a relationship based upon experimental data. Field studies were also conducted in the Trenton Channel to determine sediment resuspension parameters. These experimental data were then utilized in the numerical model to produce sediment transport calculations that were realistic and reasonably accurate.

Depositional sites and depositional rates for steady-state conditions during winter (low) and summer (high) flow conditions were first determined in all three study regions. The area over which deposition occurred was found to be greater during the winter than in summer due to the lower current velocities at that time of the year. Some of the sediment deposited during the winter months will then be resuspended and transported downstream as the flow rate increases during the summer runoff.

Maximum deposition rates along the western shore downstream of Monguagon Creek were about 2.0 gm/cm$^2$-year ($\approx$4 cm/year). The maximum deposition rate in Black Lagoon was reduced to approximately 1.5 gm/cm$^2$-year ($\approx$3 cm/year) while that in Gibraltar Bay decreased further to about 1.0 gm/cm$^2$-year ($\approx$2 cm/year). The reductions in deposition rates in Black Lagoon and Gibraltar Bay were due to upstream deposition lowering the suspended sediment load that was transported into the lagoon and the bay. The decrease in deposition rates between the study areas affects the amount of sediment available for resuspension during a storm. A lower deposition rate means that the thickness of a sediment layer deposited in one day, $T_{day}$, will also decrease. Compaction of the sediment bed increases with depth so that less sediment can be resuspended from the deeper layers. Thus, for a particular shear stress, less sediment will be resuspended for a smaller $T_{day}$ (lower deposition rate) than for a larger $T_{day}$ (higher deposition rate).

The effect of wind waves on the resuspension of sediments in the three areas was also investigated. The model storm was used to illustrate the differences between the study regions for the situation of a steady wind blowing along the maximum fetch. In this study, the probability of occurrence of this condition was not considered. However, small variations in the wind direction at Monguagon Creek (±2.5°) or Black Lagoon (±5°) from the maximum fetch axis drastically reduces the fetch in those areas. The bottom stresses due to waves will decrease sharply with reduced fetch. Sediment resuspension during a storm will thus decrease dramatically as the wind shifts away from the maximum fetch direction in Monguagon Creek and Black Lagoon. In contrast, Gibraltar Bay has a large exposure to Lake Erie with a fetch of approximately 50 km over a 60° arc. Fairly large wind direction variations could then occur during a storm without significantly affecting the wave stresses and sediment resuspension in the bay. The frequency of occurrence of resuspension due to wind waves would thus be much higher in Gibraltar Bay than in Monguagon Creek or Black Lagoon because of the greater arc of exposure in the bay.

Sediment resuspension due to wind waves was found to increase as one traveled downstream from Monguagon Creek to Black Lagoon to Gibraltar Bay. The maximum depth of resuspension in Monguagon Creek, Black Lagoon, and Gibraltar Bay was 0.036, 0.067, and 0.101 gm/cm$^2$ ($\approx$0.07, 0.13, 0.20 cm), respectively. The age of the resuspended layers at the point of maximum resuspension was 5, 8, and 14 days, with increasing age corresponding to increasing depth in the sediment bed. Bottom shear stresses due to waves and currents in the depositional areas increase approximately linearly among the three areas, from $\approx$10 dynes/cm$^2$ in Monguagon Creek to over 40 dynes/cm$^2$ in Gibraltar Bay while there is a nearly linear decrease in the sediment layer thickness, $T_{day}$, from Monguagon Creek to Gibraltar Bay. The interaction between the increasing shear stress and the decreasing layer thickness causes the maximum resuspension to increase approximately linearly as indicated above.

The total amount of resuspended sediment varied over 3 orders of magnitude for the three study areas. For winter flow conditions, 430, 6400, and 510,000 kg were resuspended in Monguagon Creek, Black Lagoon, and Gibraltar Bay, respectively. These large variations were primarily due to an increase in the area which was exposed to wave action. Less sediment was resuspended during summer storms because of a decrease in the depositional area during that season.

The above analysis was intended as a study of the relative differences in the deposition and resuspension characteristics of three regions in the Trenton Channel and, for that purpose, is quite accurate. For more extensive modeling under different conditions and in different regions, the analysis may be limited in accuracy for the following reasons. First, only an approximate method for determining the sizes of flocs was used in the present analysis. In the future, a more accurate description of the flocculation process will be included. This should be especially important at the sediment-water interface where rapid aggregation and disaggregation is expected. Second, in the present study, only one field sample to measure sediment resuspension was taken. More samples at different locations and under different environmental conditions, especially at higher shear stresses, are needed for an accurate analysis. Third, the bottom stresses due to waves, both breaking and non breaking, are not well understood at present but are obviously significant in the shallow waters that were investigated here.

The preceding analysis should be valid for studying the differences in resuspension and deposition among various sites especially under similar environmental conditions. With the improvements as described above, the validity of the model should be greatly extended. However, at the present time, no serious field verification of this model, or similar model, has been made. A systematic field test of the predictions made by using this model is urgently needed.

## Acknowledgments

This research was supported by the U.S. Environmental Protection Agency. Dr. Louis S. Swaby and Dr. Anthony Kiylauskas were the project officers.

## References

Burban, P.Y., Lick, W., and Lick, J. 1988. The flocculation of fine-grained sediments in estuarine waters. *J. Geophysical Research*.

Davies, A.G., Soulsby, R.L., and King, H.L. 1988. A numerical model of the combined wave and current bottom boundary layer. *J. Geophys. Res.*, 93:491-508.

Derecki, J.A. 1984. Detroit River physical and hydraulic characteristics. GLERL Contribution No. 417, Great Lakes Environmental Research Laboratory, Ann

Arbor, MI.

Glenn, S.M. and Grant, W.D. 1987. A suspended sediment stratification correction for combined wave and current flows. *J. Geophys. Res.*, 92:8244-8264.

Grant, W.D. and Madsen, O.S. 1979. Combined wave and current interaction with a rough bottom. *J. Geosphys. Res.*, 84:1797-1808.

Kreis, R.G., Jr., ed. 1988. *Integrated Study of Exposure and Biological Effects of In-Place Sediment Pollutants in the Upper Connecting Channels, Interim Results, Final Results.* U.S. Environmental Protection Agency, Office of Research and Development, ERL-Duluth, MN and LLRS-Grosse Ile, MI.

Lee, D.Y., Kang, S.W., and Lick, W. 1981. The entrainment and deposition of fine-grained sediment. *J. Great Lakes Res.*, 7:224-233.

Lick, W. and Kang, S.W. 1987. Entrainment of sediments and dredged materials in shallow lake waters. *J. Great Lakes Res.*, 13(4):619 627.

Lick, W. and Lick, J. 1988. On the aggregation and disaggregation of fine-grained lake sediments. *J. Great Lakes Res.* 14(4): 514-523.

MacIntyre, S., Lick, W., and Tsai, C.H. 1986. Entrainment of cohesive reverine sediments. *Biogeochemistry* 9:187-209.

Rathbun, J. 1985. Flow measurements in Monguagon Creek. Unpublished memorandum, Raytheon Service Company, Grosse Ile, MI.

Sternberg, R.W. 1972. Predicting initial motion and bedload transport of sediment particles in the shallow marine environment, in *Shelf Sediment Transport*, D.P. Swift, Dowden, Hutchinson, Stroudsburg, PA.
D.P. Swift,ed. Dowden, Hutchinson, Stroudsburg, PA.

Tsai, C.H. and Lick, W. 1986. A portable device for measuring sediment resuspension. *J. Great Lakes Res.*, 12:314-321.

Tsai, C.H. and Lick, W. 1988. Resuspension of sediments from Long Island Sound. Water Sci. Tech.
Sound. *Water Sci. Tech.*, 20: (617)155-164.

Tsai, C.H., Iacobellis, S., and Lick, W. 1987. Flocculation of fine-grained lake sediments due to a uniform shear stress. *J. Great Lakes Res.*, 13:135-146.

U.S. Army Coastal Engineering Research Center. 1977. *Shore Protection Manual*, Vol. 1, Fort Belvoir, VA.

U.S. Environmental Protection Agency, Large Lakes Research Station. 1987. *Upper Great Lakes Connecting Channel Study: Detroit River System Mass Balance.* U.S. Environmental Protection Agency, Office of Research and Development,

ERL-Duluth, MN and LLRS-Grosse Ile, MI.

Ziegler, C.K. and Lick, W. 1986. A numerical model of the resuspension, deposition, and transport of fine-grained sediments in shallow waters. U.C.S.B. Report ME-86-3.

Ziegler, C.K. and Lick, W. 1988. The transport of fine-grained sediments in shallow waters. *Environ. Geol. Water Sci.*, 11:123-132.

# CHAPTER 13

# Sediment Transport in Shallow Lakes - Two Case Studies Related to Eutrophication

Lambertus Lijklema
R. Hans Aalderink
Gerard Blom
Elisabeth H.S. VanDuin

## Introduction

Sediments and sediment transport affect eutrophication and, as a consequence, primary production in various ways. Nutrients, especially phosphate, have a strong tendency to concentrate in particulate material, and large pools frequently accumulate in lake bottoms. Recirculation into the overlying waters takes place by diffusion from pore water and through resuspension and subsequent desorption of the nutrient. Alternatively, internal sediment transport may remove phosphate from the euphotic zone into other areas such as deep pits or trenches from which recycling is obstructed. A second way in which primary production is affected by sediments is the enhanced turbidity caused by resuspension of bottom material and the concomitant reduced penetration of light. Especially in shallow wind, exposed lakes and the cycling of nutrients can be dominated by sediment transport. This is true for most Dutch lakes, which are predominantly shallow, hardly protected by hills or even trees, exposed to fairly high average wind speeds, and frequently have soft sediments as well.

Traditionally these conditions have led to much interest in the Netherlands in sediment resuspension and transport. This phenomenon is of interest because of its effect on contaminent and nutrient transport and accumulation, dredging decisions, and light penetrations. Several studies in this field of interest are presently in progress. Two of these studies will be presented here with most emphasis on the approach and modeling. Some typical results will also be presented. Both studies have been commissioned by Offices of the Ministry of Transportation and Public Works.

The first study concerns the internal transport of phosphate rich silt in Lake Veluwe. The ultimate objective is to assess the rate of accumulation of silt in deep pits in this lake, from which the material may be removed by dredging from time to time. The optimal location and size of such a pit should be established. Lake Veluwe is a small border lake remaining between the main land and a new polder constructed in Lake IJssel in 1957. This part of the lake has now been reclaimed in the interests of hydrology, shipping and recreation.

The second study addresses the role of resuspension in the control of light conditions and subsequent algal growth in Lake Marken. This is another part of Lake IJssel separated from the main lake by a new dike constructed in 1975. In contrast to other satellite lakes of Lake IJssel this lake has not suffered from troublesome blooms of blue-green algae and is considered to be light limited. New polders, dikes, and other constructions under consideration may affect the wind induced turbidity and consequently the structure of the ecosystem. The objective of this study is to develop a predictive tool for these effects.

Figure 1 and Table 1 show the situation, sampling sites, position of small islands, and some features of these lakes. Both studies concentrate on

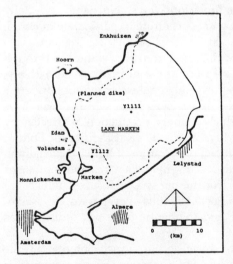 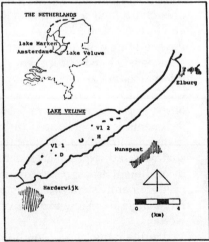

Figure 1. Location of Lake Marken and Lake Veluwe and sampling points.

Table 1

Characteristics of Lake Marken and Lake Veluwe.

|  | Lake Marken | Lake Veluwe |
|---|---|---|
| Surface area, ha | 68,800 | 3,200 |
| Depth, average, m | 3.6 | 1.3 |
| Distribution of depth and sediment composition | 8.5% : ≤2m <br> :sand <br> 8.5% : 2.0<H≤2.5 m <br> : clay <br> 10% : 2.5<H≤3.0 m <br> : clay-silt <br> 73% : ≥3.0 | 70% : ≈0.9 m <br> : sand <br> 24% : ≈2.0 m <br> : clay <br> 6% : ≈3.6 m <br> : silt <br> (pits and shipping canal) |
| Hydr. detention time | 1-2 year | summer : 0.5 year <br> winter : ≈1 month |
| External P load | 1.4 g/m², y | 1.0 g/m²,y |

resuspension and sedimentation, which are analyzed in the framework of mathematical models. This modeling framework will be discussed first, followed by a description of the characteristics of particulate material and relationships for sedimentation and resuspension. Parameter estimations and experimental results are presented and discussed. Finally, some results of simulations relevant for the two lakes are presented.

# Modeling Sediment Transport in Shallow Lakes

Sediment transport models for (shallow) lakes require a quantitative description of:

- resuspension/erosion of bottom sediment,
- sedimentation of suspended solids,
- production and decay of suspended solids,
- inputs and outputs,
- internal horizontal (and in some cases vertical) transport by advection and dispersion.

Sediment transport is essentially three dimensional but in most shallow Dutch lakes, the vertical dispersion is predominant, and vertical gradients in suspended solilds concentrations are virtually absent. Hence a two dimensional horizontal transport model may be adequate. The physical criterion for the occurrence of vertical gradients is the ratio of the characteristic times for dispersion ($t_d$) and for sedimentation ($t_s$).

$$t_d/t_s = (H^2/2D_v)/(H/w_s) = w_s H/2D_v \tag{1}$$

where:

$t_d$ : characteristic time for dispersion (s)
$t_s$ : characteristic time for sedimentation (s)
H : depth (m)
$D_v$ : vertical dispersion coefficient ($m^2 \cdot s^{-1}$)
$w_s$ : sedimentation rate ($m \cdot s^{-1}$)

If $t_d/t_s \ll 1$, dispersion will be dominating. The vertical dispersion coefficient in shallow Dutch lakes is in the order of $2\ 10^{-4}$-$10^{-3}\ m^2 \cdot s^{-1}$ (DiGiano et al., 1978). For Lake Veluwe (average depth is 1.3 m) this means that dispersion will be dominating if the sedimentation velocity is $< 3\ 10^{-4}$-$15\ 10^{-4}\ m \cdot s^{-1}$; for Lake Marken (average depth is 3.6 m) the sedimentation velocity should be less than $1\ 10^{-4}$-$6\ 10^{-4}\ m \cdot s^{-1}$. In subsequent sections it will be shown that this condition is fulfilled in both lakes.

In both studies discussed here, the transport of bottom sediment and its effects on water quality were the primary objectives. Production and decay of suspended solids were negligible and also inputs and outputs were dominated by resuspension and sedimentation. Hence the mass balance is represented by the differential equation:

$$\frac{\delta HC}{\delta t} + \frac{\delta HUC}{\delta x} + \frac{\delta HVC}{\delta y} - \frac{\delta(HD_x \cdot \delta C)}{\delta x} - \frac{\delta(HD_y \cdot \delta C)}{\delta y} - \phi_r + \phi_s = 0 \tag{2}$$

where:

C : concentration ($kg/m^3$)
Dx : dispersion coeff., x-axis ($m^2/s$)
Dy : idem, y-axis ($m^2/s$)
t : time (s)
U : water velocity along x-axis (m/s)
V : water velocity along y-axis (m/s)

Sediment Transport in Shallow Lakes    257

φr : resuspension flux (kg/m²s)
φs : sedimentation flux (kg/m²s)
U and V are depth averaged

This equation is the basis for the two-dimensional model for Sediment Transport, Resuspension and Sedimentation in Shallow lakes (STRESS-2d) developed by the authors. The model is based upon an existing two-dimensional horizontal hydrodynamic model for the calculation of water velocities and levels and the transport of conservative compounds: WAQUA. WAQUA has been developed by Rijkswaterstaat in cooperation with the Delft Hydraulic Laboratory (Stelling, 1984). This version used a rectangular grid with squared unit cells (for dimensions see applications, discussed in a subsequent section). The authors expanded the model with a wave model, terms for resuspension and sedimentation, and a mass balance for the sediment.

Flexibility was one of the objectives of the authors. The user may choose from four different descriptions of resuspension/sedimentation (Table 2). It is also possible to add other relationships to the model as was done in the Lake Veluwe study, discussed in a subsequent section. Three options for modeling the sediment are available: one single layer of infinite depth, one layer with variable volume and composition, or two layers with variable

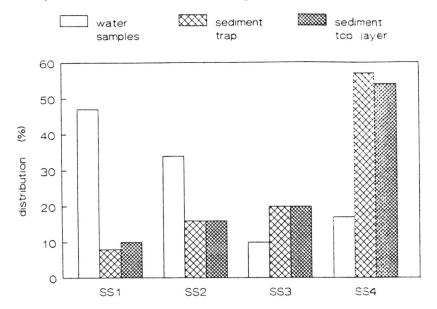

Figure 2. Distribution of material from lake water, sediment traps and sediment top layer in Lake Marken (mean values for 1988).

volume and composition.

The numerical solution of the expanded advection-diffusion equation (Eq. 2) involves an Alternating Direction Implicit (ADI) - technique. It implies central differencing and a staggered grid and provides a stable solution with limited numerical dispersion (Stelling, 1984).

The advective and dispersive transport are driven by wind induced flow and turbulence. Resuspension of bottom sediment is also induced by wind through wave action. This will be discussed in some detail later. Before that, attention will be paid to the classification of sediment in order to account for differences in settling rate, erodability, and variability in composition.

## Subdivision of Particulate Material in Fractions

For several reasons it is necessary to take into account the composition of the suspended material. These include:

- the wide range in sedimentation rates as a consequence of variations in size and density; (Blom *et al.*, 1992)
- the variation in susceptibility towards erosion of sediments for the same reasons;
- the non homogeneous distribution of phosphate over the different classes in size and composition;
- the dependency of the light extinction coefficient on the size distribution. (Van Duin *et. al.*, 1992)

Various options for the fractionation of the particles can be considered, such as size (sand, silt, clay) or composition (sand, detritus, plankton characterized by chlorophyll A, clay by Al content, etc.). Considering the objectives of the study and the model structure (Eq. 2) we have chosen an operational subdivision: the fractionation is in terms of differences in sedimentation rates. This has the advantage that experimental values can be used directly in the simulations and, from time series of suspended solids, the parameters can be estimated by application of the model. It is not easy to relate sedimentation rates and composition directly because, in natural waters, aggregation of particles occurs frequently. Empirical relationships can be assessed however, and this would be necessary, for instance, when the preference of phosphate for the fine sized fractions must be taken into account to model the selective settling of this material in deeper zones of a lake. Such relationships have been established for Lake Veluwe where these phenomena are being studied (Brinkman and Van

Raaphorst, 1986).

In both studies, the distribution of settling velocities of the solids withdrawn from water samples, from sediment traps, or from the sediment top layer, were measured in settling tubes. In these experiments, a suspension is brought in a cylindric glass tube and, after vigorous stirring, is left to settle. After a specific time, a small sample is taken at a predetermined depth with a pipette. In this sample, no particles are present with settling velocity larger than the sampling depth divided by the settling time. The concentration of the remaining particles is measured as the total dry weight after filtration. These experiments were performed on samples from surface water, sediment traps, and lake bottom sediments. By varying the sampling time and depth in the settling tables, we were able to determine the weight fractional distribution of particle sizes in the samples from various locations.

Theoretically, an infinite thin layer should be sampled from the settling tube with the pipette. Practically, a minimum quantity of suspension is needed for the analysis of total dry weight. In the Lake Veluwe study, a thin layer of about 2 cm was sampled and analyzed for total and ash free dry weight content. In the Lake Marken study, considerably more material was needed because the optical characteristics of the fraction had to be measured as well. Hence, in the Lake Marken study, a thick layer of 15 cm was sampled. Therefore, the fractionation boundaries were less sharp, but, nevertheless, the results turned out to be useful.

The fractionation bounds selected in the two lakes are different. In part this is based on field observations: in Lake Marken small particles prevail. Table 2 shows the mean values obtained as the geometric mean of the bounds for subdivisions used. In both lakes, four fractions are considered.

This settling tube method is more or less subject to variances caused by flocculation of particles in tubes, which affects the distribution. This is particularly so for the light fractions that remain long in the test tube. On the other hand, this condition to a certain extent also reflects the situation in the field, where flocculation will occur as well. The role of flocculation is discussed further in this volume by Lick and others. As a precaution, all experiments with material sampled from the field, e.g., in sediment traps, were performed in lake water, based in part on observations by Lick (1982), who found a strong influence of the composition of the medium on the distribution. Vigorous shaking of the water prior to the settling was used for disaggregation of samples.

The total suspended solids concentration, measured weekly during 8 months in 1988 in Lake Marken, varied between 10 and 200 $g \cdot m^{-3}$, with a mean value of 55 $g \cdot m^{-3}$. The total sedimentation flux, measured with sediment traps, varied between 1.6 and 41 $g \cdot m^{-2} \cdot h^{-1}$, with a mean value of

6 g•m$^{-2}$•h$^{-1}$. In Figure 2, the mean fall velocity distribution measured in Lake Marken of water samples, material collected in sediment traps, and material from the sediment top layer is presented. Most of the material in sediment traps and in the sediment top layer consists of the coarser fraction 3 and 4, whereas in the water samples around 80% of the suspended solids consists of the fine material, fractions 1 and 2.

For Lake Veluwe, where the settling process itself is a key factor in view of the objectives, some typical results obtained with material collected in sediment traps are presented in Figure 3. Such values can be compared directly with results of simulations when the net sedimentation flux for the cell representing the measuring site is integrated over the time that the sediment trap is operating (generally one week).

In the week of November 9-16, the wind speeds were high. The differences among the sampling sites can be attributed to differences in sediment composition, in fetch, and in depth. The location of the sampling sites is shown in Figure 1, the predominant wind direction in this week was from the SW-SSW, hence sampling points H and V1.2 had the highest fetch.

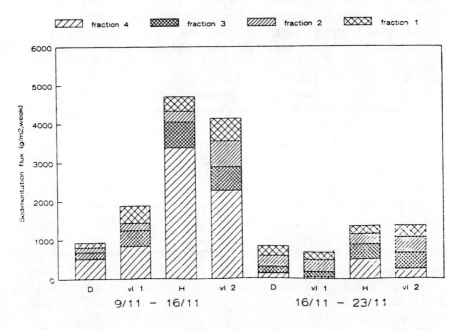

Figure 3. Quantity and distribution of settled material in sediment traps in Lake Veluwe (November, 1987).

Table 2

Frctionation of Suspended Solids According
to Mean Settling Velocity, $w_s$ (m/s)

| Fraction | Lake Marken | Lake Veluwe |
|---|---|---|
| 1 | $1.83 \times 10^{-6}$ | $0.22 \times 10^{-5}$ |
| 2 | $25.5 \times 10^{-6}$ | $0.83 \times 10^{-4}$ |
| 3 | $109.0 \times 10^{-6}$ | $2.78 \times 10^{-4}$ |
| 4 | $412.0 \times 10^{-6}$ | $100 \times 10^{-4}$ |

Further, H lies in a shallow area (0.8 m), at V1.2 the depth is 2 m. The difference between V1.1 and D can be attributed to a difference in fetch.

# Description of Resuspension

## Wave Model

The description of the generation of waves by wind is based upon the wave forecasting curves for shallow waters (Bretsneider and Reid, 1953) as described by CERC (1979). In this approach, deep water forecasting relationships are used to assess the increase in energy due to wind stress. In these equations, wind waves are related to water depth, fetch, and wind speed. An adjusted wind speed according to Bouws (1986) is used here, in which the wind speed, measured at an elevation of 10 meters, is corrected for the wind profile:

$$U_a = 0.537 * U_{10}^{1.23} \qquad (3)$$

where:

$U_a$ : adjusted wind speed (m•s$^{-1)}$)
$U_{10}$ : wind speed at 10 m (m•s$^{-1}$)

The relationships produce significant wave heights and periods. In order to validate the wave model, water level variations were measured in Lake Marken at site Y1111 in 1988 with a wave staff. From the water level variations, the significant wave height and period were computed. In Figure 4a and 4b, the measured wind speed and direction are presented for a period of 12 days in September, 1988. For the same period, the measured and simulated wave height, are compared in Figure 5. The variations in significant wave height are simulated rather well, but the absolute value tends to be systematically lower than the measured one.

In models, resuspension rates are related either directly to wind speed and direction, to the wave characteristics, or to the maximum water velocity near the bottom. The latter can be obtained from (Phillips, 1966).

$$u_b = (\pi H_s / T_s) * \sinh[2\pi H / L_d] \quad (4)$$

where:

$H_s$ : significant wave height (m)
$L_d$ : significant wave length (m)
$T_s$ : significant wave period (s)
$u_b$ : maximal orbital velocity (m•s$^{-1}$)

The shallow water wave length is related to the deep water wave length according to (CERC, 1977):

$$L_d = L * \tanh [2\pi H / L_d] \quad (5)$$

where:

L : deep water wave length (m)

In certain resuspension models the bottom shear stress is used as the forcing function for resuspension. The bottom shear stress can be calculated from:

$$\tau_b : 0.5 * d_w * C_f * U_b^2 \quad (6)$$

where:

$C_f$ : friction factor
$d_w$ : density of water (kg•m$^{-3}$)
$\tau_b$ : bottom shear stress (N•m$^{-2}$)

# Relationships for Resuspension and Sedimentation

Several relationships for $\phi_r$ and $\phi_s$ have been proposed and tested previously see, e.g., Aalderink et al. (1984). All these relationships are based upon assumed complete vertical mixing. The four models tested in this study are presented in Table 3.

Table 3

Models for Resuspension and Sedimentation

| Author(s) | Resuspension Flux $\phi_r$ | | Sedimentation Flux $\phi_s$ |
|---|---|---|---|
| Luettich, 1987 | $\phi_r = w_s \cdot \theta \, (H_s - H_{cr})$ $\phi_s = 0$ | $H_s \geq H_{cr}$ $H_s \leq H_{cr}$ | $\phi_s = w_s \, (C - C_b)$ |
| | H : wave height (cm) $H_{cr}$ : critical wave height $\theta$ : constant | | $w_s$ : sedimentation rate in m/s $C_b$ : background conc. in g/m³ |
| Lam and Jacquet, 1976 | $\phi_r = 0$ for $u_b < u_{b,cr}$ $d_s$ : density sediment $u_{b,cr}$ : critical orbital velocity k : constant | | $\phi_s = w_s \, (C - C_b)$ |
| Sheng and Lick, 1979 | $\phi_r = c_1(\tau_b - \tau_{b,cr1})$ for $\tau_{b,cr1} \leq \tau_b \leq \tau_{b,cr2}$ $\phi_r = c_2(\tau_b - \tau_{b,cr2})$ for $\tau_b \geq \tau_{b,cr2}$ $\phi_r = 0$ for $\tau b < \tau_{b,cr1}$ $\tau_{b,cr}$ : critical shear stress $c_1, c_2$ : constants | | $\phi_s = w_s \, (C - C_b)$ |
| Partheniades and Krone | $\phi_r = N(\tau_b - \tau_{b,cr})/\tau_{b,cr}$ for $\tau_b \geq \tau_{b,cr}$ $\phi_r = 0$ for $\tau_b < \tau_{b,cr}$ $C_f = 0.4 \, (A/k_n)^{1/4}$ $A = (H/2)\sinh(2\pi H/L)$ | | $\phi_s = 0$ for $\tau_b \leq \tau_{b,min}$ $\phi_s = (1 - \tau_b/\tau_{b,min})w_s \cdot 2C$ for $\tau_b \geq \tau_{b,min}$ |
| | A : orbital amplitude (m) $k_n$ : bottom roughness (m) | | $\tau_{b,min}$ = limiting shear for settling settling |

Luettich (1987) related the resuspension directly with wave height, which can be calculated or measured directly. However, this resuspension flux is derived as an approximation of:

$$\phi_r = k(\tau_b - \tau_{b,cr})^N$$

and hence based on shear stresses as in the models of Sheng and Lick and Partheniades and Krone.

In the Sheng and Lick model two trajectories for resuspension are discerned, which provides more opportunity for fine-tuning the fluxes. As the shear stress is proportional to the squared bottom velocity, this model is also more dynamic than the model of Lam and Jacquet, which is linear in this variable. The model of Partheniades and Krone is unique in the sense that no settling will occur over a certain minimum shear stress, which may lead to unrealistic accumulations in the water phase.

These 4 models were tested and compared for their usefulness in describing the dynamics of the suspended solids concentration in the lake water.

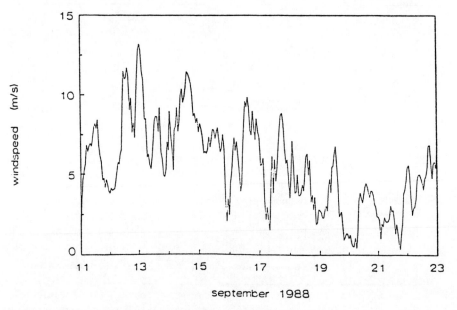

Figure 4a. Wind Speed; Lake Marken, September 1988.

## Parameter Estimation

A preliminary testing for Lake Veluwe involved several simplifications. The mass balance used for calibration of the models is:

$$dC/dt = 1/H \, (\phi_r - \phi_s)$$

which means that dispersion and advection (horizontal transport), primary production, and mineralization have not been taken into account and that the amount of resuspendable material in the top layer is assumed to be unlimited. Primary production was low during the calibration period

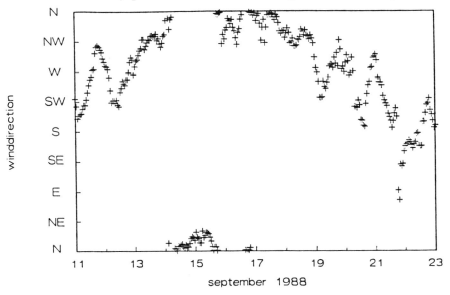

Figure 4b. Wind Direction; Lake Marken, September 1988.

(150hr) and also mineralization was of minor significance. Also, initially in the calibrationno fractionation of the suspended solids has been included. Parameter values were estimated with the methods of least squares. The least square sum was minimized using the Simplex method according to Nelder and Mead (1965). Parameter values obtained with Lake Veluwe data are presented and compared with literature values in Table 4. In Lake Veluwe, the wave characteristics were not measured. The wave model as tested in Lake Marken (see Figure 5) has been used. The errors in the forecasting of the waves will affect the estimates of the parameters in the resuspension models. Although these estimated parameters may be somewhat biased for that reason, it is still possible to compare the efficiency

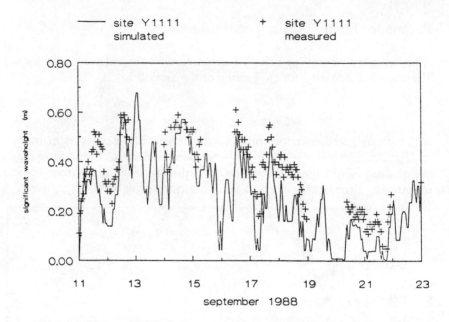

Figure 5. Simulated and measured wave height in Lake Marken.

of the different models, which was the main purpose of this exercise.

Figure 6 shows the measured (at sampling station V1.2) and the simulated data. The data of the first 150 hours have been used for the calibration.

With the parameter set obtained for the sampling site V1.2 the SS concentration at site V1.1 also was calculated. The results are shown in Figure 7, together with the limited number of measured data. Table 5 shows the standard deviation between the model results and the measured SS concentration for both the calibrated and noncalibrated period for both locations.

All models produce a reasonable reconstruction of the SS concentration (Figure 6), including the noncalibrated period ($\geq 150$ hours) and the extension to the other sampling site (Figure 7). The standard deviation was least for the model of Sheng and Lick, 1979 for the calibration period but not for the other period. Table 4 shows that two of the critical shear stresses were 0, which is not realistic. It is due to the fact that not all weather conditions were represented properly in the calibration period. A much longer calibration period is needed to obtain a good estimate of such parameters. The model of Partheniades and Krone produced the highest

Table 4

Parameters for Resuspension and Sedimentation in Lake Veluwe,
Estimated from Field Data Covering the Period of
November 9 to November 16, 1987.

| Model | Estimated Parameters | | | Literature Values | |
|---|---|---|---|---|---|
| Luettich, 1987 | $w_s$ : | $1.8*10^{-5}$ | m/s | $2.2*10^{-4}$ m/s | |
| | $H_{cr}$ : | 10 cm | | 0-16.75 cm | |
| | θ : | 5.5 | | 8.27 | |
| | $C_b$ : | 5 g/m³ | | 15 | |
| | | | (selected) | (Lake Balaton, Luettich, 1987) | |
| Lam and Jacquet, 1976 | $w_s$ : | $1.7*10^{-5}$ | m/s | $2.9*10^{-6}$ m/s | |
| | $u_{b,cr}$ : | $1.0*10^{-4}$ | m/s | 0.03 (for 4-63 μm) | m/s |
| | k : | $8.9*10^{-13}$ | m/s | $6.11*10^{-16}$ | m/s |
| | | | | (Lake Erie) | |
| Sheng and Lick, 1979 | $w_s$ : | $1.8*10^{-5}$ | m/s | $5*10^{-4}$ m/s | |
| | $c_1$ : | 0.183 | s/m | $1.33*10^{-4}$ s/m | |
| | $c_2$ : | 0.012 | s/m | $4.12*10^{-4}$ s/m | |
| | $\tau_{b,cr1}$ : | 0 | N/m² | 0.05 N/m² | |
| | $\tau_{b,cr2}$ : | 0 | N/m² | 0.015 N/m² | |
| | $\tau_{b,cr3}$ : | 0.07 | N/m² | 0.2 N/m² | |
| | | | | (Lake Erie + laboratory experiments, Sheng & Lick, 1979) | |
| Partheniades and Krone | $w_s$ : | $3.7*10^{-5}$ | m/s | | |
| | $\tau_{b,min}$ : | 0.02 | N/m² | | |
| | $\tau_c$ : | 0.01 | N/m² | | |
| | N : | $4.9*10^{-5}$ | g•m/m³•s | | |

standard deviation in the calibration period.

The settling velocity $w_s$ yielded approximately the same value in all models except that of Partheniades and Krone. The composition and fractional distribution during this period was not constant, so this estimate is very rough. The average value from the parameter estimation lies still somewhat below the settling velocity of the finest fraction as defined in Table 2.

Most of the resuspension parameters deviate considerably from the values in the literature, which is an indication of their site specific and empirical nature. From another study in the same Lake Veluwe, but at a different location (Aalderink et al., 1984), a proportionality constant k in the Lam and Jacquet, model was found, which was 40 times lower. We contribute this to the differences in local sediment composition: silt and clay at the present sampling site, sand in the former study. This introduces the question of how to deal with parameter estimation in nonhomogeneous systems where horizontal transport may be important.

## Role of Horizontal Transport

When horizontal transport cannot be neglected, Eq. 8 is not appropriate for the parameter estimation, and, instead, the full mass balance should be used, including two-dimensional transport and taking into account the

Table 5

Standard Deviation Between Simulated and
Measured SS Concentration (mg/L)

| Model | Calibrated Period V1.2 | Noncalibrated Period V1.1 | V1.2 |
|---|---|---|---|
| Luettich | 17.0 | 19.4 | 28.7 |
| Lam and Jacquet | 17.6 | 21.6 | 31.3 |
| Sheng and Lick | 13.8 | 45.9 | 27.3 |
| Partheniades & Krone | 20.3 | 22.5 | 30.6 |

variation in depth and bottom properties. This is very complex and not amenable to any formal parameter estimation method. More field data, either from direct measurement of resuspension parameters for well defined sediments or from parameter estimations in systems homogeneous with respect to bottom composition and water depth, are needed for reliable extrapolations to other systems.

An impression of the role of horizontal transport could be inferred from a storm period with the wind rising in a direction along which samples were being taken during that period. Assuming that dispersion is small compared to advection, the mass balance at any location would be:

$$dC/dt = (\phi_r - \phi_s)/H - U \cdot dC/dx$$

and advection can be neglected if $(U \cdot dC/dx):(dC/dt)$ is much smaller than 1. As an approximation $dC/dt=DC_1/T_1$ and $T_1$ are the local variation in SS concentration and the time scale for this variation. The term $U \cdot dC/dx$ can be approximated by $U \cdot DC_s/L_x$ with $DC_s$ the spatial variation in SS and $L_x$ the concomitant distance. The calculated water velocity in the x-direction was approximately 3 cm/s during the episode at a wind speed of 10 m/s. The sediment along the distance L of 2000 m was constant in composition, $DC_s$ along the streamline was 50 mg/L and the concentration along this line increased about 110 mg/L in 40 hr. From this, a value of 0.24 for the

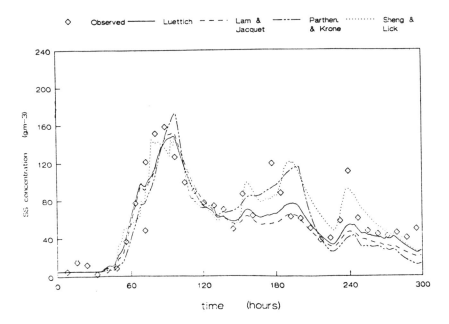

Figure 6. Simulated and measured total SS in Lake Veluwe at V1.2.

criterion quotient can be calculated, hence advection in this case cannot be neglected.

The procedure to be followed further will imply an interaction of parameter estimation based upon Eq. 9 with simulations to assess the effects of horizontal transport and inspection of gradients in time and space obtained from field data. In Lake Marken we expect a somewhat easier task because there is less variation in water depth and sediment texture. Also in this lake there is less interest in horizontal transport itself, and the periods of high wind velocities are less important for the overall primary production than the moderate wind periods.

## Calibration of a Resuspension Model Based on Fractions

The introduction of fractions in the resuspension/sedimentation model implies the extension of the number of model parameters to be estimated. When data on the concentrations for individual fractions are available, these can be used for calibration of the resuspension parameters of each fraction. However, a frequent measurement of the fractional concentrations is difficult for practical reasons. An alternative to tune the model is to

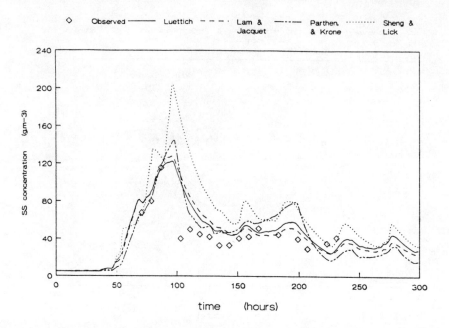

Figure 7. Simulated and measured total SS in Lake Veluwe at V1.1.

Table 6

Estimated Resuspension Parameters of the Sediment Fractions Used in the Lake Veluwe Study

| Fraction | 1 | | 2 | | 3 | | 4 |
|---|---|---|---|---|---|---|---|
| K (kg•m$^{-3}$) | 9.4 | $10^{-6}$ | 7.2 | $10^{-6}$ | 5.2 | $10^{-6}$ | --- |
| $u_{b,cr}$ (m•s$^{-1}$) | 0.0 | | 7.1 | $10^{-4}$ | 4.8 | $10^{-3}$ | --- |

relate the chemical composition of the suspended solids with the fractional compositions; Brinkman and van Raaphorst (1986) have presented data for Lake Veluwe indicating the feasibility of this approach. We used their calibration results, especially the ratio of the resuspension parameters in the model, taking into account fraction, 1, 2, and 3 (Table 6). The concentration of fraction 4 in the water compartment could be neglected.

As shown before, the Lam and Jacquet(1976) relationship resulted in a reasonable reconstruction of the total suspended solids concentration. The results of Brinkman and Van Raaphorst(1986), however, indicated that the critical orbital velocity for the lightest fraction was zero. Because the critical orbital velocity is also a term in the denominator in the Lam and Jacquet formula (Table 3), the application of this relationship is not possible. Hence for calibration and simulation a modified relationship for the resuspension was used:

$$\phi_r = K(u_b - u_{b,cr}) \qquad (10)$$

In Table 6 the estimated parameters for the fractions are summarized. Figure 8 shows the results of a simulation with the model for these three fractions.

The results show, compared to a simulation with the Lam and Jacquet(1976) relationship and only one fraction (Figure 6), a slight improvement of the simulation results in the second half of the simulation period (150-300 hours). In the first half of the period, however, the SS concentrations are somewhat overestimated. The most important advantage of subdivision of the particulate material into fractions is that the model describes the resuspension and sedimentation of individual fractions, with different chemical characteristics and phosphate contents (Table 7).

## Composition of the Sediment Top Layer

When internal transport is important, resuspension/erosion and settling will gradually lead to a change in the local composition of the sediment top layer. In turn the composition of the sediment top layer will influence the resuspension/erosion of the bottom sediments. For instance, locally, the amount of resuspendable material in the top layer may become depleted and therefore limit the resuspension flux. In order to account for these changes, a sediment mass balance is required.

In principle, a mass balance for the sediment top layer includes sedimentation, resuspension, consolidation, bioturbation and decay of organic material. Figure 9 is a schematic representation of the most important processes controlling the sediment top layer composition.

A characteristic value for sedimenation and erosion in Lake Veluwe and Lake Marken is about $4.10^{-6}$ kg/m²•s, which is equivalent to about 0.2 mm per day. Hence, at locations where erosion or sedimentation occurs with the exclusion of the other process, within a few weeks the composition of the top layer may have changed considerably.

The mixing by bioturbation can be described in terms of dispersion with a dispersion coefficient, depending on the number of organisms in the sediments and their temperature dependent activiy, and lies in the range of $10^{-10}$ to $10^{-12}$ m$^2$/s (Lake Veluwe, Brinkman and Van Raaphorst, 1986). A characteristic mixing time is 2 $(L_b)^2/2 \bullet D_b$ with $L_b$ the mixed layer of the sediment - about 5 cm - and $D_b$ the dispersion coefficient. This results in a mixing time of about 100 days or longer. Such differences in time scales should be considered when selecting a modeling strategy for the sediment.

Depending on the research objectives, system characteristics, etc., a more or less complex sediment model should be used. A relatively simple approach will be sufficient in simulations where only limited changes in sediment composition will occur. This is, for instance, the case in short term (a few hours) simulations and for longer periods when bioturbation predominates over the net resuspension/sedimentation. In these cases the sediment can be modeled as one single layer of infinite depth and constant composition.

When changes in the sediment composition cannot be neglected, a more complex sediment model has to be incorporated. At least a mass balance for the sediment top layer should be included. When bioturbation is relatively important with respect to the net resuspension/erosion a single completely mixed layer of a few (for instance 5) centimeters will be appropriate.

In many cases, however, a sediment top layer with distinct properties and a thickness of only a few millimeters can be discerned. The top layer contains a limited amount of material and has been observed visually in Lake Marken and other lakes as well.

In Luettich's work, for instance, a modeling concept with a surficial layer on top of the more consolidated sediment is presented. Such a concept can explain the leveling off of the increase in SS concentration after an initially high erosion rate during prolonged periods of strong and more or less constant winds. Also hysteresis under variable wind conditions can be explained in this way.

The two-layer sediment model developed by the authors is based on this concept. This model implies two sediment layers, of which the top layer is relatively thin (Figure 9). The mass balance for the top layer includes terms for the sediment-water interactions: resuspension, erosion, sedimentation and for interactions between the two sediment layers: consolidation, bioturbation.

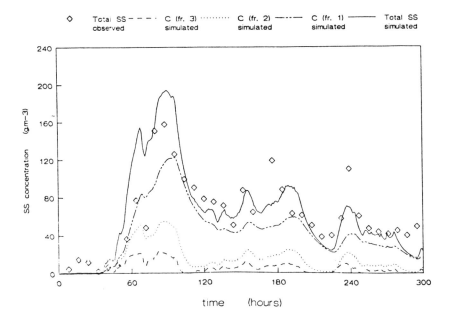

Figure 8. Simulation of the SS concentration subdivided into three fractions (Lake Veluwe, Nov. 9-16, 1987, station v1.2.

## Net Accumulation and Erosion Areas in Lake Veluwe

An illustration of the application of a sediment mass balance is presented in Figure 10. The model STRESS-2d was applied to Lake Veluwe in order to simulate the effects of sediment transport on net accumulation (or erosion). Four sediment fractions were taken into account (Table 2) and one single well mixed top layer of - initially - 5 cm. In the simulation, the modified Lam and Jacquet relationship (Eq. 10) and estimated parameters for resuspension (Table 6) were used. The grid size used in the simulations was 200 by 200 meters. The time step was 10 minutes.

The net accumulation over the period of one year (1987) was simulated. Wind speed and direction were averaged over 1 hour. The initial sediment composition was taken from Brinkman and Van Raaphorst (1986). The water content of the sediment toplayer was assumed to be 90 (volume %).

Areas of net erosion shown in Figure 10 are located in the shallow parts of the lake, especially in the southwestern region. Net accumulation occurs in relatively deep parts of the lake, especially in channels and pits.

The location of these areas with net accumulation is confirmed by the experiences with dredging.

## Light Extinction in Lake Marken

Due to absorption and scattering of the solar flux by dissolved and suspended substances, the downward irradiance, $E_d$, diminishes with depth, in an approximately exponential way described by Lambert-Beer:

$$\ln E_d(z) = -K_d z + \ln E_d(0) \qquad (11)$$

where:

$E_d(z)$ : downward irradiance at depth z (W•m$^{-2}$)
$E_d(0)$ : idem, just below the surface (W•m$^{-2}$)
$K_d$ : vertical attenuation coefficient for downward irradiance (m$^{-1}$)
z : depth (m)

$K_d$ can be computed if $E_d$ is measured at different depths with a submersible irradiance sensor. In this study, the underwater irradiance was measured with a Licor underwater quantum sensor, measuring the full 400-700 nm waveband, which covers the spectrum of Photosynthetically

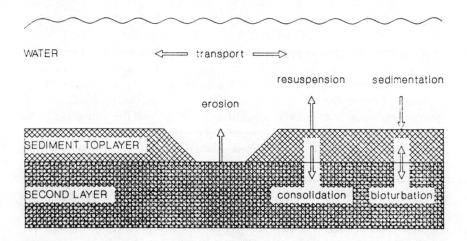

Figure 9. Schematic representation of transport processes in the sediment top layer.

Availble Radiation (PAR).

According to Kirk (1963), in any medium containing a number of absorbing and scattering components, the overall attenuation coefficient is a linear combination of the contributions of the individual components:

$$K_d = K_{d,1} + \ldots + K_{d,i} \qquad (12)$$

where:

$K_{d,1}$, $K_{d,2}$, and $K_{d,i}$ : the partial vertical attenuation coefficients corresponding to the different components ($m^{-1}$)

Table 7

Characteristics of the Sediment Fractions in Lake Veluwe

|  | 1981-1982[1] | | Autumn 1987[2] | |
| --- | --- | --- | --- | --- |
|  | % Org. Matter | P (g/kg) | % Org. Matter | P (g/kg) |
| Fraction 1 | - | - | 41 | 1.6 |
| Fraction 2 | 20 | 1.4 | 17 | 1.3 |
| Fraction 3 | 6 | 0.6 | 14 | 0.7 |
| Fraction 4 | <1 | 0.08 | - | - |

[1]Data Brinkman and Van Raaphorst (1986)
[2]Present Study

Generally, the concentration of dissolved substances has a major effect on the attenuation coefficient in fresh water system. The concentration of these substances is rather constant in the lakes considered. Hence, the variation in suspended solids concentration is the major factor determining the variation in the vertical attenuation coefficient. If the contribution per unit of suspended solids on the attenuation coefficient can be considered constant, the vertical attenuation coefficient can be written as:

$$K_d = K_{d,b} + k_{d,SS} * C \qquad (13)$$

where:

$K_{d,b}$: background attenuation of water and the dissolved substances ($m^{-1}$)

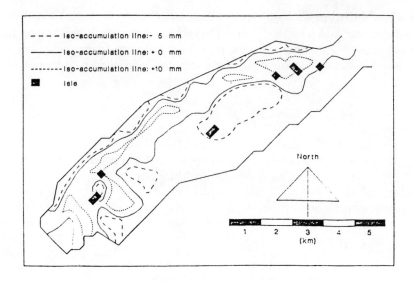

Figure 10. Simulated net accumulation of sediment depth in Lake Veluwe during 1987. Iso-accumulation lines at -5, 0, and 10 mm.

$k_{d,ss}$: specific vertical attenuation coefficient for suspended solids ($m^2 \cdot g^{-1}$)

For both studies, these parameters were obtained by linear regression on field data of $K_d$ and C. The results are presented in Figure 11. The regression equations are:

$$\text{Lake Marken} : K_d = 1.21 + 0.053*C \quad n=115 \quad r=0.91 \qquad (14)$$

$$\text{Lake Veluwe} : K_d = 1.23 + 0.021*C \quad n=23 \quad r=0.85 \qquad (15)$$

The steeper slope for Lark Marken can be attributed to the predominance of finely grained material in this lake. Because the contribution to the light attenuation is not proportional to the volume ($r_{ss}^3$) of the particles, but to the cross-sectional area ($r_{ss}^2$), it is important to distinguish the contribution of the individual fractions. For Lake Marken, the specific vertical attenuation coefficient of each fraction was obtained by measuring the attenuation of the samples from the settling experiments with a spectrophotometer and correcting them for the difference with measurements in the field. This difference is due to the fact that in a spectrophotometer only light scattered

# Sediment Transport in Shallow Lakes

at a very small angle contributes to the signal, but a very high correlation between field and laboratory data was established (r=0.94). The resulting equation is:

$$K_d = 0.64 + 0.070*C_{SS1} + 0.060*C_{SS2} + 0.045*C_{SS3} \\ + 0.037*C_{SS4} \quad n=46, \quad r=0.91 \quad (16)$$

in which:

$C_{SS1}$ ~-~ $C_{SS4}$ : concentration of fraction 1-4 (g•m$^{-3}$)

Equation 16 corroborates the theory that predicts that smaller particles have a greater contribution per weight unit material. The light attenuation as described by Eq. 16 can now be simulated using the developed routine CLEAR; Combined Light Energy Attenuation Routine, in combination with the STRESS-2d model. The attenuation in Lake Marken has been simulated for a 12-day period in 1988 and compared with field data; see Figure 12. From the results it is clear that the combination of the sediment transport model for four fractions and the equation for overall light attenuation provides a powerful tool for the calculation of the light climate. This can be used in an ecological model describing phytoplankton growth.

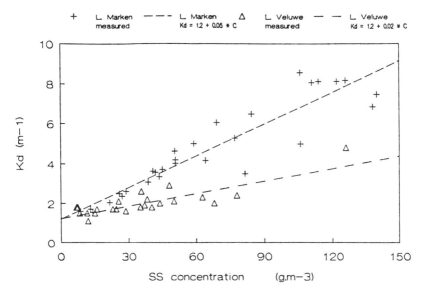

Figure 11. Relationship between extinction and SS concentration in Lake Marken and Lake Veluwe.

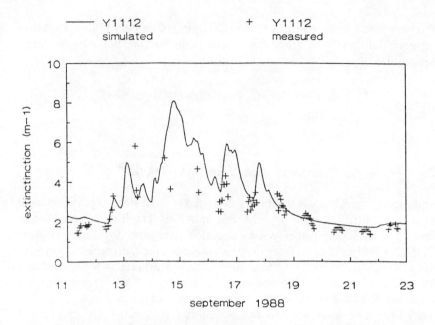

Figure 12. Measured and simulated extinction coefficients in Lake Marken (Sept. 1988).

## Conclusions

The development and testing of a model for resuspension, sedimentation, and transport of solids in shallow lakes has been described. It can be concluded that the theoretical basis and experimental calibration and validation have reached the point where this instrument becomes applicable as a water quality management tool. In Lake Marken the prediction of the light conditions is already adequate for scenario studies including algal growth. Also, in Lake Veluwe, suspended solids concentrations have been simulated with a reasonable precision. Hence the internal transport and location of areas of sediment erosion and sediment accumulation probably can be identified as well.

However, a more extensive testing of individual and combined processes is still desirable for these water systems with a rather variable spatial distribution of water depth and bottom composition. Especially the parameters characterizing the resuspension rates for spatially and temporally variable composed sediments require much attention and extensive data sets for calibration and valildation. A straight forward, reasonably simple, *in situ* field testing method to assess resuspension rates as a function of shear stress would provide a powerful tool.

# References

Aalderink, R.H., L. Likjlema, J. Breukelman, W. van Raaphorst, and A.G. Brinkman. 1984. Quantification of wind induced resuspension in a shallow lake. *Water Sci. Tech.*, Vol. 17, pp. 903-914.

Ambrose, R.B., T.A. Wool, J.P. Connolly, and R.W. Schanz. 1988. WASP4A, Hydrodynamic and water quality model, model theory, user's manual and programmer's guide. U.S. Environmental Protection Agency, Report number 600/3-87/039, Environmental Research Laboratory, Athens, Georgia.

Blom, G., E.H.S. VanDuin, R.H. Aalderink, L. Lyhlema, and C. Toet, 1992. Modelling sediment transport in shallow lakes - interactions between sediment transport and sediment composition. In: *Sediment/Water Interactions*. B.T. Hart and P.G. Sly (Eds.). Hydrobiologia 235/236 :153-166.

Bouws, E. 1986. Verwachting van zeegang door middel groeicurves:; bevindingen verekregen aan de hand van de Markermeer dataset. Koninklijk Nederlands Meteorologisch Institut, Memo-00-86-33, De Bilt (The Netherlands) (in Dutch).

Bretsneider, C.L. and R.O. Reid. 1953. Change in wave height due to bottom friction, percolation and refraction. 34th Annual Meeting of the American Geophysical Union.

Brinkman, A.G. and W. van Raaphorst. 1986. De fosfaathuishouding in het Veluewmeer. Thesis, Techn. Univ. of Twente, Enschede (The Netherlands) (in Dutch).

CERC. 1977. Shore protection manual. United States Army Coastal Engineering Center, Washington, D.C.

CERC. 1984. Shore protection manual. United States Army Coastal Engineering Center, Washington, D.C.

DiGiano, F.A., L. Lijklema, and G. Van Straten. 1978. Wind induced dispersion and algal growth in shallow lakes. *Ecological Modeling*, 4:237-252.

Kamphuis, J.W. 1975. Friction factor under oscillatory waves. *J. Waterways and Harbours Coastal Eng. Div.*, ASCE, nr. 101, pp. 135-144.

Kirk, J.T.O. *Light and photosynthesis in aquatic ecosystems*. 1963. Cambridge University Press, Cambridge.

Lam, D.C.L. and J.H. Jacquet. 1976. Computations of physical transport and regeneration of phosporus in Lake Erie. *J. Fish. Res. Board Can.*, nr. 33, pp. 550-563.

Lick, W. 1982. Entrainment, deposition, and transport of fine-graind sediments in lakes. *Hydrobiologia*, 91:31-40.

Luettich, R.A. 1987. Sediment resuspension in a shallow lake. Thesis, Massachusetts Institute of Technology, Cambridge, MA.

Nelder, J.A. and R. Mead. 1965. A simplex method for function minimization. *The Computer J*, Vol. 7, January.

Partheniades, E. 1986. The present state of knowledge and needs for further research on cohesive sediment dynamics. Third International Symposium on River Sedimentation, University of Mississippi.

Phillips, O.M. 1966. *The Dynamics of the Upper Ocean*. Cambridge University Press, Cambridge.

Sheng, Y.P. and W. Lick. 1979. The transport and resuspension of sediments in a shallow lake. *J. of Geophysi. Res.*, Vol. 84, No. C4, pp. 1809-1826.

Stelling, G.S. 1984. On the construction of computational methods for shallow water flow problems. *Rijkswaterstaat*, Rijkswaterstaat communications, nr. 35, The Hague (The Netherlands).

Van Duin, E.H.S., G. Blom, L. Lighlema, and M.J.M.Schollen, 1992. Aspects of modelling sediment transport and light attenuation in Lake Marken. In: *Sediment Water Interaction*. B. T. Hart and P. G. Sly (Eds.). *Hydrobiologia* 235/236 167-176.

# CHAPTER 14

# Parameterizing Models for Contaminated Sediment Transport

James S. Bonner
Andrew N. Ernest
Robin L. Autenrieth
Sharon L. Ducharme

## Introduction

Sediment transport in aquatic systems is governed by particle characteristics and hydrodynamics. Sediments may be introduced to a waterway via resuspension of sediment material caused by dredging operations, scouring events, erosion, or other natural processes. In addition, disposal operations of sludges, dredge material, and contaminated soils result in suspended particulate material. Regardless of the method of introduction, the transport of the particulate material will be determined by the characteristics of the waterbody and the nature of the material. Further, contaminants associated with the particulate material will sorb, desorb, or react in suspension depending on their chemical characteristics, the nature of the particulate material, and the physical - chemical characteristics of the surrounding medium, including the hydrodynamics.

Vertical transport of particulate material depends primarily on the particle density and size and the dispersive characteristics of the aquatic system. Transport mechanisms in natural systems may take the form of: fluid convection, Stokes settling, Brownian or eddy diffusion, perikinetic and orthokinetic flocculation, and differential settling (Russel, 1981; Dhamotharan et al., 1981; O'Melia, 1972; Fair et al., 1968). It has been shown (Lick, 1982) that vertical convection is only significant in large water bodies where strong upwelling and downwelling occur, (e.g., in nearshore regions) while settling and diffusion are ubiquitous phenomena. Small particles can be hydrodynamically dispersed and advected throughout the system. Conversely, large particles and small flocculating particles form a predominantly settlable material. Many studies concerning particle transport in marine environments indicate that a significant fraction of particulate material will settle via discrete or flocculent transport mechanisms (Faisst, 1976, Hunt and Pandya, 1984, Morel and Schiff, 1980). However, it is uncertain what percentage of this material will penetrate the thermocline to settle on the sediments. It is therefore necessary to make numerous assumptions when analyzing transport mechanisms. For example, with stable particle suspensions, the particles are treated as non interactive, thereby eliminating flocculation and differential settling as significant transport mechanisms. Stable particle suspensions occur when interparticle repulsive energies become sufficient enough to eliminate effective particle-particle contacts. The interparticle repulsive energies are large in a system of homogeneous particles that have high charge densities (i.e., large double layer thickness) (Shaw, 1969). Under these conditions the particles do not aggregate; therefore, there is no increased vertical transport due to changes in particle size. From a hydrodynamic point of view, a non interactive particle suspension is described in terms of dynamic fluid interactions. Under the conditions of low particle concentration, the particle is vertically transported in a free fall trajectory (i.e., no interaction across flow lines). This is termed non hindered transport and involves discrete particle settling. Particulate systems classified as destabilized colloidal suspensions or as non discrete suspensions occur when there are particle-particle interactions. It is the purpose of this study to analyze physical processes in a way so as to estimate parameters that control particle transport.

A model incorporating particle transport, contaminant sorption/desorption, and other pertinent processes can be used to evaluate contaminant fate. Using a calibrated model, data acquired from monitoring activities can be synthesized into the modeling framework to assess the potential impact of suspended contaminated particles. Determination of the controlling transport parameters coupled with hydrodynamic information are required to evaluate the rate at which the material will settle and the extent of transport from the point of initial suspension. Once the contaminant constituents have been identified and quantified, an evaluation of the toxicity can be performed. This type of model should be capable of predicting the movement of particulate material with associated contaminants and therefore could be used as a tool to assess water quality management strategies.

The development, calibration, and field validation of a model describing the particle mediated transport of contaminated material is the ultimate objective of our research. This model would consider hydrodynamic transport, convective transport of heterogeneous particles, dynamic size distribution involving particle flocculation and dynamic sorption-desorption of contaminants. This paper provides a narrower focus with the following specific objectives:

- Characterize particulate waste in terms of parameters affecting fate and transport of contaminants.

- Develop submodels that can be applied at a laboratory scale to characterize contaminant transport processes.

- Estimate model parameters by using an optimized calibration strategy that minimizes the difference between experimental data and mechanistic mathematical expressions.

- Test the model and parameter estimation strategy by comparing calibrated model predictions with observed data.

This paper describes the design, testing, and application of particle transport experiments coupled to a parameter estimation strategy used to calibrate laboratory scale research models describing particle transport. The model development section describes the model conceptualization and formulation. It involved a phased approach. The first phase focused on the development and calibration of a settling column dispersion model. This model was used to characterize the hydrodynamics in the experimental settling column. The second level concerned a model that represents a discrete, non interactive, homogeneous particle suspension. The results presented, however, are also concerned with the application of parameter estimation involving non linear optimization for calibration of homogeneous particle transport models. Of course, this approach involved various simplifying assumptions. These assumptions and other particulars are discussed below.

# Approach

## Experimental

### Particle Characteristics

The experimental approach involved the characterization of the particulate material and the measurement of its transport characteristics. The particulate material was characterized in terms of parameters that affected its transport. These included particle specific gravity, size distribution, and particle flocculation potential. The particle transport characteristics were determined by measuring vertical transport rates in a mixed settling column as a function of hydrodynamic power dissipation, random velocity gradient, vertical dispersion, and ionic strength.

## Particle Specific Gravity

Particle density was measured using three methods. For low density particles ranging between 1.0 to 1.2 g/ml (e.g., New York City sewage sludge), calibrated linear density gradients were used as described in (Bonner, 1985). For particles falling in the intermediate density range of 1.2 to 1.8 g/ml (e.g., New Bedford Harbor sediments), an electronic densiometer was used based on a modified procedure of Mettler/Paar (Paar, 1984). For high density particles (1.8 to 3.0 g/ml), standard picnometric methods were used.

## Particle Size Distribution

A Coulter Electronic Inc. (Luton, England) Multisizer was used to measure particle size distribution. Standard counting and sizing procedures were used as outlined in Treweek and Morgan (1977). Modifications to the standard procedures involving adjustment of the electrolyte strength (0.6 - 6%) and use of an electrolyte thickener were also used as outlined in Bonner (1985) and Rodgers (1983). The modified procedures were used when characterizing particle flocculation potentials as discussed in Bonner (1985).

## Particle Flocculation Potential

Particle collision efficiency, $\alpha$, here referred to as the flocculation potential, was measured using standard jar test apparatus equipped with a computer interfaced power transducer to determine hydrodynamic power dissipation and the Multisizer to determine the particle number as a function of time. To determine $\alpha$, a first order orthokinetic flocculation rate law was used of the form:

$$\frac{dN}{dt} = kN \tag{1}$$

where:

$$k = \frac{\alpha G \varphi}{\pi} \tag{2}$$

$k$ is the first order rate, t is time, N is the particle number concentration, and $\varphi$ is the particle volume fraction measured using the Multisizer. The parameter G is the root mean square of the velocity gradient represented as:

$$G = \left(\frac{P}{\mu V}\right)^{0.5} \tag{3}$$

where P is the hydrodynamic power dissipation, $\mu$ is fluid viscosity, and V is the flocculator reactor volume. Integration of this equation yields:

$$\ln\frac{N}{N_o} = kt \tag{4}$$

where $N_o$ is the initial particle number concentration. The first order rate k is found by regressing the natural log of the relative number concentration measured with the Multisizer versus time. The dimensionless collision efficiency coefficient $\alpha$ can be solved for directly:

$$\alpha = \frac{\pi k}{G\varphi} \tag{5}$$

## Particle Transport Experiments

An experimental settling column (Figure 1) was constructed as discussed in Bonner *et al.* (1990) using a 2 m PVC pipe with 1 cm wall thickness and a diameter of 38 cm. The dimensions of the column were sufficient to minimize boundary and wall effects. A loading platform was mounted at the upper boundary so that dye or particulate material could be loaded into the column as an impulse function or completely mixed initial condition. Between experiments, the column was flushed through bottom drain ports. These ports were also used to supply air for vigorous mixing used in experiments involving a uniform particle concentration initial condition or to measure completely mixed characteristic concentration at the end of appropriate experiments. The column was mixed during all transport experiments. The mixing impeller shaft passed through the center of the column and was constructed using a 2 m long optical glass shaft with 11 sets of impellers mounted at equal increments along the shaft. Each impeller was constructed using 2.54 cm PVC pipe mounted to the shaft using four-way connectors. Mixing was accomplished by rotating the impeller shaft in a sinusoidal clockwise - counterclockwise motion. The amplitude and frequency of the motion was monitored and controlled via an interfaced computer. This type of mixing minimized fluid convection per unit power dissipation. Hydrodynamic power dissipation was monitored during experiments using a computer interfaced sensor located on the impeller shaft just below the impeller

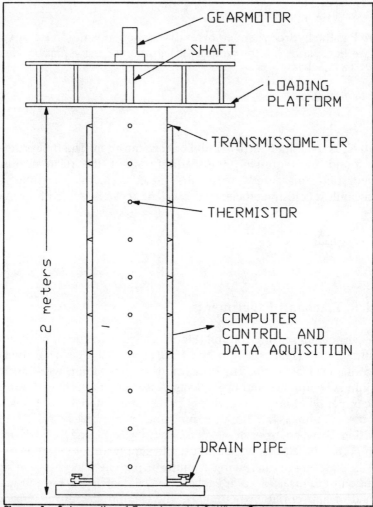
**Figure 1:** Schematic of Experimental Settling Column

drive motor. This sensor detected dynamic shaft position, angular velocity, and torque. Dynamic power dissipation was calculated by:

$$P = \tau \omega \qquad (6)$$

where P is the power dissipation (dyne-cm/sec), $\tau$ is torque (dyne-cm) and $\omega$ is the angular velocity (radians/sec). Ten transmissometers, each consisting of a computer monitored photocell and light emitting diode

(LED) pair, were mounted 180° apart at equal depths on the column. The transmissometers detected the concentration of material passing through the light path in accordance to Beer's law in the linear range. Each sensor was tested for linearity (Bonner et al., 1990) with respect to response per unit concentration for dye or particles.

## Modeling

### Dispersion

The dispersive transport model was developed to simulate one dimensional, time dependent, dispersion in our experimental settling column, assuming reflective boundaries:

$$\frac{\partial C}{\partial t} = D_z \frac{\partial^2 C}{\partial z^2} \tag{7}$$

where C is the mass concentration of the dispersing species (in this case of a conservative dye), t is time, z is the distance in the direction of the transport axis, and $D_z$ is the sum of Brownian and eddy diffusion in the direction of the transport axis. Calibration data were obtained for this model by conducting dye tracer studies at various mixing impeller angular velocities. This approach assumed that lateral variation of dye was negligible (i.e., variation of the dye concentration perpendicular to the transport axis was zero). Therefore, the initial conditions for each model calibration run assumed that all dye was located in an infinitesimal slice perpendicular to the transport axis at the surface of the water column (impulse function). The model state variable was the characteristic concentration (relative concentration) defined as the time and space variant concentration divided by the steady-state concentration.

### Homogeneous Particles

Particulate matter is transported vertically through the water column by a combination of convection, settling, and diffusion (both turbulent and Brownian). This modeling approach assumes that vertical fluid convection is not a significant transport process. Our one-dimensional, time dependent, convective diffusive transport involved two vertical transport terms. The settling term for non interacting monodispersed particles was the mean settling velocity of the material relative to the fluid (i.e., the velocity of the fluid $v_f$ minus the particle fall velocity $v_p$, where $v_f = 0$) multiplied by the concentration gradient. The diffusive transport term was the sum of the vertical eddy diffusivity coefficient and the average Brownian diffusion coefficient multiplied by the gradient of the concentration gradient as in equation (7). The state equation for the first level transport model is expressed as:

$$\frac{\partial C}{\partial t} = D_z \frac{\partial^2 C}{\partial z^2} - v_z \frac{\partial C}{\partial z} \tag{8}$$

where C is again the mass concentration of the transported species (in this case of particles) expressed as a function of depth and time, and $v_z$ is the average Stokes settling velocity. The underlying assumptions that govern the use of state equation (8) include:

- The net flux at the upper boundary is zero.

- The horizontal variation in concentration is unimportant.

- Mixing is accounted for by a spatially and temporally constant eddy diffusivity.

- The Stokes settling velocity ($v_z$) is a function of particle characteristics. There is no hindered or non discrete settling due to hydrodynamic influences.

- The initial condition is that the concentration of particles is uniform in the settling column or the impulse function (i.e., $C/C_o$ = f(z,0) = 1.0 or $\delta(z-z_o)$ = 0, where $z_o$ is the location of the upper boundary).

- The fluid convective transport is insignificant ($v_f = 0$).

- The particle size distribution can be characterized as monodispersed with a representative average particle size.

- The colloidal particles form a stable suspension; there is no flocculation or differential settling.

## Heterogeneous Particles

Most particulate systems are not monodispersed suspensions. Under conditions where the particle size distribution is such that a monodispersed assumption is invalid, a system of equations similar to equation (8) must be derived. This approach assumes that several state variables, one for each representative particle type, can be used to describe the vertical transport of a polydispersed non interactive particle suspension and can be represented by a system of partial differential equations. These equations are expressed as:

$$\frac{\partial C_2}{\partial t} = D_z \frac{\partial^2 C_2}{\partial z^2} - v_{z_2} \frac{\partial C_2}{\partial z}$$

$$\frac{\partial C_1}{\partial t} = D_z \frac{\partial^2 C_1}{\partial z^2} - v_{z_1} \frac{\partial C_1}{\partial z}$$
(9)

$$\vdots \qquad \vdots \qquad \vdots$$

$$\frac{\partial C_n}{\partial t} = D_z \frac{\partial^2 C_n}{\partial z^2} - v_{z_n} \frac{\partial C_n}{\partial z}$$

where $C_i$ is the mass concentration of particles in category i and n is the total number of particle categories. The total particulate mass concentration is equal to the sum of the solutions of the individual state equations at a given location and time.

## Parameter Estimation

Measurement of dynamic particle transport characteristics in a mixed laboratory setting is dependent on transport characteristics of the experimental apparatus. For this reason, the first step of the experimental procedure was to estimate parameters that control transport in the experimental reactor, namely the axial dispersion measured as a function of impeller angular velocity. The approach we used to measure particle transport coefficients involved a phased experimental approach as outlined above. The data generated during these experiments was synthesized using an optimizing parameter estimation routine. This procedure was successfully applied to determine parameters controlling the transport and fate of contaminated particles in aqueous environments.

The objective of the parameter estimation procedure is to determine the optimal values for the controlling transport parameters that minimize the difference between experimental data and the theoretical model. This results in a residual function expressed as:

$$S_r(P_1, P_2, P_3, ..., P_m) = \sum_{i=1}^{n} (C_{pred_i} - C_{obs_i})^2 \qquad (10)$$

where $C_{obsi}$ is the observed concentration of a specific state variable at a particular point i in space and time, $C_{predi}$ is the predicted concentration of the same state variable (model prediction) calculated at the same point in space and time, and n is the total number of observations. The residual function $S_r$ describes the deviation between experiment and theory, and is a function of m parameters $P_1$-$P_m$.

Minimization of this residual function necessarily requires the best estimate of the parameter value. This is accomplished by taking the derivative of the residual function with respect to each model parameter. The parameter (e.g., a particular rate coefficient) value that allows

$$\frac{\partial S_r}{\partial (parameter)} = 0 \qquad (11)$$

is the optimal value assuming that the predictive model is correct. This provides an optimizing statistical basis for the choice of parameter values.

It should be noted that this basic approach can be utilized in multiple parameter extraction from experimental data. The number of parameters to be extracted only increases the number of optimization submodels to be solved. For example, for m estimated parameters, equation (11) becomes a system of equations represented as:

$$\{S_r'\} = \begin{Bmatrix} \frac{\partial(S_r)}{\partial(P_1)} \\ \frac{\partial(S_r)}{\partial(P_2)} \\ \vdots \\ \vdots \\ \frac{\partial(S_r)}{\partial(P_m)} \end{Bmatrix} = \begin{Bmatrix} f_1(P_1, P_2, \ldots, P_m) \\ f_2(P_1, P_2, \ldots, P_m) \\ \\ \\ f_m(P_1, P_2, \ldots, P_m) \end{Bmatrix} = \{0\} \qquad (12)$$

where $P_1$-$P_m$ are the parameters being quantified. This is a system of non linear equations that can be solved for all parameters. To do this, each equation is expanded using Taylor expansion and then truncated after the first derivative terms. An iterative numerical technique can then be derived involving the multiequation Newton-Raphson method expressed as:

$$\{P^{k+1}\} = \{P^k\} + \{S_r'^k\}[J^k]^{-1} \qquad (13)$$

where $\{P^{k+1}\}$ and $\{P^k\}$ are the parameter vectors at the $k^{th+1}$ and $k^{th}$ iteration, respectively. The Jacobian matrix $[J^k]$ represents the matrix of partial derivative terms evaluated at the $k^{th}$ iteration:

$$[J] = \begin{bmatrix} \dfrac{\partial^2 S_r}{\partial P_1^2} & \dfrac{\partial^2 S_r}{\partial P_1 \partial P_2} & \cdots & \dfrac{\partial^2 S_r}{\partial P_1 \partial P_m} \\ \vdots & & & \vdots \\ \vdots & & & \vdots \\ \dfrac{\partial^2 S_r}{\partial P_m \partial P_1} & \cdots & \cdots & \dfrac{\partial^2 S_r}{\partial P_m^2} \end{bmatrix} \qquad (14)$$

Further, explicit analytical predictive solutions of the governing mathematical model (i.e., $C_{predi}$) are unnecessary. The mathematical model can be of any form, analytical or numerical, conducive to optimization of the unknown parameters subject to each data point constraint. This approach will be applied to determine the unknown parameters in equation (9).

## Results and Discussion

### Settling Column Characteristics

The experimentally determined relationship between power dissipation and angular velocity is displayed in Figure 2. The operating range for the settling column ranged from 0 to $2.6 \times 10^6$ dyne-cm/sec for angular velocities from 0 to 20 rpm. Power dissipation increased logarithmically with angular velocity as is shown in the plot. This relationship was strongly dependent on the impeller configuration in the column. This explains the discrepancies with other similar studies such as Lai et al., (1975) because the impeller shaft design used in these experiments was unique to this settling column. With impellers evenly spaced down the column, a more homogeneous system is acquired.

Velocity gradients are an important transport mechanism for orthokinetic flocculation. The mean velocity gradient, or G, affects the settling velocities of particles by changing their collision rate. A strong fluid shear, approximately 4.0 to 30.0 $sec^{-1}$, will promote flocculation resulting in larger and denser flocs. Higher settling velocities coincide with these conditions. Figure 3 shows a linear relationship between velocity gradient and angular velocity. For an angular velocity range of 0 to 20 rpm, the velocity gradient varied from 0 to 35 $sec^{-1}$, respectively. These values concur with the expected velocity gradient values in an estuarine system (Sverdrup et al., 1942).

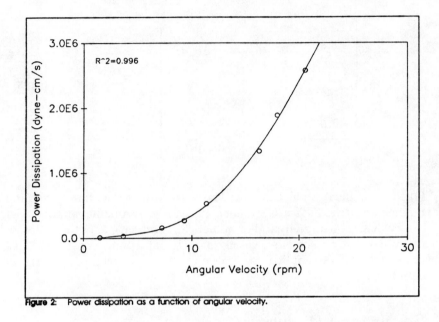

**Figure 2:** Power dissipation as a function of angular velocity.

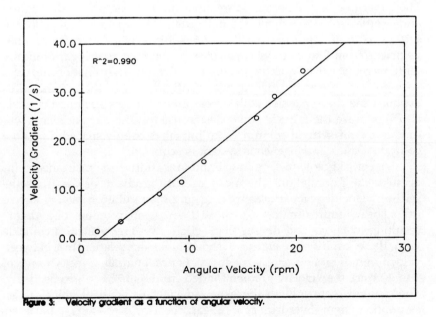

**Figure 3:** Velocity gradient as a function of angular velocity.

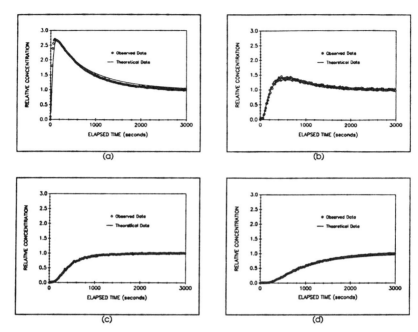

Figure 4: Observed and estimated relative dye concentrations at four characteristic locations in the column. Study run at 20.7 rpm. Representative characteristic distances: (a) 18%, (b) 36%, (c) 54%, and (d) 72%.

The dispersive nature of aquatic environments affects the settling of particulate matter. Small particles may be dispersed and advected throughout a system before they settle. Dye studies conducted with this settling column to obtain dispersion coefficients were designed to minimize the advective component in the system. This design is based on the assumption that the hydrodynamic components are separable.

Figure 4 shows a typical set of results for the dye studies. These plots depict the observed temporal dye concentration profiles at four characteristic locations within the column. Overlaid on these plots are the theoretical curves representing the dispersive transport model predictions using the $D_z$ estimated using the observed data. It is evident that the model predictions accurately reproduce the estimated data. This particular study was run with an impeller angular velocity of 20.7 rpm, corresponding to an estimated dispersion of 4.5 cm$^2$/sec. Over ten such studies were conducted over the entire working range of the impeller angular velocity to characterize completely the hydrodynamics of the column. These results are summarized in Figure 5, which portrays dispersion as a linear function of angular velocity. The range for dispersion developed in the settling column varied from 0 to 6 cm$^2$/sec for the angular velocity range of 0 to 25 rpm. Further, the relationship between dispersion and velocity gradient is shown in Figure 6. The plot is linear, having a correlation coefficient of greater than 0.90.

These results are significant in that they show that we were able both to control and quantify the hydrodynamic characteristics of the column. We were then able to obtain meaningful results concerning particle behavior in natural aquatic systems from the column.

## Model Behavior

Derivation of a dispersive transport model for the hydrodynamic characterization of the column from the basic governing equation (equation 7) was analytical. The homogeneous particle transport model (equation 8) was solved using a one-dimensional, finite difference approximation over space, with Crank Nicholson iteration over time. The model was run with sufficiently high tolerances effectively to negate numerical instability and bias. For low Peclet numbers, indicating a highly dispersive system, the column remains completely mixed over time. This results in a series of successively lower uniform concentration profiles over time. The rate at which the uniform concentration values decrease over time is equal to the advective flux at the bottom boundary.

As advection becomes dominant, mass gradually lost from the system through the lower boundary is replaced through settling from the top of the column. As a result, the column deviates further from a completely mixed condition as time progresses, with higher concentrations occurring lower in the column.

## Parameter Estimation Algorithm Behavior

The parameter estimation algorithm is an iterative numerical procedure. The basis of the method is the iterative solution of systems of non linear algebraic equations using the Newton-Raphson method.

The solution of the governing differential equation, in this case a finite difference derivation, is in itself an approximation. Further, the systems of non linear equations to be solved are derived by differencing the residual function with respect to the parameter in question. With the predicted concentration being obtained via a numerical model, the only way to represent these equations is by numerical differentiation. The heart of the Newton-Raphson method for systems of equations lies in the assembly of the Jacobian of the system of equations. This means that a matrix of the second derivatives of the residual function must be evaluated. It is evident that this method depends on the efficient and accurate evaluation of a series of numerical approximations. This can be

Contaminated Sediment Transport 295

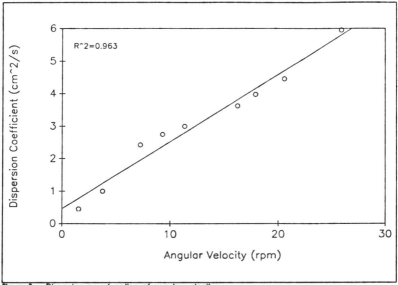

**Figure 5:** Dispersion as a function of angular velocity.

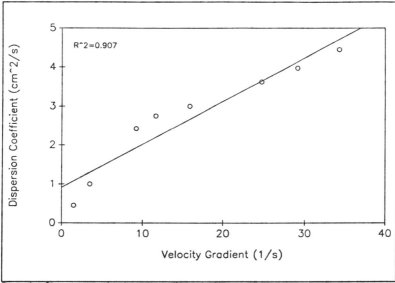

**Figure 6:** Dispersion as a function of velocity gradient.

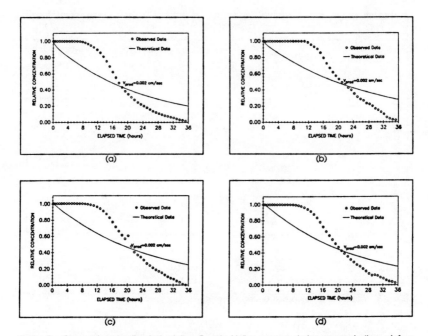

**Figure 7:** Observed and estimated relative Passaic Valley sewage sludge concentrations at four characteristic locations in the column. Study run at 8.2 rpm in freshwater. Representative characteristic distances: (a) 18%, (b) 27%, (c) 54%, and (d) 72%.

ensured by the application of rigorous tolerances and stability criteria, which unfortunately most often results in increased computational effort.

Almost all iterative equation-solving numerical techniques require an initial estimate of the solution to begin computations. The accuracy of this initial guess will likely define the convergence, or non convergence, of the method.

The dynamics of the advective-dispersive transport equation with respect to its two parameters, dispersion coefficient and velocity, differ widely. Over numerous estimation runs, it became apparent that the rate of convergence to the optimum dispersion coefficient is markedly faster than that towards the optimum velocity.

Having mentioned the possible impediments to convergence of the estimation scheme, it should be noted that by minimizing the roundoff error introduced by the numerical approximations, and by starting with a reasonable estimate of all the desired parameters, the technique will converge to a solution.

Preliminary work has been conducted that indicates that numerical

# Contaminated Sediment Transport

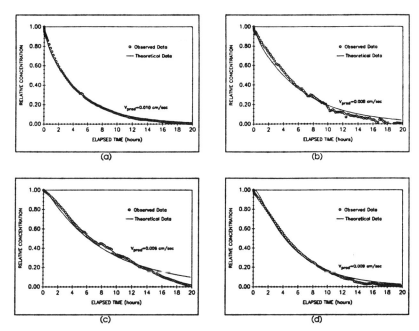

**Figure 8:** Observed and estimated relative Passaic Valley sewage sludge concentrations at four characteristic locations in the column. Study run at 8.2 rpm in saltwater. Representative characteristic distances: (a) 18%, (b) 27%, (c) 54%, and (d) 72%.

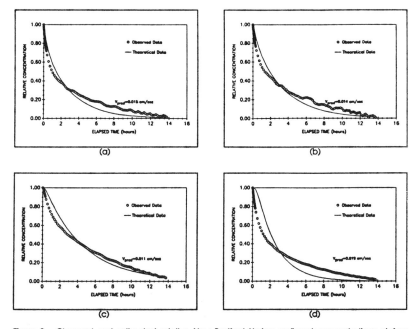

**Figure 9:** Observed and estimated relative New Bedford Harbor sediment concentrations at four characteristic locations in the column. Study run at 7.4 rpm in freshwater. Representative characteristic distances: (a) 18%, (b) 27%, (c) 54%, and (d) 72%.

degree of conformity between the mathematical model and the experimental setup. This means that the greater the amount of noise in the data that the parameter estimation algorithm has to process, the greater the chance of non-convergence. This is being further investigated (Ernest, *et al.* 1989).

## Particle Transport Experiments

Two types of particles were examined in both freshwater and saltwater. The particle types represented both highly flocculent - Passaic Valley sewage sludge, and slightly flocculent material - New Bedford Harbor sediments. Figures 7 and 8 depict the temporal concentration profile at four characteristic depths within the column for Passaic Valley sludge in freshwater and 30 ppt saltwater, respectively. Figures 9 and 10 display the same information for New Bedford Harbor sediments. The fact that these profiles are almost identical at four separate locations for each particle type indicates that the system is dispersion dominant.

Particle characterization of Passaic Valley sewage sludge in freshwater show that the particles do not have any appreciable flocculation potential (Table 1) while having the highest particle size standard deviation of all the samples studied. These characteristics indicate that the suspension, once introduced to the settling column, would be continuously polydisperse throughout the duration of the

Table 1. Summary of particle flocculation potential

| Particle Type | Experimental Conditions | Equilibrium Size (um) | Standard Deviation | Skewness | Flocculation Potential ($\alpha$) |
|---|---|---|---|---|---|
| Sewage Sludge | | | | | |
| Passaic Valley | Freshwater/G = 100 | 21.6 | 6.7 | +1.0 | 0.00 |
| | Saltwater/G = 100 | 21.9 | 4.9 | +4.0 | 0.67 |
| Sediment | | | | | |
| New Bedford Harbor | Freshwater/G = 100 | 20.7 | 2.8 | +0.1 | $6.9 \times 10^{-4}$ |
| | Saltwater/G = 100 | 19.2 | 4.7 | +6.2 | $2.0 \times 10^{-2}$ |

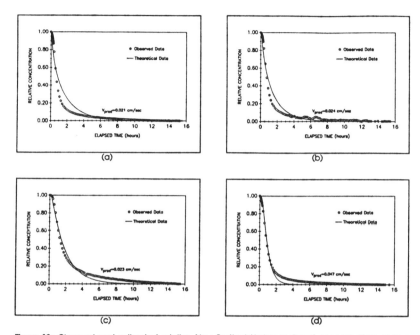

Figure 10: Observed and estimated relative New Bedford Harbor sediment concentrations at four characteristic locations in the column. Study run at 7.4 rpm in saltwater. Representative characteristic distances: (a) 18%, (b) 27%, (c) 54%, and (d) 72%.

experiment. The theoretical lines shown in figure 7 are the prediction of the homogeneous transport model (equation 8) using the optimum dispersion coefficient and settling velocity obtained using the parameter estimation procedure. The high degree of disparity between observation and model predictions indicate the inadequacy of the monodisperse transport model in predicting a polydisperse phenomena.

Studies conducted on Passaic Valley sewage sludge in saltwater show that the particles have a very high flocculation potential (0.67, Table 1) and a particle size standard deviation of 4.9 μm. This indicates that the particles flocculated very rapidly and reached a uniform narrow size distribution quickly, resulting in the monodisperse behaviour depicted in figure 8.

Conversely, New Bedford Harbor sediments, with a lower particle size standard deviation (2.8 μm in freshwater and 4.7 μm in saltwater) but a lower flocculation potential ($6.9 \times 10^{-4}$ in freshwater and $2.0 \times 10^{-2}$ in saltwater), display consistent discrepancies with the monodisperse particle transport model (figures 9 and 10). In the early stages of the study, the theoretical model consistently overpredicted the amount of mass in the column, whereas in the later stages, the model underpredicted. This phenomenon appears to a lesser extent in the studies conducted with New Bedford Harbor sediments in saltwater. The flocculation potential here is significantly higher than the freshwater case. This would indicate that the particles, as in the case of the Passaic Valley sewage sludge, flocculated rapidly in the initial period of the experiment to form a distribution with a small standard size deviation. It is interesting to note that while the standard deviation of the Passaic Valley sewage sludge and the New Bedford Harbor sediments in saltwater were relatively close (4.9 and 4.7 µm, respectively), the flocculation potential of the New Bedford Harbor sediments was over one order of magnitude less than the Passaic Valley sludge ($2.0 \times 10^{-2}$ vs. 0.67). Comparison of Figures 8 and 10 shows that New Bedford Harbor sediments in saltwater behave less like a monodisperse suspension than Passaic Valley sludge under similar conditions. This can be attributed to the fact that the New Bedford Harbor sediments flocculate slower, and so tend toward the monodisperse approximation slower than Passaic Valley sewage sludge.

The deviation of the suspensions from the monodisperse approximation may be as a result of the existence of polydispersity, heterogeneity, flocculation processes, or a combination thereof, within the column. These are unaccounted for in the present application of the theoretical model. It is conceivable, therefore, that this discrepancy may be eliminated by the application of a complex model describing the physical transport of heterogeneous particles. Such a model would account for variations in transport characteristics with respect to organic content, specific gravity, contaminant load, and other parameters affecting the heterogeneity of the system.

This paper represents the preliminary results of an ongoing study. An extension of this work, currently under progress, is the monitoring of the particle size distribution during the experiments. Knowledge of the particle size dynamics will enable more accurate perception of the controlling mechanisms and a more detailed analysis of the observed phenomena. In order to support this experimental work, particle interaction mathematical models are being developed to fully utilize the size distribution data. Coupling between the experimental work and the modeling effort is again accomplished via the parameter estimation framework for the evaluation of the particle interaction parameters. Mathematically, inclusion of these phenomena is accomplished by deriving a system of coupled partial differential equations. The equations are coupled via particle interaction terms ($\theta_k$):

$$\frac{\partial C_1}{\partial t} = D_z \frac{\partial^2 C_1}{\partial z^2} - v_{z_1} \frac{\partial C_1}{\partial z} + \theta_1$$

$$\frac{\partial C_2}{\partial t} = D_z \frac{\partial^2 C_2}{\partial z^2} - v_{z_2} \frac{\partial C_2}{\partial z} + \theta_2 \qquad (15)$$

$$\vdots \qquad \vdots \qquad \vdots$$

$$\frac{\partial C_n}{\partial t} = D_z \frac{\partial^2 C_n}{\partial z^2} - v_{z_n} \frac{\partial C_n}{\partial z} + \theta_n$$

where $\theta_k$ is the time rate of change of mass concentration of particles in category k due to particle interaction. $\theta_k$ is a function of the time rate of change in particle number concentration ($dN_k/dt$) (Smoluchowski, 1917; Friedlander, 1977) presented briefly in figure 11.

Analysis of preliminary data (Tables 1 and 2), resulted in estimated settling velocities that follow the same trend as the measured density and flocculation potential of the samples. Currently under investigation is an

Table 2. Summary of particle transport characteristics

| Particle Type | Experimental Conditions | | Equilibrium Size (um) | Density (g/ml) | Velocity Stokes (cm/s) | Velocity Estimated (cm/s) |
|---|---|---|---|---|---|---|
| Sewage Sludge | | | | | | |
| Passaic Valley | Freshwater = 100 | G | 21.6 | 1.23 | 0.0057 | 0.0022 |
| | Saltwater = 100 | G | 21.9 | | 0.0059 | 0.0068 |
| Sediment | | | | | | |
| New Bedford Harbor | Freshwater = 100 | G | 20.7 | 1.70 | 0.0160 | 0.0124 |
| | Saltwater = 100 | G | 19.2 | | 0.0138 | 0.0276 |

$$\frac{dN_k}{dt} = 0.5\alpha\sum_{i=1}^{k}\sum_{j=1}^{i}\gamma(d_i,d_j)N_iN_j - \alpha N_k\sum_{i=1}^{n}\gamma(d_i,d_j)N_i$$

where

$$\gamma(d_i,d_j) = \beta(d_i,d_j) + \phi(d_i,d_j) + \sigma(d_i,d_j)$$

and

**PERIKINETIC FLOCCULATION**
(Brownian contacts)

$$\beta(d_i,d_j) = \frac{2kT}{3\mu}\left(\frac{1}{d_i}+\frac{1}{d_j}\right)(d_i+d_j)$$

**ORTHOKINETIC FLOCCULATION**
(Fluid shear)

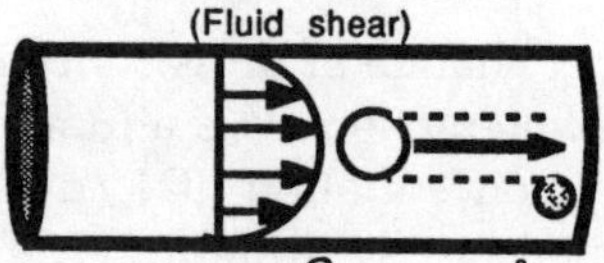

$$\phi(d_i,d_j) = \frac{G}{6}(d_i+d_j)^3$$

**DIFFERENTIAL SETTLING**

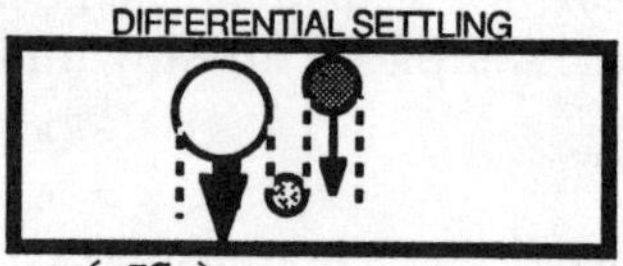

$$\sigma(d_i,d_j) = \left(\frac{\pi g}{72\mu}\right)(\rho_p - \rho_l)(d_i+d_j)^3(d_i-d_j)$$

**Figure 11:** Particle interaction terms.

extension of the existing algorithm used in the transformation and normalization of the raw experimental data. This final step in the testing of the methodology would result in greater efficiency in the actual parameter estimation.

Figure 12 shows the degree of correlation attained between the experimental data and the model predictions using estimated parameters. The degree of normal deviation of the data points from the 45° line indicates the amount of noise in the observed data. Specific trends, such as those showing generally higher predicted concentrations than those observed (Figure 12), are indicative of predictive model inadequacies.

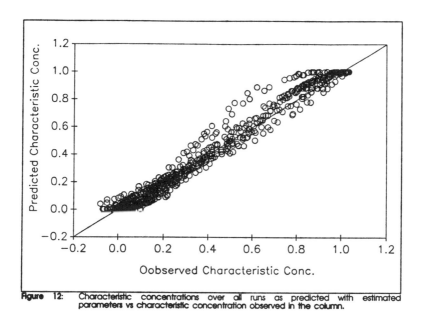

Figure 12: Characteristic concentrations over all runs as predicted with estimated parameters vs characteristic concentration observed in the column.

## Conclusions

Experiments were conducted to characterize particulate waste in terms of size, density, and flocculation potential. In addition, dynamic settling velocity was measured in a mixed laboratory settling column. The data derived from the experiments are synthesized into a modeling framework using a parameter estimation method as a model calibration tool. The following conclusions may be drawn from our study:

- Hydrodynamic characteristics in our experimental settling column can be measured, monitored, and controlled.

- In our experimental column, the random velocity gradient and vertical dispersion showed a linear relationship to impeller angular velocity.

- The parameter estimation procedure was used successfully to determine coefficients necessary to calibrate dispersive and advective-dispersive models.

- In all cases, particulate waste material was destabilized by seawater.

- Vertical transport was dispersion dominated.

- Loss of particles from the water column was directly related to settling velocity and α.

## Acknowledgments

This research was conducted under USEPA Grant #CR814257-01, "Study of Processes Affecting Particle Mediated Transport of Hazardous Materials in Marine Systems." We would like to acknowledge various people for their assistance and contributions throughout this work. Specifically, our project officer, Dr. J.F. Paul, for supporting this research. We would also like thank Stephanie Sanders and Ravindra Malali for their technical assistance. Finally, we thank Dr. Bill Batchelor for his valuable comments and suggestions.

## References

Bonner, J. S., Hunt, C. D., Paul, J. F., and Biermam, V. J. 1986. Prediction of vertical transport of low level radioactive middlesex soil at a deep ocean disposal site. Technical Report EPA 520/1-86-016, U.S. Environmental Protection Agency, Environmental Research Laboratory, Narragansett, Rhode Island.

Bonner, J. S. 1985. The Vertical Transport and Aggregation Tendency of Freshwater Phytoplankton. Ph.D. dissertation, Department of Civil and Environmental Engineering, Clarkson University, Potsdam, New York.

Cornwell, D., and M. Bishop. 1983. Determining velocity gradients in laboratory and full scale systems. *J. of the Amer. Water Works Assoc.* 9:470-475.

Dhamotharan, S., J. Gulliver, and H. Stephan. 1981. Unsteady one-dimensional settling of suspended sediment. *Water Resources Res.* 17(4):1125-1132.

Ducharme, S.L. 1989. Design and Validation of a Settling Column for Particle Transport Studies. M.S. Thesis, Department of Civil Engineering, Texas A&M University, College Station, Texas.

Ernest, A. N., Bonner, J. S., and Autenrieth, R. L. 1991. Model parameter estimation for particle transport. American Society of Civil Engineers, *J. of Environmental Eng.* 117(5): 573-594.

Fair, G., J. Geyer, and D. Okum. 1968. Water and Wastewater Engineering, Vol. 2. John Wiley and Sons, New York.

Faisst, W. 1976. EQL Report 13, Environmental Quality Laboratory, California Institute of Technology, Pasadena, California.

Friedlander, S. 1977. Smoke, Dust and Haze. John Wiley and Sons, New York.

Hunt, J. R., and J. Pandya. 1984. Sewage sludge coagulation and settling in seawater. *Environmental Sci. and Technol.*, 18(2):119-121.

Lai, R., H. Hudson, Jr., and J. Singley. 1975. Velocity gradient calibration of jar-test equipment. *J. Amer. Water Works Assoc.*, 10:553-557.

Lick, W. 1982. Entrainment, deposition and transport of fine grain sediments in lakes. *Hydrobiologia*, 91(1):31-40.

Morel, F. M. and S. Schiff. 1980. Ocean disposal of municipal wastewater: The impact on estuary and coastal waters. Cambridge, Massachussetts. MIT Sea Grant College Program.

O'Melia, C. R. 1972. Coagulation and flocculation. In: *Physico-Chemical Processes for Water Quality Control*. Wiley Interscience, New York.

Paar, A. 1984. DMA External Measuring Cell. Graz, Austria. A 8054.

Rodgers, P. 1983. Model simulation of PCB dynamics in Lake Michigan. In: Mackay, D., Paterson, S., Eisenich, S., and Simmons, M., eds., Physical Behaviour of PCBs in the Great Lakes. International Association of Great Lakes, Michigan. pages 311-328.

Russel, W. 1981. Brownian motion of small particles suspended in liquids. *Ann. Rev. Fluid Mech.*, 13:425-455.

Shaw, D. 1969. *Electrophoresis*. Academic Press. Butterworths & Co., New York.

Smoluchowski, M. 1917. Versuch einer mathematiischer theorie der koagulation kinetik kolloides lasungen. *Z. Physic. Chem.*, 92:128-168.

Sverdrup, H., M. Johnson, and R. Fleming. 1942. *The Oceans*. Prentice Hall, Englewood Cliff, New Jersey.

Treweek, G. and J. Morgan. 1977. Size distribution of flocculation particles: Application of electronic particle counters. *Environ. Sci. Tech.*, 11(7):707-714.

# CHAPTER 15

# The Effects of Sediment Mixing on the Long-Term Behavior of Pollutants in Lakes

David N. Edgington

## Introduction

Over the last 150 years the waters of the Great Lakes have been subjected to many different chemical insults as a result of human activity. The majority of these chemicals have been more or less strongly adsorbed onto particles, and, as a result, the sediments provide a record of the history of the changes in the magnitude of the inputs of pollutants to the lakes (Edgington and Robbins, 1976; Christensen and Chien, 1981). In the last four decades a large number of xenobiotic organic compounds have been synthesized and widely distributed in the environment. The possible toxic effects of these compounds, as well as of increasing loadings of inorganic ions, particularly many metals used in a variety of metallurgical processes, have led to concerns regarding their long-term behavior in the Great Lakes and the efficacy of reductions in input fluxes. In fact, the amelioration of problems related to the persistence of these in-place pollutants in the sediments is now regarded to be of prime importance to meet the water quality objectives of the International Joint Commission (IJC) (1987).

Effective enforcement of the discharge of pollutants in the Great Lakes region will reduce the potential for further degradation of this environment. The question remains, for how long will the effects of pollutants, presently in the water and sediments, persist after effective control of new inputs has been achieved? In spite of the large interest in the long-term problems of pollutant control in the Great Lakes and the measurement of thousands of samples for a wide variety of inorganic and organic pollutants, there are very few data sets available that have been accumulated consistently over a time period sufficiently long to provide an assessment of long-term changes in the concentration of any particular pollutant.

It is fortuitous that the accumulation of one of these data sets, measurements of the long-lived radionuclides ($^{90}$Sr, $^{137}$Cs, and $^{239-240}$Pu), initiated by Argonne National Laboratory in the early 1970s, provides a long-term record of the decrease in concentration of these pollutants in the water, primarily of Lake Michigan, following the almost total cessation of new inputs (see Wahlgren et al., 1980 and references therein; Nelson and Metta, 1983; Orlandini, 1988). At the same time, Robbins and Edgington (1975) initiated a detailed study of the rates of sedimentation in Lake Michigan using one of these radionuclides, $^{137}$Cs, and the naturally occurring radionuclide $^{210}$Pb. They found that these two methods of dating sediments, one based on the identification of two horizons and the other on radioactive decay, gave self-consistent results provided that an allowance was made for the complete mixing of the upper few centimeters of the sediment. Later, Robbins et al. (1977) showed that this mixing was due to the activity of benthic organisms.

In Wahlgren et al. (1980) and N.C.R.P. (1984), it has been shown that it was possible to predict the changes in concentration of both $^{90}$Sr and $^{239-240}$Pu in the waters of the Great Lakes using a simple time-dependent model similar to that described first by Lerman (1972). In this model it was assumed that the transfer of a particle reactive radionuclide, such as plutonium, to the sediment could be expressed in terms of a sediment (particle) residence time. This simple model:

$$dC_n/dt = (1/V_n)[A_n\phi(t) + \alpha_n W_n + Q_{n-1}C_{n-1}] - C_n(1/T_n + 1/T_n' \tag{1}$$

predicts within ±10% the changes in the concentration of the non-particle-reactive radionuclide $^{90}$Sr, $C_n$, with time in each of the Great Lakes between 1970 and 1983 (N.C.R.P., 1984). In Eq. (1), $\phi(t)$ is the input flux from the atmosphere, and, for each lake, n, $\alpha_n W_n$ is the input from the watershed (where $\alpha_n$ is the fractional loss, and $W_n$ is the total pollutant stored), $A_n$ is the surface area, $V_n$ is the volume of water, and $Q_n$ is the flow rate, where the water residence time $T_n = V_n/Q_n$. Since $^{90}$Sr is not strongly bound by sediment particles, $T_n^1$ (= $V_n/k$, where k is a constant), the loss of the pollutant from the water to the sediments ($1/T_n'$) was set to zero.

For these calculations, the source terms, the time variant annual atmospheric flux to the lakes and their watersheds, $\phi(T)$ and $\alpha_n W_n(t)$

[= $\alpha_n A'_n \cdot \Sigma\phi(t)$ where $A'_n$ is the area of the watershed], were estimated using the monthly values of $^{90}$Sr in fallout measured at Argonne National Laboratory (Krey et al., 1974; Nelson, D.M., 1988). (The ratio of $^{239,240}$Pu/$^{90}$Sr was taken to be ≈0.02 (Hardy et al., 1973). The best fit to the experimental data was obtained using a value of $\alpha_n = 0.005$ for erosion of $^{90}$Sr from the watershed (N.C.R.P., 1984). Measurements of the concentrations of plutonium in river waters entering the lake have confirmed that fractional losses of $^{90}$Sr of this magnitude from midwestern watersheds are not unreasonable (Sprugel and Bartelt, 1975).

In order to apply this simple model to predict the behavior of plutonium in the Great Lakes is was necessary to set values for the pollutant/sediment resident times, $T'_n$, that would reproduce the earliest available measured concentration for each lake (Wahlgren et al., 1980). However, it was found that the measured concentrations of $^{239-240}$Pu remaining in the water did not decrease as rapidly as predicted by the model. It was at first assumed that this may have resulted from additional inputs of plutonium from the watershed as was needed to predict the concentrations of $^{90}$Sr correctly. It was shown, however, that even when using the same value of $\alpha_n = 0.005$ as used in the calculations for $^{90}$Sr (a much more mobile radionuclide because of its greater solubility in water), the added inputs to the system were inadequate to maintain the measured concentrations of plutonium in the water. More recent direct measurements of plutonium in river water have shown that a more reasonable value of $\alpha$, based on these rivers entering the Great Lakes is ≈0.00002 (Nelson et al., 1984). Therefore, it must be concluded that, as the influence of new inputs decreased, the effects of desorption from the sediments must become dominant in controlling the concentration in the water.

Independent experiments have shown that the adsorption/desorption of plutonium to sediments is rapid and can be described in terms of the formation of complexes of Pu(IV) with, most probably, hydrated iron oxides at the sediment surface (Edgington et al., 1979; Edgington and Nelson, 1984; Balistrieri and Murray, 1983).

These studies of radionuclides in the water and sediments of the Great Lakes have provided a basic understanding of the time scales of the processes that control sedimentation and the long-term persistence of pollutants as controlled by sediment-water interactions. It is the purpose of this paper to show how it is possible to predict the change in concentration of a particle reactive pollutant such as $^{239-240}$Pu in the water of Lake Michigan as a function of time using a simple time-dependent model similar to that developed by Lerman (1972). This model has been modified by the inclusion of a sediment-water interaction term controlled by the sedimentation rate, the thickness of the mixed sediment layer, and an equilibrium constant, $(K_d)$, for the reaction of the pollutant between water and sediment, where

$$K_d = (\text{conc. X in sediment})/(\text{conc. X in water})$$

and usually is expressed in units of ml·g$^{-1}$ or l·kg$^{-1}$.

## Model Development

In this model, it is assumed that the lake is well mixed both in terms of the concentration of plutonium in the water and on the suspended solids. One of the major surprises that arose from the initial surveys of plutonium in Lake Michigan was that, if the water was sampled before stratification, there was little variation in the measured concentrations both areally and with depth (Wahlgren and Nelson, 1975; Wahlgren et al., 1975; 1976). It was also found that the concentration of plutonium measured on suspended particles was the same as that found in the fine-grained surface sediments (Alberts et al., 1976). Even though the concentration of plutonium decreases to zero in the epilimnion during the summer months due to scavenging by particles, in particular by the *in situ* precipitation of calcite, and settling through the thermocline, the well-mixed state is rapidly restored after the fall turn-over (Wahlgren et al., 1976; 1980). Hence, the assumption of complete mixing is reasonable since the time-scale for the sedimentation process will be shown to be of the order of years to decades.

Clearly, the simple irreversible term involving the pollutant/sediment particle residence time in Eq. (1), $1/T'/_{rv}$ must be modified to permit the transfer of pollutant to and from the sediment. The combination of the processes implied in the steady-state sedimentation rate - rapid mixing model that Robbins and Edgington (1975) constructed to describe the observed profiles of radionuclides (such as $^{210}$Pb, $^{137}$Cs, and $^{239,240}$Pu) with depth in the sediments and the reversible formation of surface complexes as formulated by Balistrieri and Murray (1983) provides a simple parameterization for the sediment interaction term for a mass balance equation and inclusion in the simple time-dependent concentration model. The sediment part of the model is illustrated in Figure 1.

Implicit in this model is the assumption that, over the time scales involved in the processes of sedimentation and the ultimate burial of the pollutant beneath the mixed layer, the pollutant adsorbed onto the sediment in the mixed layer essentially remains in contact, and in equilibrium, with the overlying water as a result of a combination of bioturbation and physical disturbance. The thickness (depth) of this

# Sediment Mixing in Lakes

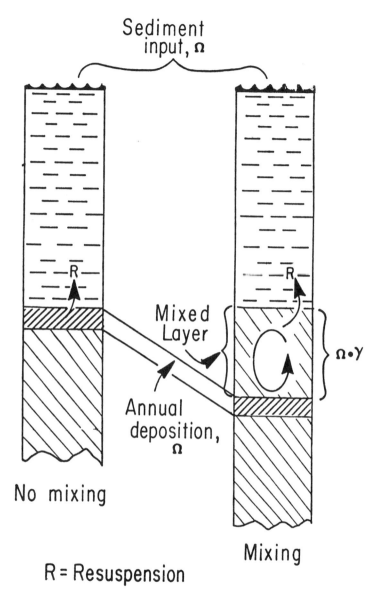

Figure 1. Diagramatic illustration of the sediment interaction term used in the model for no-mixing and mixing. The sediments within the mixed layer are assumed to be in active equilibrium with the overlying water. The historical record of inputs is maintained below this active sediment layer.

mixed layer is determined by the nature of the sediment, in particular its organic carbon content, and the species and densities of benthic organisms present (Robbins, 1982, Robbins et al., 1977 and Fisher et al., 1980). This mixed layer will act as a capacitor or buffer: when $\phi(t)>0$ there is a net transfer of the pollutant into the mixed layer by exchange onto the sediments, and a fraction of the total pollutant stored in the mixed layer will be lost to the undisturbed layer below. When $\phi(t) \approx 0$ and the pollutant flow is out of the lake, $Q_n C_n > \phi(t)$, then a fraction of the pollutant will be lost from the active mixed layer to both the overlying water, by desorption to maintain equilibrium between the sediment and water, and to the undisturbed layer below. The sediment adsorption/desorption process effectively occurs within the water column and is rapid in relation to the time scale of the sedimentation process so that the interaction can be represented by a simple distribution coefficient $[K_d, (ml \cdot g^{-1})]$.

The desorption from the sediments in the mixed layer will maintain an elevated concentration in the overlying water for a period of time, dependent, to a large degree, on the time constant for the transfer of pollutant from the active mixed layer to the undisturbed layer. If the sedimentation rate is $\Omega$ (g·cm$^{-2}$·yr$^{-1}$) and the mixed layer thickness is S (g·cm$^{-2}$), then the time constant for loss from the sediments, $\gamma$ (yr), is given by $S/\Omega$. Since benthic organisms have been shown to mix sediments to a depth of several centimeters and sedimentation rates are usually of the order of millimeters per year, $\Omega \ll S$, then the value of $\gamma$ is measured in decades (Robbins et al., 1977; Edgington and Robbins, 1975).

If the concentration of particles in the water is $Cp_n$ then the total mass of sediment in contact with the water is

$$A_n \Omega_n (1 + \gamma_n) + V_n Cp_n \tag{2}$$

Since the units for $K_d$ are expressed as $1 \cdot kg^{-1}$ or $ml \cdot g^{-1}$, the total equivalent water volume of sediment is

$$A_n \Omega_n K_d (1 + \gamma_n) + V_n K_d Cp_n . \tag{3}$$

# Sediment Mixing in Lakes

Therefore, constructing a mass balance for the pollutant in lake n

$$\begin{aligned}
dC_n/dt = {} & (1/V_n)(A_n\phi_n + \alpha_n W_n) && \text{(inputs from atmosphere and watershed)} \\
& + Q_{N-1}C_{n-1}(1 + K_d Cp_{n-1}) && \text{(inputs from lake upstream)} \\
& + A_n \Omega_n K_d C_n(1 + \gamma_N) && \text{(storage in sediments)} \\
& - A_n \Omega_n K_d C_n && \text{(loss to historical sediments)} \\
& - Q_n C_n(1 + K_d Cp_n) && \text{(loss to next lake downstream)}
\end{aligned} \quad (4)$$

Equation (4) can be solved using simple numerical integration techniques. Let

$$dC/dt = f(C) \tag{5}$$

where $f(C)$ = right hand side of Eq. (4). Since for the special case of the fallout radionuclides (or any other non natural pollutant), $C=0$ at $t=0$, then

$$(C_{i+1} - C_i)/dt - f(C_i) = 0 \tag{6}$$

where $i = 0, 1, 2 \ldots$ and $dt=1$. In practice for Lake Michigan where there is only outflow, equations (4) and (6) simplify to

$$C_{i+1} = B_{i+1}/[V(1 + K_d Cp) + \Omega K_d(1 + \gamma)] \tag{7}$$

where

$$B_{i+1} = B_i + \phi_{i+1} + \alpha W_{i+1} - C_i[\Omega K_d + Q(1 + K_d Cp)] \tag{8}$$

and $V = V_n/A_n$ and $Q = Q_n/A_n$.

Since Lake Michigan has an average depth of 100 m and a water residence time of 85 years, than V=10 liters and Q=10/85 = 0.118 $1 \cdot yr^{-1}$. The concentration of particles in the water column, $Cp \approx 10^{-3}$ $g \cdot l^{-1}$. For plutonium, the fractional annual loss from the watershed, $\alpha$, has been estimated to be $2 \times 10^{-5}$ (Nelson et al., 1984). The average sedimentation rate, $\Omega$, is assumed to be 0.020 $g \cdot cm^{-2} \cdot y^{-1}$.

The curves presented in Figure 2 demonstrate the effect of varying $K_d$ between 1 and $10^5$ on the relative concentration of pollutant in water as a function of time after an instantaneous input and assuming no mixing below the sediment/water interface. The curves presented in Figure 3 demonstrate the effect of varying the depth of the mixed layer (time constant), $\gamma$, between 0 and 20 years on the relative concentration in water as a function of time for a pollutant that has a $K_d=10^5$. It is clear that the decrease in concentration in the water has a half-life of $\gamma$ years.

In a lake as deep as Lake Michigan, pollutants that have a low to moderate tendency to be adsorbed to sediments, with $K_d<10^3$, behave as if they remain completely in solution, and there would be a decrease of a factor of 2 in concentration in solution (in the absence of radioactive decay) in 85 years, governed by the water residence time. For values of $K_d>2 \times 10^3$ the sediment interaction term becomes more important and there is a dramatic change in behavior when the values of $K_d$ increase to between $10^4$ and $10^5$. At $K_d=10^5$, a common value for many organic and inorganic pollutants, the initial concentration will decrease by a factor of 100 in 25 years.

The effect of the active mixed layer acting as a buffer or capacitor is very clear from Figure 3. For pollutants with a $K_d=10^5$, adsorption to the sediments in the mixed layer results in an initial decrease of a factor of four to five in concentration over the situation where there is no mixing. After 8 to 10 years, however, the situation is reversed, and, after 20 years, the difference in concentration rapidly increases because of the buffering by the mixed layer.

## Application to Lake Michigan

There are very few data sets for pollutants that can match the extensive set of measurements of plutonium and other radionuclides such as $^{137}Cs$ and $^{210}Pb$ that were made by scientists at Argonne National Laboratory, on samples of water and sediments that were collected more or less systematically in Lake Michigan between 1970 and the present (Wahlgren and Nelson, 1975; Wahlgren et al., 1975; 1980; Edgington and Robbins, 1975; 1976; Orlandini, 1988). As such, these data provide the best opportunity to test a model such as this.

In order to apply this model to predict the change in concentration of plutonium in the water of Lake Michigan with time, it is necessary to provide reliable estimates of three parameters: (1) average sedimentation rate, $\Omega$ (g·cm$^{-2}$·y$^{-1}$); (2) average mixing depth, $\gamma$ (yr); and, (3) sediment distribution coefficient, $K_d$.

The calculation of average sedimentation rates and mixing depths for Lake Michigan is made difficult by the highly variable patterns of sedimentation. The early seismic studies by Lineback and Gross (1972) indicated that, in the southern half of the lake, large areas of the bottom of the central and western parts of the basin were essentially bare of

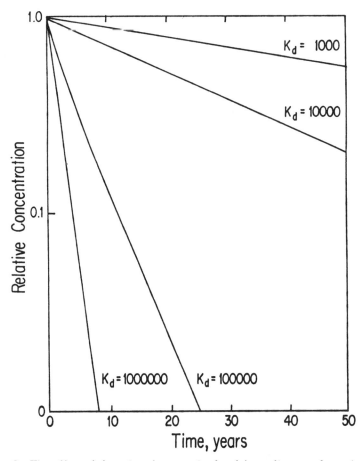

Figure 2a. The effect of changing the magnitude of the sediment adsorption term, $K_D$, in equations (7 and 8) on the variation of the concentration of a pollutant in water with time after a single pulsed input.

sediment and that the largest deposition occurred in a relatively narrow band 20 km off the eastern shore of the lake. The initial studies of Edgington and Robbins (1975) showed that the largest deposition of $^{137}$Cs and $^{239-240}$Pu from weapons testing was coincident with the zones of greatest historical deposition as defined by Lineback and Gross: in other words, the focusing of pollutants was occurring in the same manner as the focusing of sediments in the basin. Quantification of the $^{137}$Cs and $^{239-240}$Pu in the sediment cores indicated that many of the cores examined contained a considerable excess of the radionuclide over that delivered

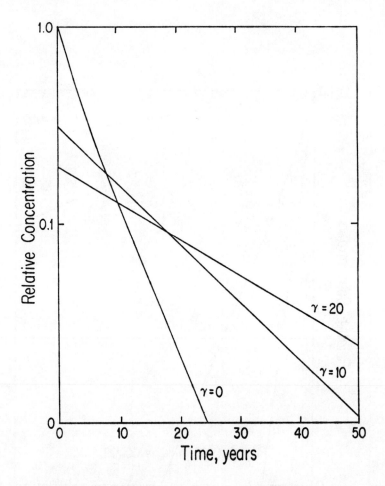

Figure 2b. The effect of changing the depth of the mixing layer, expressed in years of sedimentation, $\gamma$, in equations (7 and 8) on the variation of the concentration in water as a function of time of a pollutant that has a $K_D$ of $10^5$.

Sediment Mixing in Lakes 317

to the surface of the lake (Edgington *et al.*, 1976). From detailed studies of the distribution of $^{137}$Cs and $^{210}$Pb in a series of sediment cores, collected in 1972 and 1982 to be representative of the depositional area, it was possible to calculate values of the sedimentation rate and mixed layer depth for each core (see Robbins and Edgington, 1975; Edgington and Robbins, 1975). A comparison of the data obtained in 1972 and 1982

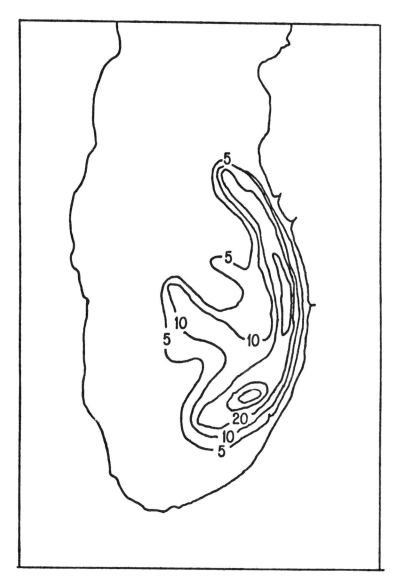

Figure 3a. Areal patterns of the variation in sedimentation rate expressed as mg cm$^{-2}$y$^{-1}$.

indicated that the average sedimentation rate had not changed, thus confirming the steady-state sedimentation assumptions made by Robbins and Edgington (1975). However, it was found that, while the total loading of $^{137}$Cs, corrected for radioactive decay and new inputs, had remained constant, there was a redistribution of this radionuclide from the edges to areas of higher sedimentation rate in the center of the depositional area (Benante, 1984). There was also an apparent increase between 1972 and 1982 in the thickness of the mixed layer that would be expected from the long-term focusing implied by the redistribution of $^{137}$Cs in the sediments (Benante, 1984).

The areal variations in sedimentation rate (mg·cm$^{-2}$·y$^{-1}$) and mixing depth (expressed as mg·cm$^{-2}$) are shown in Figures 4 and 5. In order to calculate lake wide averages, the fractional areas within isopleths shown in Figures 4 and 5 were measured. The relative contributions of the sedimentation, which in each of these isoplethsis equal to the average sedimentation rate and mixing depth in Lake Michigan, are summarized in Table 1. Since the areas associated with the <5 mg·cm$^{-2}$·y$^{-1}$ and <100 mg·cm$^{-2}$ isopleths represent greater than 60% of the total area, average sedimentation rates and mixing depths were calculated on the basis of both the areas of the total lake and the depositional zone. While the range of sedimentation rates is relatively large from 4.3 to 24.0 mg·cm$^{-2}$·y$^{-1}$, the range of mixing depths expressed in terms of years of sedimentation (S/$\Omega$) is small, between 17.0 and 19.0 years.

The interaction of plutonium with sediments has been the subject of a large number of studies. A large proportion of the data indicates that, for both the Great Lakes and oceans, the measured values of $K_d$, from both field and laboratory measurements are remarkably close, between $10^5$ and $10^6$ (Edgington, 1981). It has been shown that plutonium exists in two oxidation states, Pu(V) and Pu(IV), in the water of Lake Michigan (Nelson et al., 1980; Nelson and Orlandini, 1979) and that the measured $K_d$s for these two oxidation states are 250 and 2.5x10$^6$, respectively (Nelson and Lovett, 1978). Since the measured Pu(V)/Pu(IV) ratio in Lake Michigan water is 4.0, the value for the $K_d$ of plutonium onto the sediments of Lake Michigan would be predicted to be 5x10$^5$. One of the most recent values measured by Nelson et al. (1984) is 5.8x10$^5$, which is in good agreement with the value predicted above. Other experiments reported by Edgington et al. (1979) have shown that the interaction of plutonium with Lake Michigan sediments is reversible with equilibrium being established for both adsorption and desorption within 24 hours.

The changes in the concentration of plutonium in the water of Lake Michigan between 1970 and 1985 (Wahlgren et al., 1980; Nelson and Metta, 1983; Orlandini, 1988) are plotted in Figure 4a. A linear least squares fit to these data indicates that the concentration of plutonium in Lake Michigan water is decreasing with a half-life of 10.0 years ($r^2$=0.94).

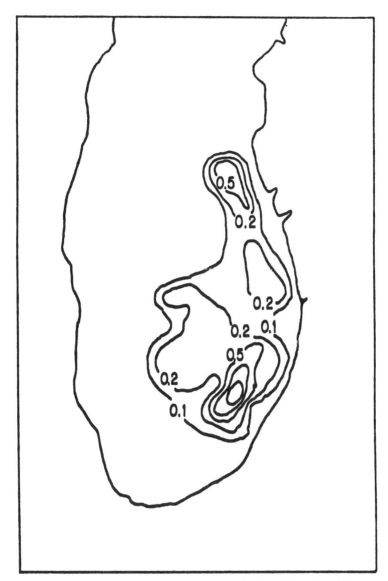

**Figure 3b.** Areal patterns on the variation in depth of the mixed layer in g·cm$^{-2}$ in the southern two-thirds of Lake Michigan.

The additional curves plotted in Figure 4a illustrate the expected changes in the concentration with time for the following three cases: (1) the absence of mixing, $\gamma=0$, using values of $K_d=5.8\times10^5$, fractional erosional loss from the watershed, $\alpha=2\times10^{-5}$, and sedimentation rate, $\Omega=0.024$ g·cm$^{-2}$·y$^{-1}$; (2) using the same values of $K_d$, $\alpha$ and $\Omega$ with a mixed layer depth, $\gamma=20$ years; and (3) no mixing, same values of $K_d$ and a more soluble pollutant, $\alpha=0.001$. A closer fit to the observed data is shown in Figure 4b. For this calculation, the parameter values were as follows: $K_d=5.8 \times 10^5$, $\Omega=0.024$ g·cm$^{-2}$·y$^{-1}$, $\gamma=10$ years, and $\alpha=2 \times 10^{-5}$.

## Discussion

The observed decrease in the concentration of plutonium in the water is faster than would be expected from the magnitude of the average mixed layer depth, $\gamma \approx 20$ years (Table 1), that was estimated from the analysis of sediment cores for $^{137}$Cs and was used for the calculation of the dashed curve in Figure 6. The rate of decrease is far slower, however, than would be predicted from a model that does not take account of sediment mixing, solid curves.

This difference could be explained in two ways. In the first

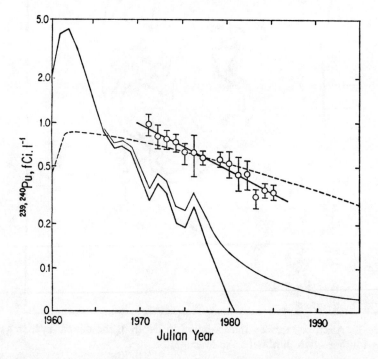

Figure 4a. The variation of the concentration of plutonium in water of Lake Michigan as a function of time (-0-). The line drawn through the data if the best least squares fit. (a) The other two solid lines represent the predicted change in concentration with time for no association with sediment and the inclusion of continued leakage from the watershed. The dashed line represents the change in concentration predicted using equations (7 and 8) with the best estimates of $K_D$,

# Sediment Mixing in Lakes

Table 1

Estimation of the Average Simulation Rate, $\Omega$, and Mixing Depth, $\gamma$, for Lake Michigan

(A) Sedimentation Rate

| Isopleth (A) (mg·cm$^{-2}$·y$^{-1}$) | Normalized Area (B) | | Effective Sedimentation Rate (= A x B) | | |
|---|---|---|---|---|---|
| < 5 | 0.671 | | 0.0 | - | 3.36 |
| 5 - 10 | 0.123 | | 0.62 | - | 1.13 |
| 10 - 20 | 0.115 | | 1.15 | - | 2.30 |
| 20 - 40 | 0.064 | | 1.28 | - | 2.56 |
| 40 - 60 | 0.019 | | 0.76 | - | 1.13 |
| 60 - 80 | 0.008 | | 0.48 | - | 0.64 |
| Whole Lake | 1.000 | $\Omega =$ | 4.30 | - | 11.10 |
| Depositional Zone | 0.329 | $\Omega =$ | 13.0 | - | 24.00 |

(B) Mixing Depth

| Isopleth (A) (mg·cm$^{-2}$) | Normalized Area (B) | | Effective Mixing Depth (= A x B) | | |
|---|---|---|---|---|---|
| < 100 | 0.677 | | 0.0 | - | 67.7 |
| 100 - 200 | 0.135 | | 13.5 | - | 27.0 |
| 200 - 500 | 0.141 | | 28.2 | - | 70.5 |
| 500 - 1000 | 0.33 | | 16.5 | - | 33.0 |
| 1000 - 1500 | 0.011 | | 11.0 | - | 16.5 |
| 1500 - 2000 | 0.002 | | 3.0 | - | 4.0 |
| Whole Lake | 1.000 | $S =$ | 72.0 | - | 219.0 |
| | $\gamma$ (years) = | $S/\Omega =$ | 17.0 | - | 20.0 |
| Depositional Zone | 0.323 | $S =$ | 223.0 | - | 467.0 |
| | $\gamma$ (years) = | $S/\Omega =$ | 17.0 | - | 19.0 |

instance, it could be postulated that the measured rate of erosion from the watershed, $\alpha=2\times10^{-5}$, is too small. It can be readily shown that, even if the value of the leakage rate from the watershed is increased to a value greater than that estimated for $^{90}$Sr, a far more hydrophilic radionuclide having a $K_d<1000$, this will not provide a large enough continuous input to maintain the observed elevated concentration of plutonium in the water [curve (3), Figure 4a]. Even with this scenario, the observed concentration in 1980 is still a factor of four larger than the predicted value.

The more likely explanation is that the steady-state sedimentation rate model of Robbins and Edgington (1975) has overestimated the magnitude of the depth of the mixed layer and, therefore, the value of $\gamma$. As mentioned earlier in this paper, a comparison of the results for the geochronology of the same series of stations in Lake Michigan, sampled ten years apart, led to the realization that continued focusing of sediments within even the major areas of deposition is leading to changes in the distribution of the $^{137}$Cs inventory and an apparent increase in the depth of activity by benthic organisms (increased thickness of the mixed

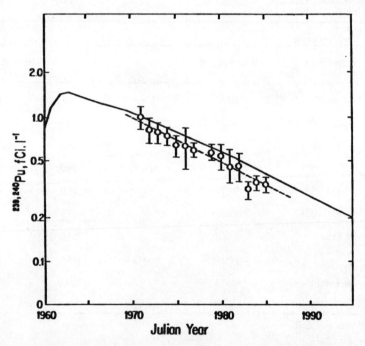

Figure 4b. (b) The best fit to the experimentally observed data using $K_D$, $\Omega$, $\gamma$ and that are within reasonable bounds for the errors associated with these parameters.

layer depth) (Edgington and Robbins, 1988). This change in the thickness of the mixed layer depth is illustrated in Figure 7. If it is remembered that there had been considerable focusing of these pollutants during the period between the initial major inputs (1959 to 1965) and the first measurements in 1972, with inventories in the center of the depositional zone being up to three times the measured integrated atmospheric input, a further focusing during the next decade is not surprising.

This continued focusing of the sediments cannot be handled satisfactorily by the model developed by Robbins and Edgington (1975). In that, and similar models, it was assumed that the new sediment depositing at any location was uncontaminated with $^{137}$Cs. With no new inputs from the atmosphere, i.e., $\phi(t)=0$, the least squares model interprets the continued redeposition of old resuspended sediment as an increase in the depth of the mixed layer. In future surveys, a more

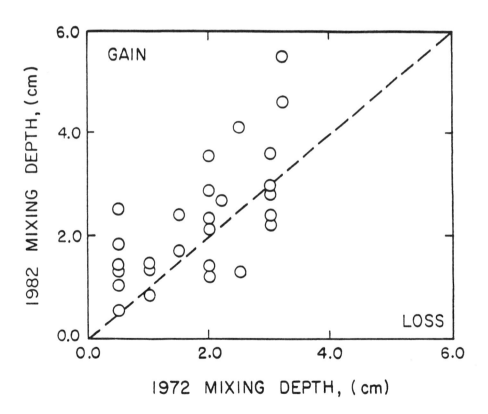

Figure 5. Comparison of mixed layer depth for several sites in Lake Michigan between 1972 and 1982.

increase in the depth of the mixed layer. In future surveys, a more accurate assessment of the sedimentation rate and mixed layer depth will have to be made using the $^{210}$Pb profiles rather than the more easily measured $^{137}$Cs profiles. When using $^{210}$Pb for geochronology, the problems of resuspension and continued focusing of old sediment are minimized because the inputs of this natural radionuclide are continuous relatively constant. Furthermore, the sedimentation parameters are calculated on the basis of radioactive decay rather than the position of horizon corresponding to initial and maximum inputs.

The best fit to the experimental data is plotted in Figure 4b. The maximum value of the average sedimentation rate, $\Omega$, from Table 1 was used in this calculation. Smaller values of the sedimentation rate when combined with the chosen value of $K_d$ do not permit storage of sufficient plutonium in the sediment mixed layer to maintain the observed concentrations with time in the water. Examination of the sediment storage term from Eq. (4), $\Omega K_d(1+\gamma)$, shows that, in order to maintain or increase the fraction of the plutonium stored in the sediments, an increase in any of the three parameters is required. Therefore, if the value of $\Omega$ is decreased by a factor of 2, then the value of either $K_d$ or $\gamma$ must be increased by the same amount to maintain the same loading of plutonium in the mixed layer. Since the estimated value of $\gamma$ is already too large, it would be necessary to increase the value of $K_d$. This is also not reasonable because the chosen value of $5.8 \times 10^5$ is at the high end of the range of values that has been observed from either field or laboratory measurements (Edgington, 1980; Watters *et al.*, 1980). Again, a careful examination of the steady-state sedimentation model of Robbins and Edgington, will show that, in the analysis of a $^{137}$Cs profile, if the value of the mixing depth, S, is overestimated, then the value of the sedimentation rate, $\Omega$, must be underestimated because of the finite depth of penetration of this radionuclide into the sediments. These problems will be resolved in future studies when the estimates are made using the $^{210}$Pb rather than the $^{137}$Cs profiles. This problem becomes critical in shallow areas, such as Green Bay, where sedimentation rates are large, and resuspension events predominate. Here, contrary to the results obtained using $^{210}$Pb, the measurements of $^{137}$Cs lead to predictions of very low sedimentation rates as a result of the massive resuspension and redeposition of old sediment (Deering, 1985).

## Conclusions

A simple model has been developed that couples a time-dependent concentration model for the transport and fate of pollutants in the Great Lakes developed by Lerman (1972) with a sediment complexing model suggested by Balistrieri and Murray (1983).

This model assumes that the interaction of the pollutant with the sediment is a versible reaction and that sediment contained within the layer of mixed sediment immediately below the sediment/water interface is essentially in contact with, and as a result of biological or physical movement, remains in equilibrium with, the overlying water.

For inputs, this model requires a knowledge of the inputs of the pollutant to the lake from the atmosphere and watershed, the average sedimentation rate and mixing depth in the lake, and the distribution coefficient for the pollutant.

This model has been used to predict the behavior of fallout plutonium in Lake Michigan using as inputs the monthly fallout data, an average sedimentation rate of 0.024 $g \cdot cm^{-2} \cdot y^{-1}$ and mixing depth of 20 years calculated from measurements of $^{137}Cs$ in sediment cores collected in 1982. The observed concentrations of plutonium in the water indicated that the mixing depth should be closer to 10 years.

It is suggested that this twofold difference in the value of the time constant for the loss of plutonium from the mixed layer, as well as the need to use a value for the sedimentation rate that is higher than the calculated average may arise from problems related to the continued resuspension and deposition of old sediment affecting the reliability of the $^{137}Cs$ method of dating sediments.

## Acknowledgments

This work was supported in part by the U.S. Environmental Protection Agency, Office of Research and Development, Grants Program, Grant #R809344010, and the Wisconsin Sea Grant Program. The author is indebted to D. Schwetz, J.A. Benante, L.G. Deering, P. Anderson, and R. Paddock for their valuable contributions to this study. In particular, he thanks D.M. Nelson and K.A. Orlandini of Argonne National Laboratory for providing unpublished data that were critical for the success of this paper.

## References

Balistrieri, L.S. and Murray, J.W. 1983. Metal solid interactions in the marine environment: estimating apparent binding constants. *Geochim. Acta.* 47. 1091-1098.

Benante, J.M. 1984. Studies of natural and artificial radionuclides in recent Lake Michigan sediments. M. S. Thesis, University of Wisconsin, Milwaukee.

Christensen, E. R. and Chien, N.K. 1981. Fluxes of arsenic, lead, zinc, and cadium to Green Bay and Lake Michigan sediments. *Envr. Sci. Technol.* 15.

553-557.

Deering, L. G. 1985. The uses and limitations of $^{210}$Pb and $^{137}$Cs geochronology to date recent sediments in Green Bay, Lake Michigan. M.S. Thesis, University of Wisconsin, Milwaukee.

Edgington, D. N. 1981. A review of the persistence of long-lived radionuclides in the marine environment -- sediment/water interactions. Impacts of Radionuclide Releases in to the Marine Environment. IAEA. Vienna. 67-91.

Edgington, D. N. and Nelson, D.M. 1984. The chemical behavior of long-lived radionuclides in the marine environment. In: Proceedings of the International Symposium on the Behaviour of Long-Lived Radionuclides in the Marine Environment. Commission of European Communities. EUR 9214. 19 - 68.

Edgington, D. N. and Robbins, J. A. 1975. The behavior of plutonium and other long-lived radionuclides in Lake Michigan. II. Patterns of deposition in the sediments. Impacts of Nuclear Releases into Aquatic Environment. IAEA. Vienna. 245-260.

Edgington, D. N. and Robbins, J.A. 1976. Records of lead deposition in Lake Michigan sediments since 1800. *Envr. Sci. and Technol.* 10: 266-274.

Edgington, D. N. and Robbins, J. A. 1988. Time scales of focusing in large lakes as revealed by measurements of fallout Cs-137. Proc. SIL Conf. on Large Lakes. In press.

Edgington, D. N., Kartunnen, J.O., Nelson, D.M., and Larsen, R.P. 1979. Plutonium concentration in natural waters - its relationship to sediment adsorption and desorption. Radiological and Environmental Research Division Annual Report. Argonne National Laboratory. ANL-79-65 Part III. 54-56.

Edgington, D.N., Alberts, J.J., Wahlgren, M.A., Karttunen, J.O., and Reeve, C.A. 1976. Plutonium and americium in Lake Michigan sediments. Transuranium Nuclides in the Environment. IAEA. Vienna. 493-516.

Fisher, J. B., Lick. W.J., McCall, P.L., and Robbins, J.A. 1980. Vertical mixing of lake sediments by tubificid oligochaetes. *J. Geophys. Res.* 85: 3337-4006.

Hardy, E. P., Krey, E.P., and Volchok, H. L. 1973. Global inventory and distribution of fallout plutonium. *Nature.* 241: 444-445.

I.J.C. 1987. Great Lakes Water Quality Agreemnt between the United States and Canada. International Joint Commission

Krey, P. W., Schonberg, M., and Toonkel, L., 1974. Updating stratospheric inventories -- January 1974. Fallout program Quarterly Summary Report. U. S. Atomic Energy Commission Report. HASL-281.

Lerman, A. 1972. Strontium-90 in the Great Lakes: Concentration-time model. *J. Geophys. Res.* 77: 3256-3264.

Lineback, J.A. and Gross, D.L. 1972. Depositional patterns, facies, and trace element accumulation in the Waukegan member of the Lake Pleistocene-Lake Michigan formation in southern Lake Michigan. Illinois State Geol. Survey. Environ. Geol. Notes. EGN-58.

N.C.R.P. 1984. Assessment of radionuclides released to surface waters. Radiological assessment: predicting the transport, bioaccumuation, and uptake by man of radionuclides released to the environment. N.C.R.P. Report No. 76 96-136.

Nelson, D. M. 1988. Argonne National Laboratory, Personal Communication.

Nelson, D.M. and Lovett, M.B. 1978. Oxidation State of Plutonium in the Irish Sea. 276: 599-601.

Nelson, D.M. and Metta, D.N. 1983. Plutonium concentrations in the Great Lakes -- an update. Radiological and Environmental Research Division Annual Report. Argonne National Laboratory. ANL82-65. Part III. 45-48.

Nelson, D.M. and Orlandini, K.A. 1979. Identification of Pu (V) in natural waters. Radiological and Environmental Research Division Annual Report. Argonne National Laboratory. AN79-65 Part III. 19-25.

Nelson, D.M., Metta, D.N., and Kartunnen, J.O. 1984. $^{239\text{-}240}$Pu, $^{137}$Cs and $^{90}$Sr in Tributaries of Lake Michigan. Radiological and Environmental Research Division Annual Report. Argonne National Laboratory. ANL-83-100 Part III. 45-54.

Nelson, D.M., Metta, D.N. and Larsen, R.P. 1980. Oxidation state distibution of plutonium in marine waters. Radiological and Environmental Research Division Annual Report. Argonne National Laboratory. ANL-80-115 Part III. 26-28.

Orlandini, K. A. 1988. Argonne National Laboratory, Personal Communication.

Robbins, J.A. 1982. Stratigraphic and dynamic effects of sediment reworking by Great Lakes zoobenthos. *Hydrobiologia*. 92: 611-622.

Robbins, J.A. and Edgington, D.N. 1975. Determination of recent sedimentation rates in Lake Michigan using Pb-210 and Cs-137. *Geochim. Cosmochim. Acta*. 39: 285-304.

Robbins, J.A., Krezoski, J.R., and Mozley, S.C. 1977. Radioactivity in sediments of the Great Lakes: Postdepositional redistribution by deposit feeding organisms. Earth planet. *Sci. Lett.* 36: 325-333.

Sprugel, D.G. and Bartelt, G.E. 1975. Preliminary mass balance of plutonium in a watershed near Sidney, Ohio. Radiological and Environmental Research Division Annual Report. Argonne National Laboratory. ANL-8060. Part III. 20-22.

Wahlgren, M. A. and Nelson, D. M. 1975. Plutonium in the Laurentian Great Lakes: comparison of surface waters. *Verh. Int. Ver. Theor. Angew. Limnol.* 19: 317-322.

Wahlgren, M.A., Alberts, J.J., Nelson, D.M. and Orlandini, K.A. 1976. Study of the behavior of transuranic and possible chemical homologues in Lake Michigan water and biota. Transuranium Nuclides in the Environment. IAEA. Vienna. 9-24.

Wahlgren, M.A., Robbins, J.A., and Edgington, D.N. 1980. Plutonium the Great Lakes. In: Transuranic Elements in the Environment (Hanson, W. and Watters, R. A., Eds.). U. S. Dept. of Energy. 659-683.

# CHAPTER 16
# Research Needs and Summary

John F. Paul
Joseph V. DePinto
Wilbert Lick

## Environmental Management Perspective

The point of view that has been taken in the preparation of the material in this volume is that we wanted to show to what extent the research community has been able to provide necessary tools to environmental managers for use in decision making. After all, in addition to providing a better understanding of the world in which we live, we would like to provide environmental decision-makers with the best scientific information available for use in their deliberations. We must all recognize, however, that decisions are not made solely on the basis of the scientific information presented. Economics, societal values, legislative mandates and regulations, politics, and public perceptions all play a role in the ultimate settlement of environmental issues. But this reality should not deter us from providing sound technical advice.

The environmental problem that we have been focusing on is "how do we determine the significance of in-place pollutants?' Are they hazardous enough that they are a major cause of degradation to the ecosystem and/or are they a significant human health risk? Is it better to leave the contaminated sediments where they are? How do we determine which remediation options would provide the most effective clean-up, in ecological and human health terms? The goal of the work that has been presented in this volume and of research needs that are provided later in this chapter is to provide scientifically credible information to environmental managers for use in addressing these types of questions. In a sense, this type of research, called directed research, is different from the narrowly-directed, highly-specialized work that most researchers are trained to conduct. Of necessity, this directed research is multidisciplinary, and requires coordination among many investigators if the end result is to be credible and useful. Chapter 2 provides examples of this type of work being conducted in EPA's Sediment

Quality Criteria Research Program.

To typify this type of research, consider the distinction between models used to describe physical, chemical, and biological processes and those used as management tools. Researchers traditionally use process models, whether they be conceptual or mathematical, to form a picture of the domain they are investigating and for use in organizing and presenting the results of their investigations. These types of models are generally not of use for managers because they are (1) difficult to understand, (2) narrowly focused in problem specification, (3) have not been constructed to address specific management questions, and (4) do not provide answers in a form that managers can translate into action. Models as management tools are developed to provide answers to specific management questions in a form that are intelligible and useful to mangers who need the information. Typically, management models are comprised of multiple process submodels that have been constructed by research studies; however, they often contain simplifying and site-specific assumptions that facilitate their application and interpretation with limited data.

Traditionally, researchers have viewed management models as "back-of-the-envelop" calculations, at best. Chapter 3 illustrates the types of models that have been used by environmental managers in decisions associated with siting of the Boston Harbor sewage outfall. Obviously, some of these models are quite simplistic in their structure. Were these the best models available at the time? The answer is no! But, were these the best models that could have been used in a timely fashion to address the environmental decision? The answer to this question is probably yes. This simple example emphasizes the conflict, or rather the gap, that exists between models which provide the best understanding of a detailed process and models which managers can effectively use. We do not mean to paint a grim picture, but rather want to emphasize the distinction between what a researcher uses and what a manager can use. This gap will always exist, but we need to make sure that as our scientific understanding improves the management tools derived from this improved understanding keep pace.

The next section provides a technical summary of the work presented in this volume. It gives a perspective on where we are in our technical understanding of transport and transformation of contaminants near the sediment-water interface. The final section presents our perspective on the knowledge gaps that exist and where research needs to be focused to provide better management tools for addressing the contaminated sediment problem.

# Technical Summary

## Physical Processes

In chapter 4 by Lick, recent work on the flocculation, deposition, and resuspension of fine-grained sediments is discussed. There is particular emphasis on a quantitative description of particle aggregation and disaggregation because of the effect on floc sizes, effective densities, settling speeds, and adsorption/desorption of contaminants from these flocs. Recent experimental work on settling speeds of aggregated particles as well as experimental and field measurements of sediment resuspension demonstrates the need to understand flocculation in order to properly compute particle dynamics in aquatic systems. Because of this and related work, it is becoming evident that the qualitative behavior of fine-grained sediments is becoming relatively well-understood, at least for certain ranges of the governing parameters.

Bedford (Chapter 5) extensively reviews and evaluates possible methods for determining vertical sediment fluxes at the sediment-water interface. The emphasis is on field measruement procedures and the risks, assumptions, and errors involved in using a particular method for determining resuspension rates. It is suggested that one of the most significant and challanging problems in this area is how to adequately sample the wave-current boundary layers and near-bottom sediments with sufficient precision and without disruption of the interface. Newly developed remote acoustic devices have been used for this purpose and show great promise.

In chapter 6, Sly uses particle size data from the bottom sediments of the Great Lakes in an attempt to understand the resuspension, deposition, and transport of different size particles as a function of the local water depth and hydraulic energy. From this information, a general qualitative understanding of particle transport throughout a lake as a function of these parameters is being obtained. Aspects of sediment geochemistry and biological productivity are also linked to the physical processes active at the sediment-water interface.

## Chemical/Biological Processes

Because of its importance in governing toxicity to or bioaccumulation by aquatic organisms, it is extremely important to have a good quantitative understanding of the distribution of a contaminant among its various phases and among potential species within a phase. This applies to both in-place sediment systems as well as those introduced into the water column either externally or by resuspension of bottom sediments. Chapter 7 reviewed approaches for quantifying the distribution of heavy metals in bottom sediments where the potential sorbents are very heterogeneous in nature. In this chapter, Allen documented that good models exist for computing metal distribution in the aqueous porewater phase of sediments provided good thermodynamic information exists on the various ligands in the system. The main problem arises in dealing with natural organic complexing agents (such as humic and fulvic acids) found in sediments. Because there are a variety of binding sites on these compounds, their metal binding properties are normally determined empirically (by titration on a case-by-case basis) and fit to either a discrete multi-ligand model or a continuous distribution ("affinity spectrum") ligand model. In either case we need a great deal more data on the binding properties of natural organic compounds and on the general applicability of these data.

With respect to the partitioning of metals between aqueous and solid phases in sediments, Allen again emphasizes the differences between understanding sorption to "pure" sorbent phases and predicting partitioning for naturally occurring mixtures of sorbent binding sites. While mechanistic models operate quite well for "model" systems, the additivity of sorption onto mixtures of discrete sorbent phases (clay, organic, iron oxides and manganese oxides) in natural sediments is still an open question. Also, as in the aqueous phase the problem of characterizing the type of organics associated with the solid phase in sediments and their effect on metal sorption is still unresolved. In summary, Allen states and demonstrates that we can model metal distribution in sediments if we can determine three quantities: (1) the total concentration of <u>exchangeable</u> metal in the system; (2) the values of stability constants (intrinsic or conditional) for all metal binding reactions important for the system; and (3) the concentration of each of the available binding sites for these reactions. Further progress will require more measurement and better measurement techniques for these quantities.

Tessier, *et al.* (Chapter 8) studied the importance of pH in governing the processes (sorption, precipitation, and fluxes of dissolved species across the sediment-water interface) that influence metals behavior in lakes by making extensive measurements of six different metals in porewater and sediments of lakes covering a wide range of pH values (4-8). In an attempt to simplify sediment metal equilibrium modeling, they focused only on what was felt to be the dominant sorbent solid phase and analyzed their field data in terms of a surface complexation reaction obeying the following mass action equation:

$$K_M = \frac{(Fe-M)}{(Fe-ox)\,[M^{z+}]} \tag{1}$$

where,

# Research Needs and Summary 333

$K_M$ = binding constant for cationic metal to sedimentary iron oxyhydroxides
(Fe-ox) = total concentration of iron oxyhydroxide sites
(Fe-M) = concentration of occupied iron oxyhydroxide sites
$M^{z+}$ = free metal concentration in porewater.

They found that these field-derived Log $K_M$ values increased linearly with pH and, in accordance with theory, followed the sequence Pb > Cu > Zn > Ni ≈ Cd. Thus, at low pH they predict the major portion of these metals in the water column of the lake, while at higher pH values (pH > 7 for Cd, Ni, and Zn, and > 6 for Pb) most of the metal pool would be associated with bottom sediments. They also demonstrated that depth profiles of metals in the upper layers of sediments were strongly dependent on the redox conditions as they influenced the distribution of Fe(III) and Fe(II). Upward diffusion of Fe(II) from anoxic layers and formation of reactive iron oxyhydroxides upon reoxidation creates strong binding sites for cationic metals. In spite of these accomplishments, the actual values of the field-derived $K_M$'s differed from laboratory-determined values, due to such hypothesized explanations as the failure to properly account for the effect of natural organic complexes in estimating $[M^{z+}]$ (especially for Pb and Cu).

While Tessier, et al. overcame the need for measurements of sediment porewater metal concentrations by employing a sediment "peeper" sampler, there remains a paucity of direct measurements of organic chemical concentrations in interstitial waters. Bierman (Chapter 9) used existing field data for body burdens in benthic invertebrates in combination with total sediment chemical concentrations to estimate interstitial dissolved chemical levels and sediment partition coefficients. By assuming that partitioning of hydrophobic organic chemicals could be completely described in terms of organic carbon normalization and octanol-water partition coefficients, he found a three-phase (truly dissolved, colloidal/DOC-bound, and solid phase-bound) model to be consistent with the available data for PCBs but not necessarily for PAHs. Bierman also found that in-place sediment partition coefficients were not controlled by particle concentrations, a finding consistent with the theory that particle collisions in agitated suspensions are responsible for the "solids effect." One of the biggest hindrances in the application of the three-phase model is the fact that the value of $K_{oc}^{nsp}$ (organic carbon normalized partition coefficient for non separable particles and DOC) is not independent of geographic location or season. In fact, it is not even clear whether measurement of interstitial DOC is an adequate surrogate for the quantity of "third phase" in the system.

While equilibrium partitioning models seem to be quite adequate for assessing the distribution of toxicant within in-place bottom sediments, DePinto, et al. (Chapter 10) offered evidence from field measurements that a kinetic approach may be more appropriate for modeling metals partitioning during the highly dynamic process of sediment resuspension. In an effort to understand field data of metals partitioning during resuspension experiments, they undertook a preliminary application of a time-dependent sorption-desorption model. The model coupled process descriptions of mass transport within the bulk solution, mass transport across the liquid film separating the bulk solution from the particle surface, radial diffusive transport within the porous structure of the sorbent, and reversible adsorption/desorption kinetics at the particle surface as well as at surface sites within the pores. This application provided evidence for film transport rate limitation for systems exposed to low shear stress and reaction rate limitation at higher shear stresses. Also, the model demonstrated how pore diffusion limitation could explain the slower desorption kinetics observed in many systems. As with all metal partitioning studies, pH was an important variable in determining whether net metal flux was out of the bulk solution and onto the particles or the reverse. Although this model could explain trends and relative rate limitations, considerable model development, parameterization, and field testing for the partitioning behavior of both metals and toxic organics is required.

We have already seen that heavy metals transport and fate in bottom sediments can be influenced by the redox status of the system. In Chapter 11 Myers and Nealson investigate how microbial-mediated processes, especially manganese oxide reduction, can affect sediment geochemistry. While relatively much is known about the oxidation of organic matter in sediments coupled to the reduction of nitrate and sulfate and to the production of carbon dioxide and methane, the authors claim that Mn and Fe oxides can serve as important electron acceptors in many situations. They have isolated and characterized a facultative, nonfermentative bacterium, *Alteromonas putrifaciens* strain MR-1, from anaerobic freshwater sediments which can reduce $MnO_2$ either directly (resulting from respiratory electron transport-linked Mn reductase systems) or indirectly (resulting from the bacterial production of products that can chemically reduce Mn). Understanding microbial-mediated processes in sediments and how they are linked to physical and chemical processes is a desperately needed area of investigation.

## Modeling/Synthesis

In Chapter 12, Ziegler, *et al.* use a numerical model to simulate the resuspension, transport, and deposition of fine-grained sediments in the Trenton Channel of the Detroit River. Recent process experimental results were included in the calculations. The modeling analysis focused on three areas of interest in the channel: Monguagon Creek, Black Lagoon, and Gibraltar Bay. The model was capable of identifying

# Research Needs and Summary 335

depositional sites, along with the corresponding deposition rates, as a function of channel flow and wind waves. It was felt that the analysis was reasonably accurate for purposes of comparing deposition and resuspension characteristics of these sites. Deficiencies in the modeling were due to inadequate data on sediment resuspension as a function of shear stress (especially high values of shear), inadequate knowledge of shear stresses due to non-breaking and breaking waves in shallow waters, and a simplified analysis of flocculation.

Lijklema, et al. (Chapter 13) hypothesized the importance of sediment transport to the phosphorus dynamics of two shallow lakes in the Netherlands. They therefore investigated sediment transport as predicted by several transport models. The authors then selected an approach that they felt reproduced field observations sufficiently well to be used within a nutrient management model. They then used this framework to study the transport of phosphorus-rich silt in Lake Veluwe with the objective of assessing the rate of accumulation of silt in the deeper portions of the lake and to study the role of resuspension in the central sediments in Lake Marken and its impact on algal growth in this system.

Recognizing the complexity particle-mediated transport of contaminants, Bonner, et al. (Chapter 14) developed a model incorporating hydrodynamic transport, convective transport of heterogeneous particles, dynamic size distribution involving particle aggregation/disaggregation, and dynamic sorption/desorption of contaminants. This model was parameterized by using a non-linear optimization strategy for calibrating transport coefficients to laboratory-scale processes experiments. It was then tested by comparing its predictions with observed particle concentrations (from two sources representing extreme and intermediate flocculation potential) in a larger-scale, mixed settling column that experimentally combines all the processes included in the model. The model was able to adequately predict particle concentrations under a range of column operating conditions, including freshwater and saltwater, $0-2.6 \times 10^6$ dyne-cm/sec hydrodynamic power dissipation, $0-35$ sec$^{-1}$ velocity gradient, and $0-6$ cm$^2$/sec vertical dispersion. In all cases particles in the column were destabilized by seawater, vertical transport within the water column was dominated by dispersion, but actual deposition loss from the column was directly related to settling velocity and flocculation collision efficiency. This study demonstrated the feasibility of integrating a variety of important mechanisms into a modeling framework that can be used for understanding the behavior of sediments in aquatic systems.

Finally, in chapter 15, Edgington describes a relatively simple model for the long-term transport and fate of radioisotopes in large lakes, with specific applications to the Great Lakes. The intent of this work is to used a radioisotope as a tracer of sediments in these systems. The model requires knowledge of atmospheric and watershed loadings, average sedimentation rate, surface sediment mixing depth, and an equilibrium distribution coefficient between particulate and water phases of the element. The model was successfully applied, with plutonium as the tracer, to describe the long-term sediment dynamics in Lake Michigan. Reasonable agreement with observations was obtained; however, some discrepancies were attributed to inadequate spatial resolution of the model.

## Research Needs

1. Resuspension properties of sediments. The qualitative transport behavior of fine-grained sediments is relatively well understood. However, for quantitative results on resuspension rates and amount of material resuspended for a particular shear stress at a given site, the sediments in question must be tested on a site-specific basis. To improve this situation, the most obvious needs are knowledge of a) resuspension rates and amount of material resuspended at high shear stresses (i.e., greater than 10 dynes/cm$^2$) when major storm events can resuspend and transport large amounts of sediments, b) effects of bioturbation, and c) effects on resuspension properties of mixtures of fine-grained and coarse-grained sediments.

2. Net resuspension and deposition of sediment bed. For contaminant flux and transport, the net gain or loss of sediment from the sediment-water interface needs to be computed on a spatial and temporal basis. This is not quantitatively understood at present and no model exists which can predict this gain or loss with sufficient confidence for a range of sediment types (i.e., a mixture of grain sizes). To accomplish this goal, there is a need for field measurements of this net sediment deposition or erosion for a variety of conditions and forcing functions. For example, accurate measurements of bathymetry are needed for a typical river-estuary system on a relatively fine spatial and temporal scale. It is especially important to attempt to obtain data such as this before, during, and after extreme events when most sediment flux occurs.

3. Flocculation. A general understanding of flocculation process in natural systems is being attained; however, this area requires considerable work before the process can be confidently incorporated into sediment transport models. Specific needs are: a) investigation of aggregation and disaggregation at low particle shears and concentrations where quantitative experiments have not been performed; b) investigation of the effect of organic matter on flocculation; c) and development and field applicability testing of

more numerically efficient submodels of flocculation in system-level sediment transport models.

4. Contaminant partitioning to aquatic particulate matter. It is extremely important to have a good quantitative understanding of the phase distribution of contaminants between the solution phase and the solid phase. This applies to both in-place sediment systems as well as those introduced into the water column either externally or by resuspension of bottom sediments. It requires the ability to quantify those characteristics of the sorbent (e.g., mineralogy, oxide coatings, organic content, size distribution, particle porosity, and sorption site density, selectivity and location) and bulk solution (e.g., pH, ionic strength, redox potential, complexing agent type and concentrations, and sorbate chemical concentration, reactivity, and other properties) that are essential in governing partitioning. We must also formulate models that can predict the phase distribution as a function of these basic properties of the system. Also, we need further research to develop scientifically credible explanations for observations that appear to conflict with existing theory, such as: decrease in measured partition coefficient with increasing sorbent solid concentration; apparent irreversibility of contaminant adsorption or at least relatively slow desorption; and inability to distinguish between truly dissolved species in the solution phase and those associated with macromolecules or colloidal-sized sorbents.

5. Interaction of physical and chemical processes in contaminant partitioning. In addition to the recognized chemical influences on the kinetics and equilibrium of contaminant partitioning to a solid phase, there are several physical processes that may also regulate the contaminant distribution. These processes -- which include mass transfer between bulk solution and the particle surface induced by fluid shear stress, particle-particle interactions that can alter the state of aggregation of sorbent particles and hence sorbent site density and porosity, and intra-particle macropore and micropore diffusion processes -- must be researched and understood in terms of their effect on the partitioning process.

6. Contaminant biotransformation and biodegradation in sediments. A potentially significant factor in governing the long-term fate of contaminants in aquatic systems is the microbial-induced transformation or degradation of these compounds in bottom sediments. A significant amount of research on the biodegradation of xenobiotics in sediments and soil systems has been conducted in recent years; however, much remains to be learned about this

important process as it occurs in natural systems. In particular, research is required on degradation rates and seasonal variation of those rates, the influence of horizontal heterogeneity and vertical stratification on degradation, the importance of these spatial scales on the sequential reactions often necessary for biochemical processes, the influence of the type and amount of contaminants and other environmental factors on the diversity and functioning of *in-situ* microbial communities.

7. <u>Appropriate physical description of the sediment-water boundary</u>. Two questions regarding the physical description of the sediment-water interface continue to perplex modelers. One is the selection of the appropriate boundary condition for models dealing with the exchange of particles between the sediment and water column. Also, selection of the appropriate upper mixed-layer depth for modeling sediment and associated contaminant transport is still an unresolved issue. These decisions are not only governed by the physical processes of deposition, entrainment and burial but one must also consider the effect of bioturbation and the characteristic reaction time of the material of interest. The sensitivity of model predictions to these formulations must be investigated and a methodology for parameterizing them on a system-specific basis must be developed.

# Conclusions

Sediments are very complicated and highly variable systems, the understanding of which requires a systematic integration of the physical sciences of hydrodynamics, rheology, and particle transport, the chemistry of phase partitioning, speciation, colloids, and surfaces, and the biology of contaminant biotransformation, bioavailability, and bioturbation. Mathematical models have the best likelihood of synthesizing these areas into management tools that can assist regulators in making informed resource management decisions. In constructing useable resource management models, modelers have many decisions to make regarding the trade-offs between scientific rigor and efficacy of addressing the particular management question at hand. These decisions involve such things as using appropriate spatial and temporal resolution, incorporating the <u>necessary</u> (so that the model will have the required sensitivity and feedback control) processes without becoming overly complex, and insuring that the model will be transferable and user-friendly enough so that it will, in fact, be used by the regulatory community.

One of the biggest impediments to achieving the goal of using models for management of systems containing contaminated sediments

# Research Needs and Summary 339

is the lack of coherent, system-wide data sets that can be used to field test models and thereby gain the necessary confidence for their application as management tools. A related question is what is an appropriate timescale for calibration of a model that must be used for making long-term predictions of a system response to regulatory or remediation actions. Too often we are asked to calibrate and/or verify a site-specific model with one or two years of field data and then make predictions of the system response thirty or forty years into the future. The only way to build confidence in making these long-term predictions is to conduct post-audits of systems for which a remedial or regulatory action have been taken; that is, we should build a long-term, follow-up monitoring schedule into all contaminated sediment management programs.

# Index

Abiotic suspended matter, 1
Acid-base titration, 124
Acid lake, 136
  littoral zone of, 130
  sediment-water interface of, 134
Acoustic scattering devices, 76
Active sediment layer, 311
Adsorption-desorption model, 180
Adsorption/desorption kinetics, 194
Adsorption isotherms, 179
AET, see apparent effects threshold
Affinity spectrum, 118, 332
Aggregation, 3
  of fine-grained sediments, 52
  rates of, 36
Agitated suspensions, 170
Algal assays, 116
Algal growth, 254, 278
Algebraic stress models (ASM), 71
Aluminum, 138, 139
Amphipods, 158, 159
Anaerobic genes, 210
Angular velocity, 286, 291, 293, 294
Animal body burdens, 154
Anoxic porewaters, 133
Antimycin A, 209
Apparent effects threshold (AET), 9
Apparent partition coefficients, 168, 169
Applied shear stress, 50
Aquatic communities, impact of toxic
    chemicals on, 177
Aquatic environments, 205, 293
Aquatic food chains, bioaccumulation in, 177
Aquatic organisms, 1
  bioaccumulation by, 331
  metal toxicity to, 115
Aquatic particulate matter, contaminant
    partitioning to, 337
Aquatic sediment, humic extracted from, 117
Aquatic systems, sediment transport in, 281
ARCS, see Assessment and Remediation of
    Contaminated Sediments
Arsenic
  concentration profiles, 131
  equilibrium constants for sorption of, 142
ASM, see algebraic stress models
Assessment and Remediation of
    Contaminated Sediments (ARCS),
    2
Atmospheric fallout, 129
Azide, 209

Bacteria
  iron-reducing, 213
  modes of manganese reduction by, 207
  sulfate-reducing, 213
Bacterial reduction, mechanisms of, 206
Balanced, indigenous population (BIP), 20,
    21
Bays, toxic contaminants in, 7
BBL, see bottom boundary layer
Bed transport regime, 65
Benthic biota, 7
Benthic boundary layer, 81
Benthic community
  chemical effects on, 11
  degraded, 12, 28
  structure, 8, 14
Benthic ecosystems
  degradation of, 10
  monitoring, 14
Benthic exchange model, 10
Benthic irrigation, 136
Benthic organisms, 1, 4, 312, 322
  modification of sediment bed by, 49
  number and types of, 50
Benthic productivity, 107
Benthic resources, 10
Bimodal sands, 102
Binary collisions, 40, 42
Bioaccumulation, 12, 13, 331
  factors, sediment-to-organism, 10
  ratio of to bioconcentration, 157
Bioavailability, 14
Biochemical markers, 14, 15
Biochemical oxygen demand (BOD), 17, 18
Bioirrigation, 136
Biological productivity, 95
Biological variability, 163
Biotic community integrity, 1
Biotic suspended matter, 1
Bioturbation, mixing by, 272
BIP, see balanced, indigenous population
BOD, see biochemical oxygen demand
Boston Harbor, 2, 18
  secondary outfall sites, 20
  sewage outfall, 3, 330
  wastewater outfalls in, 19, 29
Bottom boundary layer (BBL), 60, 62, 63,
    65, 79
Bottom feeders, 1
Bottom flow behavior, 60
Bottom sampling, 85

**341**

Bottom sediments, 1, 35, 59, 255, 337
Bottom shear stress, 71, 231, 246, 262
Bottom sitting towers, 72
Bottom stress, 48
Boundary layer, 64, 71
Brownian motion, 36, 37
Buckingham analysis, 71
Bulk equilibrium concentration, 197
Bulk solution, 183
Bulk transport, 182

Cadmium
  carbonate, 119
  ME-SORB calibration plots for, 189
  toxicity of, 115
Calcium, 134
Calcium carbonate, particulates composed of, 106
Calibration parameters, 187
Calibration values, 194
Capping, 12, 13
Carbon dioxide, conversion of organic matter to, 205
Carbonyl cyanide $m$-chlorophenyl hydrazone (CCCP), 209
Cationic metals, partition coefficient for, 196
CCCP, see carbonyl cyanide $m$-chlorophenyl hydrazone
Chemical/biological processes, 3
Chemical contaminants, 7
Chemical equilibrium computer programs, 118
Chemical reductants, 218
Chemical speciation, 115
Chesapeake Bay, 2
Chironomids, 158, 159, 162
Chlorobenzenes, 159
Chromium, ME-SORB calibration plots for, 191
Clays, 139
Clean Water Act (CWA), 2, 17
Clean Water Act Section 301(h), 7, 13
Coagulation, 30
Coagulation model, 22, 24
Coarse sediments, 96, 99
Coastal waters, discharge of wastewater into, 17
Cobalt, ME-SORB calibration plots for, 190
Cohesive sediment transport, 60
Collision
  efficiency factor, 38
  frequency functions, 41
  rate, 39
Colloidal particles, 288

Combined sewer outflows (CSOs), 19
Commonwealth of Massachusetts Metropolitan District Commission (MDC), 19, 21, 25
Consolidation, 272
Contaminant(s)
  biotransformation, 337–338
  chemical release of, 54
  deposition, 12
  exposure, frequency of, 12
  flux, 52
  long-term accumulation of, 1
  particulate-associated, 3, 4
  partitioning, interaction of physical processes in, 337
  phase redistribution, 178
  resuspension, 177, 178
Contaminated sediment transport, parameterizing models for, 281–305
  experimental approach, 283–287
    particle characteristics, 283–284
    particle flocculation potential, 284–285
    particle size distribution, 284
    particle specific gravity, 284
    particle transport experiments, 285–287
  modeling, 287–291
    dispersion, 287
    heterogeneous particles, 288–289
    homogeneous particles, 287–288
    parameter estimation, 289–291
Contaminated sediments
  biological effects of, 178
  resuspension of, 1
  in suspension, 236
Contiguous zone, discharges into waters of, 17
Control volume, 74, 83, 88
Convection-diffusion transport model, 26
Copper, 121, 133
  adsorption of onto salt marsh sediment, 122
  in overlying water, 132
  solubility diagram for, 134
  toxicity of, 116
Coprecipitation, 107, 118
Coriolis effects, 60
Critical orbital velocity, 271
Critical shear stress, 232, 266
CSOs, see combined sewer outflows
Current velocities, 22, 231, 237, 241, 244
CWA, see Clean Water Act

# Index

Deep water forecasting relationships, 261
Deep water sediments, 107
Deposition, 4, 30
  of fine-grained sediments, 52
  rates, steady-state, 238, 242, 245
  /resuspension models, 13
  sites, steady-state, 244
Depositional flux, 69
Depositional regime, breaks in, 108
Desorption kinetics, slow, 197
Destabilized colloidal suspensions, 282
Dicumarol, 209
Diethylene-triamine-penta-acetic acid (DTPA), 119
Differential settling, 36
Diffusivity coefficient, 71
Dinitrophenol, 209
Diprotic acids, 119
Disaggregation, 3, 30
  of fine-grained sediments, 52
  rate of, 36, 38
Discharge conditions, 30
Dispersion coefficient, 272
Dispersive transport model, 287
Disposal options, 12
Dissolved aqueous concentration, 166
Dissolved organic ligands, complexation by, 144
Dissolved organic matter, 153
Distribution coefficients, 138
Downward irradiance, 274
Dredging operations, 12, 281
DTPA, see diethylene-triamine-penta-acetic acid
Dutch lakes, 253
Dye studies, 293

Ecological damage, by benthic contamination, 12
Ecosystem
  recovery, 12
  structure, disruption of, 1
  vulnerability of, 11
Eddy viscosity, 241, 243
EDTA, see ethylene-diamine-tetra-acetic acid
Effluent limitations, 17
Effluent quality, 18
Electromagnetic (EM) current, 76
Electron acceptor mutants, 212
EM, see electromagnetic current
Entrainment, 4
  estimating, 59
  flux, 69, 71, 78

Environmental laws, applications of sediment criteria in implementing, 8
Environmental management decisions, 30
Environmental phase partitioning, 153
Environmental Protection Agency (EPA), 2, 17
  Office of Research and Development (ORD), 7, 9, 10, 13, 15
  Office of Water, 7
  Sediment Quality Criteria program, 3
EPA, see Environmental Protection Agency
Equilibrium, 69
  bottom boundary layer, 86
  concentration, bulk, 197
  constant, 196, 309
  partitioning, 8–10
    coefficients, 154
    model, 180, 159
    theory, 15
Erosion, 243, 272, 273, 281, 309, 322
Estuaries
  contaminant fluxes in, 54
  fine-grained sediments in, 35
Estuarine protection, 7, 14
Estuarine water, sizes of flocs in, 40
Ethylene-diamine-tetra-acetic acid (EDTA), 116, 117, 119, 132
Eutrophication, 253

Factories, wastewater drains from, 225
Farfield problem, 30
Federal Water Pollution Control Act (FWPCA), 17
Field
  measurements, 12
  resuspension experiments, 186
  tests, 54
Film
  diffusion, 197
  metal concentration, 185
  transfer coefficient, 188
  transport, 182
Fine-grained sediments, 242, 243, 331
  aggregation and disaggregation of, 228, 229
  analysis of, 225
  flocculation, deposition, and resuspension of, 35–57
  flocculation, 36–43
  *in situ* measurements of resuspension, 50–52
  laboratory measurements of resuspension, 47–50
  settling speeds, 43–47

resuspension of, 3, 229, 230
transport of in Trenton Channel of Detroit
   River, 225–252
   Black Lagoon, 241–243
   bottom stresses due to currents and
      waves, 230–232
   Gibraltar Bay, 243–248
   governing equations, 227–228
   Monguagon Creek, 236–240
   results of numerical calculations,
      232–236
   sediment dynamics, 228–230
First-order decay process, 27
First-stage desorption, 197
Fish disease, incidence of, 11
Fish tumor incidence, 11
Floc(s)
   average density of, 43
   diameters, time variations of median, 39,
      41
   settling speed of, 248
Flocculation, 3, 4, 35, 36, 336–337
   of fine-grained sediments, 52
   model of, 53
   potential, 303
Flocculator reactor volume, 285
Flocculent transport mechanisms, 282
Flow field physics, 59
Fluid mechanical processes, 59, 60
Fluid mechanics, 71
Fluid momentum, 228
Fluid shear, 36, 37, 42
Fluid shear stress, 337
Flux
   estimation, 84
   potential, 11
   terms, 68
Formaldehyde, 209
Formate, 211
Forward reaction rate, 196
Four-point sampling, 84
Fraction organic carbon, 159
Fraunhofer diffraction, 38
Free stream velocity, 64
Fulvic acid, 133, 332
Fumarate, 209
FWPCA, see Federal Water Pollution Control
      Act

Gibraltar Bay, 235, 244, 334
Glacial till, 96
Glaciolacustrine clays, 107, 109, 110
Glycine, 209
Grass shrimp, 115

Gravity waves, 81
Great Lakes, 2, 13, 95
   contaminant transport in, 35
   distribution of sediment types in, 96
   water levels in, 99
   Water Quality Board, 2

Hazardous pollutants, 225
Heavy metals
   kinetic partitioning of, 178
   transport, 334
2-heptyl-hydroxyquinolone-$N$-oxide (HQNO),
      209
Heterogeneous particles, convective transport
      of, 335
Hexachlorobiphenyl, 166, 167
Homogeneous transport model, 299
Horizontal eddy diffusivity, 228
Horizontal flux, 68, 83
Horizontal transport, 265, 267, 268
HQNO, see 2-heptyl-hydroxyquinolone-$N$-
      oxide
Human food chains, 7
Human health, impact of toxic chemicals on,
      177
Humic acid, 117, 133, 332
Hydraulic energies, 99, 100, 331
Hydraulic equilibrium, 101, 104
Hydrodynamic power dissipation, 284, 285
Hydrodynamics, 281, 283, 338
Hydrogen ion activity, 17
Hydrogen sulfide, 212
Hydrophobic organic chemicals, 153, 166,
      170, 333
   activity coefficients for, 156
   animal concentrations of, 158
   distribution of, 157
   equal partitioning of, 161
   phase partitioning of, 154, 155, 171
Hypolimnion flow velocities, 108

Ice-drop debris, 102
IJC, see International Joint Commission
Industrialization, 129
In-place bottom sediments, 333
In-place contamination, 2, 7
In-place pollutants, 1, 177, 329
Internal horizontal transport, 255
Internal sediment transport, 253
International Joint Commission (IJC), 307
Interstitial water quality, 14
Ion exchange, 118
Iron, 139
   cycle, association of arsenic with, 132

Index                                                                        345

oxide, 121, 124
oxyhydroxides
    natural, 145
    sedimentary, 146, 333
    synthetic, 146
reductase system, 215
sulfides, 133
Irradiance sensor, 274
Isothermal conditions, formation of, 106–107
Isotropic homogeneous turbulence, 61

Kurtosis, 101, 102

LACSD, see Los Angeles County Sewage District
Lactate, 211
Lag-gravel, fine sands from, 108
Lake(s)
    bottom sediments, 259
    contaminant fluxes in, 54
    dynamics, 107
    eutrophication of shallow, 4
    fine-grained sediments in, 35
    physical stress-response effects in, 98
    toxic contaminants in, 7
Lake Erie, surficial sediments in, 35
Lake Ontario, 95
Lake water
    dissolved concentrations of elements in, 146
    pH, 132, 133
    suspended load of, 105
    trace elements in, 129
Lead
    adsorption of onto salt marsh sediment, 122
    ME-SORB calibration plots for, 193
LED, see light emitting diode
Light
    climate, 277
    emitting diode (LED), 286–287
    extinction coefficient, 258
Lipid-normalized relationships, 159
Local sediment composition, differences in, 267
Long Island Sound, sediments from, 49, 50
Long-term sediment diffusion experiments, 170
Los Angeles County Sewage District (LACSD), 24, 25

Macrobenthic invertebrates, 157
Management strategies, 59
Manganese, 138, 139, 206

oxide, 124, 205
    adsorption of copper onto, 121
    reduction of, 4
oxyhydroxides, formation of, 129
reduction
    biological and chemical mechanisms of in aquatic and sediment systems, 205–223
        direct manganese reduction, 207–211
        indirect reduction, 211–215
        manganese reduction mechanisms, 206
    mechanisms of, 218, 219
Mass balance models, 153
Massachusetts Bay, 18, 21, 22
    convection-diffusion model for, 26
    secondary outfall sites, 20
Massachusetts Water Resources Authority (MWRA), 19, 21, 26
Mathematical models, 338
MDC, see Commonwealth of Massachusetts Metropolitan District Commission
Mean flow, determination of, 80
Mean settling velocity, 261
Median particle size, 53
Metal(s)
    adsorption kinetics, 187
    binding, 124
    bioavailability of, 115
    carbonates, 119
    equilibrium partitioning of, 3
    -humate interactions, 118
    -ligand-sorbent systems, 179
    partitioning behavior, 187
    phase distribution, 181
    speciation, 3
Metals partitioning, predicting during resuspension events, 177–203
    development of ME-SORB, 180–186
        ME-SORB mathematical formulation, 183–186
        model framework and assumptions, 180–183
    equilibrium metals partitioning, 178–180
    model application, 194–197
    model calibration, 186–194
Methane, production of, 205
Methanogenesis, 206
Microbial redox reactions, 205
Microcosm research, 13
Mid-basin sediments, 107
Mid-lake basin sediments, 107
Mid-lake water column, 106
Mineralogy, 50

Mine waters, 142
Mirex, distribution of in Lake Ontario sediments, 109
Mixed layer depth, comparison of, 323
Mixing depth, average, 315
Mixture density, 67
Molecular diffusion, downward, 136
Molecular viscosity, 67
Momentum boundary layer, 79, 88
Momentum flux, 70
Municipal wastewater, 17, 19
MWRA, see Massachusetts Water Resources Authority

National Environmental Policy Act (NEPA), 7, 14
National Pollution Discharge Evaluation System (NPDES), 21
Natural convection currents, 44
Natural iron oxyhydroxides, 145
Natural organic matter, decay products of, 117
Natural sediments, 35
Natural systems
  effect of sediment transport processes in, 3
  sorption of trace elements in, 139
Natural weathering, 129
Nearshore areas, fine-grained sediments in, 35
Negative depth contours, 240
NEPA, see National Environmental Policy Act
Nepheloid layer, formation of, 105
Net resuspension, 49
New Bedford Harbor, 2
Nickel
  distribution coefficients, 138
  ME-SORB calibration plots for, 192
  in overlying water, 132
Nitrate, microbial reduction of, 205
Nitrilotriacetic acid (NTA), 115
Non-cohesive sediments, 36
Nonequilibrium fluxes, 86
Nonlinear drift, 81
NPDES, see National Pollution Discharge Evaluation System
NTA, see nitrilotriacetic acid
Nutrient
  release, 13
  removal, 19

Ocean
  contaminant fluxes in, 54
  dumping, 7, 14

Octanol-water partition coefficients, 333
Old sediment, deposition of, 4
Oligochaetes, 158, 161, 162, 168
One-point measurements, 85
Orbital velocities, 81
ORD, see EPA Office of Research and Development
Organic acids, 216
Organic carbon, 312
Organic carbon normalization, 166, 333
Organic chemicals
  binding of to natural sediments, 3
  partitioning of in sediments, 4, 153–175
    conceptual framework for phase partitioning, 154–158
    methods, 158–161
      field data, 158–159
      phase partitioning models, 159–161
      simulation of $K_{OC}$ values, 159
    results, 161–169
      comparisons of phase partitioning models, 166–169
      phase distributions between animal lipid and sediment organic carbon, 161–163
      relationship between estimated $K_{OC}$ and $K_{OW}$, 163–166
Organic matter, 139
  degradation of, 106
  oxidation of, 129, 205
Organic wastes, 7
Outer surface sorbed metal, 183
Overlying water
  arsenic diffusion to, 132
  undersaturation of oxic, 136
Oxic lake waters, 133
Oxide/hydroxide solid phase, 134

PAHs, see polyaromatic hydrocarbons
PAR, see photosynthetically available radiation
Parameter estimation, 294
Particle
  characteristics, 281
  collision efficiency, 284
  density, 284
  diameter, settling speed as function of, 44–46
  flocculation potential, summary of, 298
  interaction model, 159–160, 169
  settlement, 106
  size
    characteristics, 101
    data, 95

Index 347

distribution, 30, 38, 40, 42, 50, 288
    dynamics, 198
    monitoring of, 300
  specific gravity, 283
  surface, film solution transfer to, 183
  transport, 338
    characteristics, summary of, 301
    coefficients, measurement of, 289
Particulate
  aggregation, 30
  deposition, models of, 13
  diluted wastewater, 22
  matter, resuspension flux of, 178
PCBs, see polychlorinated biphenyls
Pelagic fish, 158
Phase averaging, 78
Phase partitioning, 338
Photocell, computer monitored, 286
Photosynthetically available radiation (PAR), 274–275
Physical forcing functions, 97
Physical processes, 3
Phytoplankton growth, 277
Plankton, 147, 258
Plant nutrition, 119
Plume dispersion, 104
Plume dynamics, models of, 30
Plutonium
  behavior of in Great Lakes, 309
  concentration of in water, 320
  interaction of with sediments, 318
Point source discharge, 107
Point sources, 129
Pollutants
  effects of sediment mixing on long-term behavior of in lakes, 307–328
    application to Lake Michigan, 314–320
    model development, 310–314
  focusing of, 316
Polluted waters, heavily, 52
Pollution-sensitive species, 22
Polyaromatic hydrocarbons (PAHs), 158, 159, 162, 171, 333
Polychlorinated biphenyls (PCBs), 158, 159, 162, 164, 165, 168, 171, 333
Pore
  adsorption, 182
  diffusion, 182
  solution, 183
  surface sorbed metal, 183
Porewater
  concentrations, 4, 131
  peepers, 133, 134, 137
POTW, see publicly owned treatment works

Primary outfalls, 27
Primary treated wastewater, 19
Priority chemicals, 9
Process synthesis/modeling, 3
Publicly owned treatment works (POTW), 17
Puget Sound, 2
Pure sorbent phases, 332
Pyruvate, 211, 212

Radioactive decay, 318, 324
Radioisotopes, long-term transport of, 4, 335
Radionuclides
  hydrophilic, 322
  measurements of, 308, 309
Random velocity gradient, 303
Record length, 77
Regulatory management perspectives, 3
Remedial actions, simulation of, 13
Research needs, 329–339
  appropriate physical description of sediment-water boundary, 338
  contaminant biotransformation and biodegradation in sediments, 337–338
  contaminant partitioning to aquatic particulate matter, 337
  environmental management perspective, 329–330
  flocculation, 336–337
  interaction of physical and chemical processes in contaminant partitioning, 337
  net resuspensions and deposition of sediment bed, 336
  resuspension properties of sediments, 336
  technical summary, 331–336
    chemical/biological processes, 331–334
    modeling/synthesis, 334–336
    physical processes, 331
Residual function, 290, 294
Resource management, 4
Respiratory processes
  bacterial electron transport-linked, 206
  electron transport-linked, 211
Resuspended sediment, 240
Resuspension, 30, 272
  event, 178, 198
  of fine-grained sediments, 52
  flux, 69, 257, 264, 271
  models for, 263
  rate, 47, 278
Reverse equilibrium models, 153
Rheology, 338
Rivers, toxic contaminants in, 7

Saline estuarine waters, 17
Sampling biases, 98
Sampling frequency, 82
Sampling variability, 78
Sands, resuspension rate for, 36
Sandy silt, 97
Scouring events, 281
Sea water, flocs in, 40
Seasonal resuspension, 239
Second-stage desorption, 197
Secondary effluent treatment, 28
Secondary outfalls, 27
Secondary treated wastewater, 21
Secondary treatment, 18, 19
Sediment
  bed, net resuspension and deposition of, 336
  complexing model, 324
  composition, changes in, 272
  contamination, 198
  degradation, 10
  deposition, rate of, 12
  depth profiles, 3
  distribution coefficient, 315
  entrainment rates, 3
  flux, 36, 79, 228, 239
  fractionation, 120
  geochemistry, 4, 95
  layer, thickness of, 249
  load, suspended, 243, 245
  management programs, 339
  mitigation
    options, 14
    procedures, 12
  mixing, effects of on long-term behavior of pollutants in lakes, 307–328
    application to Lake Michigan, 314–320
    model development, 310–314
  model, two-layer, 272
  molecular diffusion coefficient, 67
  monitoring methods, 14
  organic carbon, 154
  partition coefficients, 154, 157, 166
  quality criteria (SQC), 7
    chemical-specific, 8
    contaminant-specific, 9
    effects of additivity on, 11
    regulatory applications of, 10
  research, EPA/ORD role and perspective in, 7–15
    current monitoring of sediment quality, 14
    current research in mitigation/prevention of sediment quality problems, 12–13
    current sediment quality criteria research program, 9
    current sediment quality evaluation, 10–11
    future monitoring research, 14–15
    future research in mitigation/prevention of sediment quality problems, 13–14
    future sediment quality criteria research, 9–10
    future sediment quality evaluation research, 11–12
    mitigation/prevention of sediment quality problems, 12
    monitoring of sediment quality, 14
    regulatory applications of sediment quality criteria, 7–8
    sediment quality evaluation, 10
  resuspension, 4, 240, 242, 247
    events, metals partitioning during, 180
    effect of storm duration on, 248
    measurement of, 50, 51
  samples, particle-size anomalies in, 102
  texture, variation in, 269
  top layer, change in composition of, 271
  toxicity, 10, 11
  transport
    calculations, 232
    equation, 236
    processes, 3
    traps, 259, 260
  types, distribution of in Great Lakes, 96
  wasteload allocation, 7
  -water boundary, appropriate physical description of, 338
  -water interface, *in situ* measurement of entrainment, resuspension, and related processes at, 59–94
    analytical and computational framework for entrainment flux, 69–72
    auxiliary variables, 87–88
    basic equations, 65–67
    control volume method, 83–84
    definition of flux terms, 68–69
    differential/computational method, 84–85
    empirical Buckingham $\pi$ approach, 86–87
    evaluation, 82–83
    fluid mechanical setting, 60–65

# Index

boundary layers, 63–65
scales of activity, 60–63
instrument sampling effects, 75–80
  instrument noise and calibration, 75–77
  record length, 77–80
inverse or boundary layer methods, 85–86
phenomenological turbulence model approach, 86
physical considerations of data collection, 72–75
  instrument separation, 73–75
  instrumentation size and obstruction considerations, 75
  tower tilt and settling, 73
regression and curve fitting, 80–82
  determination of mean profiles, 81–82
  determination of mean, 80–81
  mean flow conditions affected by waves, 81
sampling, averaging, and statistical considerations for *in situ* measurement, 72
setting, basic equations, and flux definitions, 59–60
-water interface, net flux at, 228
-water interface, transformation of pollutants across, 30
/water interface, impact of physical processes at in large lakes, 95–113
  basin deposits and nepheloid layer, 105–107
  depositional regime and hydraulic equilibrium, 100–103
  erosion, 108–110
  mean particle-size and silt/clay ratio, 103–105
  particle-size, geochemistry, and sediment focusing, 107–108
  particle-size/water depth relationships, 96–99
Sedimentation
  flux, 257, 259
  models for, 263
  rate, 136, 309
    average, 315, 318, 325
    variation in, 317
Sensor volume, 75
Sequential selective fractionation, 118
Settling
  columns, 4, 22, 23,
  speeds, 35, 36, 43, 45, 53, 229, 331
  tube, 259
  velocities, 23, 25, 259, 267, 288, 291
Sewage outfall, lessons learned from siting of Boston Harbor, 17–33
  existing and proposed Boston Harbor outfalls, 19–21
  methods used to evaluate sedimentation, 21–26
    MDC 301(h) application, 21–22
    MDC revised 301(h) application, 22–26
  MWRA secondary outfall siting, 26–29
Sewage sludge, 19, 298, 300
Shallow lakes, sediment transport in, 253–280
  composition of sediment top layer, 271–272
  description of resuspension, 261–271
    calibration of resuspension model based on fractions, 269–271
    parameter estimation, 265–267
    relationships for resuspension and sedimentation, 263–264
    role of horizontal transport, 268–269
    wave model, 261–262
  light extinction in Lake Marken, 274–278
  net accumulation and erosion areas in Lake Veluwe, 273–274
  subdivision of particulate material in fractions, 258–261
Shear
  stress, 82, 85, 180, 186
  turbulence resulting from, 60
  velocities, 64, 98
Silicon oxides, 139
Silt/clay ratio, 104, 106
Skewness, 101, 102
Skewness/kurtosis relationships, 103
Sludge, discharge of, 19
Solids
  accumulation, 27
  steady-state accumulation of, 22
Solid waste dumpsites, groundwater seepages from, 225
Solubility calculations, 133
Sorption-desorption hysteresis, 153
Sorption-desorption reactions, 144
Sorption-desorption reversibility, 160
Source identification, 13
Spectral spike, 61
SQC, see sediment quality criteria
Stability constant, 123
St. Lawrence River, channel outflow of, 108

Stokean settlement, 105
Storm calculations, 246
Storm sediment resuspension during, 249
Submodels, development of, 283
Sulfate, microbial reduction of, 205
Sulfide formation, 133
Sulfide production, 213
Sulfite, 209
Sulfur compounds, 212
Summer resuspension events, 21
Superfund, 7, 13
Surface
  adsorption, 182
  complexation
    model, 139, 141, 143
    theory, 179
  film layer, 183
Surficial oxidized sediments, 146
Suspended flocs, effective diameter of, 248
Suspended sediment, 234
  concentration, 228
  load, 248
Suspended solids, 17, 18
  concentration
    dynamics of, 264
    variation in, 275
  production and decay of, 256
  sedimentation of, 255
Synthetic iron oxyhydroxides, 146

Taylor's frozen turbulence hypothesis (TFTH), 75
Technical Support Document (TSD), 21–23
Terminal electron acceptors, 211, 215
Territorial sea, discharges into deep waters of, 17
Tetrathionate, 209
TFTH, see Taylor's frozen turbulence hypothesis
Thermodynamic bioaccumulation, 158
Thermodynamic bioaccumulation potential, 159, 171
Thiosulfate, 209
Thiosulfate reductase system, 215
Three-body collisions, 42, 43
Three phase model, 159, 160, 171
Tidal movement, 17
Tke, see turbulent kinetic energy
TMAO, see trimethylamine N-oxide
Tower
  settling, 72, 73
  tilt, 80, 82
Toxic chemical(s)

potential hazards of, 153
transport, 2, 19
Toxic metals, partitioning of in natural water-sediment systems, 116–127
  metal partitioning between solid and aqueous phases, 118–124
    mechanistic approaches for multiples phases, 122–124
    mechanistic approaches for single phases, 119–121
    selective fractionation approach, 118–119
  metal speciation in aqueous phase, 115–118
Toxic substances, 1
Toxic waste load allocations, 177
Toxicity
  chemicals causing, 12
  models of, 12
  monitoring of sediment, 14
  tests, partial life-cycle, 14
Trace elements, reactions of near sediment-water interface in lakes, 129–152
  anoxic pore waters, 133
  benthic fluxes, 134–136
  dissolved concentration profiles, 130–132
    arsenic, 131–132
    iron, 130–131
    trace metals, 132
  distribution coefficient, 137–139
  overlying waters, 133–134
  partitioning of labile trace elements in lakes of various pH, 144–146
  solubility equilibrium, 133
  sorption, 136–137
  surface complexation, 139–143
Trace metal(s), 115
  fixation, 136
  pollutants, 115
  rapid release of, 146
Transitional interface materials, 102
Transport/deposition models, 12
Trenton Channel, of Detroit River, 4, 180, 181, 194, 197, 226
Trimethylamine N-oxide (TMAO), 209, 212
Triple layer models, 179
TSD, see Technical Support Document
Turbidity, enhanced, 253
Turbulence, 76, 81, 104, 108
  isotropic homogeneous, 61
  models, 71, 88
  wind induced, 258

# Index

Turbulent fluctuations, 78, 79
Turbulent kinetic energy (tke), 86
Turbulent shear stress, 86
Two-body collisions, 42
Two phase model, 159, 160, 170
Two-stage sorbate release, 197

Underwater photography, 108
Undisturbed sediments, 51

Variance spectrum, 63
Vertical attenuation coefficient, 274–276
Vertical dispersion, 283
Vertical flux, 68, 83
Vertical scaling, 79
Vertical sediment flux, 3
Vertical transport, 304

Wasteload allocation strategies, 12, 13
Wastewater
  diluted, 22, 26
  outfalls, 30
  plume, trapping level of, 22
  treatment facilities, 19
Water column, 144, 240, 312
  contamination in, 7
  loss of particles from, 304
  organic carbon in, 107
  organisms, effects on, 1
  physical processes, 61
  removal of particulate material from, 27
  removal of trace elements from, 147

  resuspended sediments in, 178
  stratification, 27
  trace elements in, 129
  washout from, 2
Water quality criteria, 9
Water renewal rate, 129
Water residence time, 308
Water resource managers, 1
Wave-current boundary layer, 84, 89
Wave-current interactions, 80, 231
Wave frequency data, 98
Wave parameters, 231
Waves, generation of, 261
Wetlands, 12
  experimental, 13
  protection programs, 14
  research program, 13
Wind velocity, 242
Wind waves, 225, 227, 234, 247
  effect of, 235, 249
  generated by storms, 239
Winter flow conditions, 249

Xenobiotic organic compounds, 307
Xenobiotics, biodegradation of, 337

Zinc, 116
  concentration gradients of, 135
  distribution coefficients, 138
  equilibrium constants for sorption of, 142
  in overlying water, 132, 133
  solubility diagram for, 134